HUMAN DIMENSIONS OF WILDLIFE MANAGEMENT

HUMAN DIMENSIONS OF WILDLIFE MANAGEMENT • 2nd EDITION

EDITED BY *Daniel J. Decker*
Shawn J. Riley
William F. Siemer

Published in affiliation with The Wildlife Society

THE JOHNS HOPKINS UNIVERSITY PRESS | BALTIMORE

The Johns Hopkins University Press
2715 North Charles Street
Baltimore, Maryland 21218-4363
www.press.jhu.edu

Library of Congress Cataloging-in-Publication Data

Human dimensions of wildlife management / edited by Daniel J.
Decker, Shawn J. Riley, and William F. Siemer. — 2nd ed.
        p. cm.
    Rev. ed. of: Human dimensions of wildlife management in North
America / edited by Daniel J. Decker, Tommy L. Brown, and William F.
Siemer.
    Includes bibliographical references and index.
    ISBN 978-1-4214-0654-1 (hdbk. : alk. paper) — ISBN 978-1-4214-
0741-8 (electronic) — ISBN 1-4214-0654-3 (hdbk. : alk. paper) —
ISBN 1-4214-0741-8 (electronic)
    1. Wildlife management—North America. 2. Wildlife management—
Social aspects—North America. I. Decker, Daniel J. II. Riley, Shawn J.
(Shawn James), 1955–   III. Siemer, William F. IV. Human dimensions
of wildlife management in North America.
    SK361.H78 2012
    639.9—dc23   2012002461

A catalog record for this book is available from the British Library.

*Special discounts are available for bulk purchases of this book. For more
information, please contact Special Sales at 410-516-6936 or specialsales@
press.jhu.edu.*

The Johns Hopkins University Press uses environmentally friendly book
materials, including recycled text paper that is composed of at least 30
percent post-consumer waste, whenever possible.

CONTENTS

PREFACE

No matter where you are in the world, there is wildlife management, because someone values wildlife or is affected by wildlife in a way that matters to them. Public wildlife management formally emerged during the late nineteenth and early twentieth centuries as part of a social movement away from an era of natural resource exploitation and toward the new concept of conservation. Nongovernmental organizations were formed to promote particular interests in wildlife and to advocate for legal protection of wildlife and their habitats. Statutes, laws, and treaties between and among states, provinces, and nations were established to protect and conserve wildlife. Governmental agencies were created to ensure that societal values for wildlife resources were attended to, laws enforced, and management undertaken.

In the early years of the wildlife profession, the biological dimensions—those concerned with organisms, populations, and their habitats—were the dominant scientific concern. As species were restored, habitat protected, and populations of some species flourished, human–wildlife interactions increased and diversified, as did the stakeholders of wildlife management. The growth and diversification of stakeholders increased demand for insights into the human dimensions of wildlife management. The knowledge and application of human dimensions in wildlife management is still developing, enabling the ongoing shift toward collaborative governance of wildlife resources.

In our numerous and varied collaborations with wildlife professionals in governmental agencies and non-governmental organizations around the world, we regularly encounter a pervasive and persistent lament, "Wildlife management is 10 percent working with wildlife and 90 percent working with people!" This comment is often offered as a plea for assistance in understanding and dealing with the human dimensions of wildlife management. In response, we are pleased to present this book on the concepts and principles underlying the human dimensions of wildlife management, including applications of these ideas from across many facets of wildlife conservation throughout the world. This volume is intended to serve as a textbook to educate aspiring wildlife professionals and as a reference for those already engaged in the practice of wildlife management. The contributing authors are experts in their respective fields of study and were asked to convey the most important ideas from those fields relevant to human dimensions of wildlife management. Readers also are introduced to some novel content for a wildlife management text, even one with a human dimensions emphasis: governance, decision making, and ethical analysis and discourse. Our hope is that this book will better prepare the current generation of students for an interesting, effective, and rewarding career in wildlife management.

Human Dimensions of Wildlife Management in North America (Decker, Brown, and Siemer, eds.) was published in 2001 by The Wildlife Society; two of us (DJD and WFS) were co-editors of that book. A primary purpose of that first book was to reach an audience of practicing wildlife professionals. Three things became clear to us in the intervening decade. First, more college and university wildlife programs are offering courses focused on human dimensions (and professional societies are either encouraging human dimensions–oriented coursework or expecting such coursework for purposes of professional certification). So instructors are seeking a book written more with a student audience in mind and around which a course can be developed. Second, inquiry and application of human dimensions is advancing rapidly and results are being reported in a wide range of outlets (more than any student or practicing professional likely can read) and a need exists for continual updating and synthesis in a book-type reference. Third, human dimensions of wildlife management are being studied and the insights applied far beyond the borders of North America. In an era of globalization of knowledge, students of wildlife management will benefit from learning concepts and skills (e.g., governance, ethical analysis, structured decision-making, adaptive leadership and management) applicable to wildlife management wherever it is practiced. Thus, this volume covers a larger array of topics than the first book and is aimed more directly at students.

We have divided the book into six parts: Overview and Fundamental Concepts, Social Science Considerations, The Management Process, Human Dimensions Methods and Skills, Human Dimensions Applications, and Professional Considerations for the Future. In "Overview and Fundamental Concepts" we define what is meant by the human dimensions of wildlife management, and we introduce two fundamental components central to the purpose of wildlife management: governance and stakeholders. "Social Science Considerations" covers the foundational social sciences of social psychology, sociology, and economics. These disciplines are needed to understand how individuals and society interpret human–wildlife interactions and how wildlife is valued. Impacts—the most important effects on values resulting from human–wildlife interactions—are a central concept throughout the book. In "The Management Process," wildlife management is recognized as a cyclical and iterative process within a larger goal-seeking system; decisions are the basic task of managers within this system. "Human Dimensions Methods and Skills" includes planning a human dimensions study and common social-science methods, stakeholder engagement practices, and communication. Theory and methods are demonstrated in "Human Dimensions Applications" by examining roles for human dimensions in management of abundant wildlife, scarce wildlife, and use and users of wildlife. The last section,

"Professional Considerations for the Future," encourages readers to reflect on the need for ethics and ethical discourse in deliberations related to wildlife, as well as the need for continual learning to keep pace in an ever-changing world.

We conclude the book with consideration of the importance of taking an adaptive approach as wildlife professionals, and some challenges for improving inquiry and application of the human dimensions of wildlife management now and in the future. A valuable component to the book is the many topical boxes throughout each chapter intended to provide examples or additional discussion of selected topics. A list of suggested readings at the end of each chapter guides readers seeking either additional exploration of each chapter topic or useful references for inquiry and application in human dimensions of wildlife management.

We approach wildlife management as a complex undertaking that requires a mix of technical and value judgments, supported by social science and stakeholder engagement processes. A career in wildlife management requires knowledge about the coupled sociocultural–ecological systems in which management occurs and directs its effort. Familiarity with the human dimensions of wildlife management is a necessity. Arguably, it is impossible to possess the entire breadth and depth of knowledge needed to effectively handle all aspects of the wildlife management enterprise, and we had to leave out some areas such as anthropology, law, and wildlife criminology that are emerging as important human dimensions considerations for wildlife management. Nonetheless, the book presents the reader with a broad set of mission-critical content, which we and the chapter authors hope are useful and motivating for you. We also hope the book leads you to new areas of intellectual exploration and continual learning. We wish you success in a rewarding career as a wildlife professional who creates lasting positive impacts for present and future generations of stakeholders.

ACKNOWLEDGMENTS

We are indebted to the scores of wildlife students and professionals throughout the world with whom we have a distinct privilege to interact on a regular basis. What we are learning from these innovators as they apply human dimensions theory and processes in practice, combined with our own research, teaching, and outreach is reflected in this volume.

We thank The Wildlife Society for providing the opportunity to work with the Johns Hopkins University Press to produce this replacement for the previous Wildlife Society publication, *Human Dimensions of Wildlife Management in North America* (Decker, Brown, and Siemer, eds. 2001). We thank Vincent J. Burke and Jennifer Malat at the Johns Hopkins University Press for their help and patience in guiding us through the publication process.

Margie Peech provided valuable administrative assistance during the initial stages of the book. Meghan Baumer joined the team in spring 2011 and earned considerable recognition for keeping the three of us organized and for completing with excellence the many administrative tasks associated with readying the book manuscript and associated files for submission. Mo Viele prepared figures for all chapters.

The chapter contributors bring perspectives from many universities, state and federal agencies, and a private consulting firm. We thank them for their commitment, diligence, and patience. To make this book more of a textbook speaking from one voice, rather than the more usual book of contributed chapters, required occasional heavier-than-normal editing on our part; we are grateful for the flexibility and understanding shown by the chapter authors. A special note of thanks goes to Tommy Brown, who was a co-editor with DJD and WFS on *Human Dimensions of Wildlife Management in North America*. Much of what Tommy contributed to that book was used in some way in the current book.

In addition to chapter authors, 19 external reviewers volunteered time and expertise to critique chapters: Dale Blahna, Tommy Brown, Lisa Chase, Paul Curtis, Göran Ericsson, Anne Forstchen, Larry Gigliotti, Cynthia Jacobson, Heidi Kretzer, Bruce Lauber, Katherine McComas, Craig Miller, Charles Nilon, Perran Ross, Camilla Sandström, Tania Schusler, Chris Smith, W. Daniel Svedarsky, and James Tantillo.

An important component of this book is the text boxes, with examples of insights from the human dimensions approach on various topics. Most of the boxes were prepared by an additional group of authors: Neil H. Carter, Rebecca Christoffel, Janis Dickinson, Kipp Frohlich, Thomas Heberlein, Cynthia Jacobson, Jaime Jiménez, Bruce Lauber, Stacy Lischka, Francine Madden, Angela Mertig, Marina Michaelidou, Karen Murphy, John Organ, Daniela Raik, James Perran Ross, Camilla Sandström, and Tania Schusler.

The book contains photographs provided through the generosity of many people, organizations, and agencies. For their assistance in locating photographic material, we thank Jody Enck, Göran Ericsson, Nina Fascione, Dennis Figg, Jeffrey Flocken, the Florida Fish and Wildlife Conservation Commission, Ann Forstchen, Meredith Gore, Jaime Jiménez, David Kenyon, Kirsten Leong, Francine Madden, Matt Merchant, Marina Michaelidou, Michigan State University, Larry Nielsen, Melissa Normann, North Carolina State University, Serda Ozbenian, Joe Paulin, Daniela Raik, Shawn Riley, William Robichaud, Ed Schools, Gina Schrader, Tania Schusler, Diane Tessaglia-Hymes, Keith Tidball, and Jerry Vaske.

We gratefully acknowledge use of photographs from the following sources: Aldo Leopold Foundation (www.aldoleopold.org); Eric Andersson; M. Andreou; Liz Bihrle; Bureau of Land Management; Neil H. Carter; Jim Clayton (New York State Department of Environmental Conservation); Conservation International (Sterling Zumbrunn); Cornell Laboratory of Ornithology (Donna Salko, Tammie Sanders, Susan Spear, and Diane Tessaglia-Hymes); Paul Curtis; Emerging Wildlife Conservation Leaders Program; Ernesto C. Enkerlin Hoeflich; Florida Fish and Wildlife Conservation Commission (Cliff Leonard); Fotolia.com and contributors to Fotolia.com (Adam Golabek-Fotolia.com, Friday-Fotolia.com, Glenda Powers-Fotolia.com, Kablonk Micro-Fotolia.com, Megan Lorenz-Fotolia.com, Rafa Irusta-Fotolia.com, Tomasz Kubis-Fotolia.com); Meredith Gore; Jaime Jiménez; Tana Kappel (The Nature Conservancy); David Kenyon; Isle Royale Wolf–Moose Project; Library of Congress; Michael Lykins (Missouri Department of Conservation); Francine Madden; Silvio Marchini; Matt Merchant (New York State Department of Environmental Conservation); Marina Michaelidou; Steve Milpacher; Bret Muter; Mark Needham; Melissa Normann–Rainforest Alliance and Northern Jaguar Project; North Carolina State University; Dion Ogust; Serda Ozbenian (Saola Working Group); Joe Paulin; Ed Schools; Tania Schusler; Michael Shaffer (Michigan State University); William Siemer; Rick J. Sinnott–Alaska Department of Fish and Game; Chris Slesar (Vermont Agency of Transportation); USFWS (Steve Hillebrand); USFWS Digital Archive; Jerry Vaske; Julie Whittaker; and Roger Winstead (North Carolina State University).

The editors extend thanks to their specific supporting entities. DJD and WFS appreciate the support of the Department of Natural Resources in the College of Agriculture and Life Sciences, Cornell University. We also thank the students, staff, and other colleagues in the Human Dimensions Research Unit at Cornell University for various forms of support and encouragement for our efforts on the book. Special thanks are extended to the Michigan Department of Natural Resources (MSU) for their support of SJR through the Partnership for Ecosystem Research and Management and to Michigan State University Department of Fisheries and Wildlife and MSU AgBioResearch for additional financial support. The Swedish Fulbright Commission and the Department of Wildlife, Fish and Environmental Studies at the Agricultural University of Sweden in Umeå provided financial and logistical support for SJR to write during a sabbatical year.

We extend marked thanks to our colleagues Len Carpenter, John Organ, and others with whom we have been engaged for more than a decade on a bootleg project we call "thinking like a manager." That effort transformed into workshops and publications on management thinking and agency change that informed our work on the book. These outreach efforts have been aided and abetted by Tim Breault, Ann Forstchen, Elsa Haubold, Cynthia Jacobson, Patrick Lederle, Kirsten Leong, and others.

Finally, we each are grateful for the support and understanding of our families for the precious time we borrowed from them to complete this book project.

CONTRIBUTORS

Shorna B. Allred
Cornell University
Ithaca, New York

Alistair J. Bath
Memorial University of Newfoundland
St. John's, Newfoundland, Canada

Lisa C. Chase
Vermont Tourism Data Center
University of Vermont
Burlington, Vermont

Nancy A. Connelly
Cornell University
Ithaca, New York

Michael R. Conover
Utah State University
Logan, Utah

Daniel J. Decker
Cornell University
Ithaca, New York

Jonathan B. Dinkins
Utah State University
Logan, Utah

Jody W. Enck
Cornell University
Ithaca, New York

Meredith L. Gore
Michigan State University
East Lansing, Michigan

Robin S. Gregory
Decision Research and Value Scope
Research, Inc.
Eugene, Oregon

Barbara A. Knuth
Cornell University
Ithaca, New York

T. Bruce Lauber
Cornell University
Ithaca, New York

Kirsten M. Leong
National Park Service
Fort Collins, Colorado

Michael J. Manfredo
Colorado State University
Fort Collins, Colorado

Terry A. Messmer
Utah State University
Logan, Utah

Michael P. Nelson
Oregon State University
Corvallis, Oregon

Larry A. Nielsen
North Carolina State University
Raleigh, North Carolina

Richard C. Ready
Penn State University
University Park, Pennsylvania

Shawn J. Riley
Michigan State University
East Lansing, Michigan

Brent A. Rudolph
Michigan Department of Natural Resources
East Lansing, Michigan

Michael G. Schechter
James Madison College
Michigan State University
East Lansing, Michigan

Tania M. Schusler
Antioch University New England
Keene, New Hampshire

James E. Shanahan
Boston University
Boston, Massachusetts

William F. Siemer
Cornell University
Ithaca, New York

Richard C. Stedman
Cornell University
Ithaca, New York

Jerry J. Vaske
Colorado State University
Ft. Collins, Colorado

John A. Vucetich
Michigan Technological University
Houghton, Michigan

PART I · OVERVIEW AND FUNDAMENTAL CONCEPTS

In Part I, we define what is meant by the human dimensions of wildlife management and introduce two components central to the purpose of wildlife management: governance and stakeholders. Chapter 1 gives a brief history and outlines the breadth of human dimensions of wildlife management while introducing the chapter topics of the book. In the early years of the wildlife profession, the biological dimensions—those concerned with organisms, populations, and their habitats—were the dominant scientific concern. As species were restored, habitat protected, and populations of some species flourished, human–wildlife interactions increased and diversified, as did the stakeholders of wildlife management. The enlargement and diversification of stakeholders increased demand for human dimensions insights to support wildlife management. Human dimensions knowledge and application in wildlife management is still developing, which enables the ongoing shift toward collaborative governance of wildlife resources.

Chapter 2 traces the emergence of collaborative governance as the context within which wildlife management occurs throughout the world. Governance is affected by worldwide trends of increased democratization and globalization of transportation, trade, and information. Governance and wildlife management are focused on and benefit stakeholders, which are the topic of Chapter 3. A gradual, yet steady, trend has occurred as wildlife management has evolved; it is a shift from a view that wildlife professionals represent people who are constituents and clients toward a more inclusive perspective that focuses on all people who are beneficiaries of management. These beneficiaries are called stakeholders. The need for insights into how best to achieve good governance that serves the diverse interests of stakeholders provides the motivation for the study and practice of concepts, principles, and skills covered in this book.

1

HUMAN DIMENSIONS OF WILDLIFE MANAGEMENT

DANIEL J. DECKER, SHAWN J. RILEY,
AND WILLIAM F. SIEMER

Wildlife management conventionally is depicted as having three dimensions: humans, wildlife, and habitats. Definitions of wildlife management to date give attention to wildlife and habitats (Decker et al. in press), yet effective management addresses all three of these interdependent dimensions and the environment in which they occur (Fig. 1.1). Effective management addresses an entire management system, not just one part. In this context, the scope of human dimensions is vast; everything in the wildlife management system that is not directly about wildlife and habitat is about humans, and much of the environment in which people, wildlife, and habitats interact is influenced by humans in some way. Most concerns related to wildlife populations and habitats usually have direct or indirect human dimensions considerations, either as a cause of, or a cure for, problems. The pervasive effects of humans on wildlife management motivate managers to consider individuals, groups, social structures, cultural systems, and institutions. Professional wildlife-management work includes discovering, understanding, and applying insights about how humans value wildlife, how humans want wildlife to be managed, and how humans affect, or are affected by, wildlife and wildlife management decisions. Taken together, these components are the human dimensions of wildlife management, which is the topic of this book. There are other topic areas that could fall within human dimensions, such as human resource management, organizational behavior, or environmental law, but we do not include such functions explicitly in this book.

Human dimensions considerations, from broad social ideals (e.g., good governance, biodiversity conservation) to specific traits of individuals (e.g., motivations to hunt, photograph, or observe wildlife), are addressed in this book. In this first chapter we introduce core principles, concepts, and applications of human dimensions covered in greater depth throughout the book. The idea of "impacts" is woven throughout this chapter and the book because it is a useful concept when thinking about the human purposes of wildlife management, and impacts often are the topics of research and the impetus for stakeholder engagement (Riley et al. 2002).

1.1. FUNDAMENTAL CONCEPTS

People and wildlife affect one another within complex, coupled sociocultural–ecological systems (sometimes referred to as coupled human and natural systems; Liu et al. 2007). People interact with wildlife directly and indirectly in myriad ways. Their interactions vary in intensity and occur at many geographic and temporal scales. Wildlife management attempts to create, facilitate, regulate, or forbid various interactions that humans might have with wildlife. Outcomes of these human–wildlife interactions can be thought of as their effects. The important effects of human–wildlife interactions, those that cause strong stakeholder interest and draw management attention, are termed "impacts" (Riley et al. 2002).

Impacts may be positive or negative and take a variety of forms (e.g., economic benefits or costs; threats to or enhancement of human health and safety; ecological services wildlife provide; physical, mental, and social benefits produced by recreational enjoyment of wildlife; and many others). In addition to being produced by interactions between humans and wildlife, impacts can arise from interactions among humans where the referent is wildlife. One reason for human dimensions research and stakeholder engagement practices is to contribute knowledge about the nature, extent, and outcomes of interactions and to thereby enhance the process of making informed choices about management. The purpose of wildlife management is to influence these interactions. The responsibility of the wildlife manager is to ensure that relevant insights from human dimensions inquiry inform management decisions and actions.

Wildlife management from this perspective can be defined as:

> The guidance of decision-making processes and implementation of practices to influence interactions among people, and between people, wildlife, and wildlife habitats, to achieve impacts valued by stakeholders. (Adapted from Riley et al. 2002)

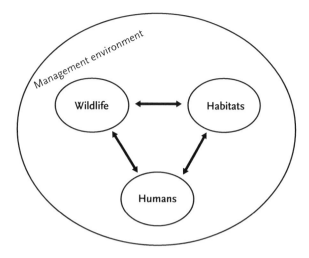

Figure 1.1. Wildlife management triad

The emphasis of wildlife management indicated in this definition is on production of value as defined by society, where values or benefits are the outcomes (impacts created, avoided, or mitigated) that are experienced by stakeholders as the result of wildlife management (e.g., values associated with biodiversity, wildlife-associated recreation, economic activity, and others). Wildlife management, therefore, is one of a set of processes that produce ecosystem goods and services (de Groot et al. 2002). Consistent with the definition above, human dimensions of wildlife management are the reasons for public and private interest in management of wildlife and wildlife habitat. Such considerations emphasize the fundamentally human purpose of wildlife management, which, like any form of management, is to turn complexity, information, and specialized activity into benefits at individual, group, and societal levels (Decker et al. 2009b).

An impacts approach to wildlife management (Riley et al. 2003) links social and ecological systems. Consequently, whether actively managed or not, social and ecological systems interact to produce diverse outputs, some desirable and some not. Understanding how coupled systems produce human–wildlife interactions, how various interactions create different effects, how those effects are perceived by people (i.e., whether they are good or bad), and whether these are important to them, allows managers to identify impacts for management attention.

The need to balance positive and negative impacts of wildlife is supported by numerous studies in which stakeholders report mixed reactions to the presence of wildlife (e.g., Decker and Gavin 1987, Chase et al. 1999b, Riley and Decker 2000, Ericsson and Heberlein 2003, Decker et al. 2006a). Conflicting beliefs about wildlife occur not only between different people but within an individual who experiences both positive and negative impacts from wildlife. For example, research has described the influence of both positive and negative impacts of white-tailed deer on residents in southern Michigan, where rural residents were conflicted in their desire to derive benefits from viewing and hunting deer (which were among

their stated reasons for living in a rural environment) and their concern about risks of deer–vehicle collisions (nearly one-third of the respondents reported being involved in a recent deer–vehicle collision; Lischka et al. 2008). To identify and understand the effects and impacts produced by this coupled sociocultural–ecological system, one has to comprehend the human as well as non-human components. This calls for a high degree of public participation in deliberations about wildlife management.

Regardless of form, the approach taken to wildlife management reflects the governance structure that has been adopted by the appropriate authority (e.g., a state wildlife management agency) or been mandated by law. As detailed in Chapter 2, governance refers to mechanisms whereby governments and other organizations direct their activities. Governance is the way things get done, and it includes the processes that implement laws, rules, and policies that collectively guide decisions. In western democracies, good governance is usually characterized as participatory, transparent, and accountable. Agencies and others involved in wildlife management rely on human dimensions insights to understand and meet citizens' expectations of the institutions and processes of government. Thinking about wildlife management in the context of governance is an alternative to a more bureaucratic, top-down view of management. The meaning of management as we think about it in the United States, however, is not applied uniformly throughout the world (Box 1.1). In many places the approach to wildlife management takes on forms adapted to, and reflective of, the culture.

Because wildlife is a public trust resource in the United States, various levels of governments are responsible for public wildlife management, a responsibility that is typically accomplished through state or federal wildlife agencies. Historically, wildlife management was perceived as the exclusive purview of state, tribal, and federal governments because they played the dominant role in policy and decision making related to wildlife management. In the past 20 to 30 years, however, partnerships among multiple levels of government and non-governmental organizations (NGOs) and across administrative or geographical boundaries are fashioned through collaborative forms of governance (Lauber et al. 2011).

A desire to govern well in the public interest conveys a fundamental purpose for gaining insights about the human dimensions of wildlife management. Significantly, the human dimensions of wildlife management refer to a body of knowledge, knowledge-generating processes, and knowledge-application activities that enable governance of wildlife resources. As contexts—especially the number and diversity of wildlife stakeholder interests seeking management consideration—become more complex, the demand for human dimensions insight grows and the sophistication required of human dimensions research and application to support good governance increases.

Attention to the public interest is reflected in the legal basis for American wildlife management, referred to as the Public Trust Doctrine (Batcheller et al. 2010), which is case law that

Box 1.1 *MANAGING, CARETAKING, OR BROTHERHOOD?*

Although its roots are European, the term "scientific management" emerged as an American concept that fueled the Industrial Revolution and the Progressive era. We extended this thinking to management of our forests and wildlife as if they were machines for producing goods and services. In other parts of the world, "management" may seem a term ill-suited to link with wildlife and nature. For example, Swedes use words with roots in nursing or caretaking when talking about wildlife (Heberlein 2005). Although they have mined minerals for more than 500 years and forestry is their country's second-largest industry, the Swedes' orientation toward wildlife management is less focused on manipulating land to produce wildlife for human uses than in the United States. This orientation leads one to think about partnering with nature rather than controlling it; the orientation is found even in the language.

When talking about moose populations and goals, Swedes often use the word "älgskötsel." Älg is Swedish for "moose," and "skötsel" translates into English as "care." It does not stop there. Another word associated with management is "vård." When you feel ill in Sweden, you seek help at a Vårdcentrum, or "care center." The Swedish Environmental Protection Agency, which oversees the country's wildlife management, is called "Naturevårdsverket" or "nature's care administration."

Of course, as North American influences extend to Europe and other places, Swedes adopted the following tools and processes: radiocollars, habitat manipulation, and population goals achieved through tightly controlled hunting. They are managing more than caretaking (and thus more often use the word "förvaltning"—"administration"). Nonetheless, "skötsel" is still commonly used when Swedes talk with each other about interacting with nature,

A biologist collects data from an immobilized moose near Umeå, Sweden (courtesy Eric Andersson)

which signals a fundamental difference from how Americans perceive nature.

The Swedes are not alone. When we discuss wolf management with the Anishinabe in northern Wisconsin, they repeatedly tell us "The wolves are our brothers. If our brothers thrive, we thrive." When asked how many wolves are enough, they look puzzled. "How many brothers are enough? Will wolves be there to the seventh generation? If we protect them and care for nature, our brothers will be there for us." The Anishinabe think they have done fine for hundreds of years on the landscape. Is it not time for scientific management to step back and to learn lessons from others? Caring for our brothers may be the true step forward toward sustainability.

THOMAS A. HEBERLEIN

---

defines ownership of wildlife and empowers the government (especially states) to be the trustee of wildlife as a public trust resource (Box 1.2). One of the obligations of a trustee (typically state governments in the United States) is to understand needs of the beneficiaries of the trust (citizens of the states in which wildlife resides) in determining goals and objectives for the trust. In most cases, state wildlife agencies serve as agents of the trustees (legislatures); technically, agency personnel are not the trustees or decision makers with respect to wildlife resource management (Smith 2011). Though they often are decision makers by default, officially agency personnel are not the primary decision makers with respect to public-trust wildlife resources. Those decisions are the responsibility of wildlife commissions, commissioners, and directors. Although few likely view themselves this way, wildlife professionals in state government are mainly trust managers (Smith 2011) in roles

of technical analysts, process managers, and implementers of management actions ranging from habitat manipulation on public lands to review of permits for land-use actions that disturb habitats on private lands, and from game and furbearer harvest regulation to public education about wildlife and conservation of threatened and endangered species. This arrangement gives rise to one of the basic functions of human dimensions inquiry and stakeholder engagement—seeking better understanding of the outcomes (impacts) of management that are desired by stakeholders (beneficiaries) within the context of existing policies and laws and incorporating such information into analyses and recommendations offered to decision makers.

A stakeholder is any person who is significantly affected by, or significantly affects, wildlife or wildlife management decisions or actions (Decker et al. 1996). Stakeholders are

Box 1.2 *THE PUBLIC TRUST DOCTRINE*
Why Human Dimensions Is Essential to Wildlife Management in North America

Who owns wildlife? The answer could vary considerably depending on what part of the world is referenced. In the United States no one owns wildlife. Wildlife is held in trust by the government for the benefit of present and future generations of Americans. The common-law basis for this is a U.S. Supreme Court decision from 1842 (*Martin v. Waddell*, 41 U.S. 367) known as the Public Trust Doctrine (PTD; Sax 1970, Bean and Rowland 1997, Geist and Organ 2004, Batcheller et al. 2010). The case centered on a landowner's claim to the right to exclude others from harvesting oysters on tidal flats because the property was granted by King Charles II in 1664 to his brother, the Duke of York, and subsequently transferred to the landowner's family. Chief Justice Roger Taney researched what authority the king had to grant such rights, which he ultimately traced back to the Magna Carta (1215 A.D.). Taney concluded that the king held such rights in trust for the nation, and no personal rights for access to fisheries could have been granted after the Magna Carta. Essentially, they "were intended to be a trust for the common use" (*Martin v. Waddell*, 41 U.S. 367). The PTD was extended to all wildlife in a subsequent Supreme Court decision in 1894 (*Geer v. Connecticut*, 161 U.S. 519) stating "the common property in game. . . . a trust for the benefit of all people" (Horner 2000:40).

What relevance does this have to inquiry into human dimensions of wildlife management? The PTD indicates the importance of human dimensions considerations by

- affirming that wildlife is for all citizens of both present and future generations;
- establishing government as the trustee, responsible for oversight on decisions affecting wildlife that are in the best interest of the beneficiaries (i.e., all citizens); and
- making government accountable to citizens for those decisions. (Organ and Batcheller 2009)

Human dimensions inquiry reveals insights about the "public interest" essential for decision making by trustees. In addition, stakeholder engagement by wildlife agencies can help citizens understand the trustees' role and responsibilities and the boundaries and limits that constrain decisions. The essence of the relationship between the PTD and human dimensions of wildlife management is evident in the following aspects of PTD and wildlife management in the United States:

- The PTD was established "by the people" and is the *raison d'etre* for governmental wildlife management agencies.
- Trustees (Government) oversee and manage wildlife "for the people."
- Trustees allow people (the Beneficiaries) latitude for wildlife use within legal limits established "by the people" but are responsible for holding them within those limits.
- The decision space available relative to those limits may be broader than that traditionally assumed or preferred

---

people with various kinds of interests or stakes in wildlife, human–wildlife interactions, and management interventions. Stakeholders are not just groups with special interest in wildlife, nor are they equivalent to lobbyists for these special interests (although lobbyists in various forms may represent some stakeholders for wildlife management). As discussed in Chapter 3, stakeholders may indeed be individuals who are well-organized into formal interest groups, but they also may be individuals joined in ad hoc, situation-specific, grassroots groups or simply individuals who are unaffiliated with (and perhaps even unknown to) one another, yet who have a similar interest or stake in a management issue. Wildlife management informed by knowledge of stakeholder needs and interests contributes to good governance practices.

Stakes (i.e., impacts of interest to some segment of the public) can take the form of recreational, cultural, psychological, social, economic, ecological, or health and safety impacts (Siemer and Decker 2006). Impacts commonly of interest in wildlife management include recreational benefits, aesthetic benefits related to quality of life, economic costs and benefits, or a species' contribution to biological diversity. Any particular

wildlife issue may entail a range of stakeholder concerns and interests, and a variety of these and other factors (e.g., past experience with wildlife managers, perceptions of agency authority) influence stakeholders' expectations of management. Sometimes stakeholders themselves may not even be aware of their stakes in wildlife management decisions (i.e., unaware of the impacts they will experience as a consequence of management), especially if impacts they never previously experienced arise from management actions. In addition, public wildlife managers have an obligation to consider future generations, consistent with the public trust responsibility of government.

Regardless of the degree or formality of organization around stakeholders' interests, wildlife management often occurs because stakeholders express a need for an agency or NGO to influence impacts associated with wildlife. Stakeholders often experience, or anticipate experiencing, undesirable impacts arising from management actions. In some cases, this can produce new stakeholders focused on management actions rather than the outcomes that are the intended purpose of the actions. This frequently leads to resistance and controversy common in response to wildlife management.

by the trustees, but it could also be narrower than that desired by some beneficiaries (e.g., wildlife managers may have preferred outcomes or management options based on their training or values, and some stakeholders may want complete elimination of a species).

- Human dimensions inquiry can provide trustees with knowledge about beneficiaries' attitudes, values, normative beliefs, and acceptability of wildlife species and management options.
- Stakeholder management can facilitate greater beneficiary understanding of roles, responsibilities, and limits, and integrate them into the decision-making process in a manner that fosters greater ownership and durability of those decisions.

Fulfillment of the PTD is a responsibility shared by both government and the people, because both entities have roles in assuring its effectiveness (Smith 1980).

JOHN F. ORGAN

A family poses for a picture during a hike near Flattop Mountain in Chugach State Park, Anchorage, Alaska. In the United States, public lands and the wildlife they support are held in trust for current and future generations. (courtesy USFWS)

A dall sheep ram in Denali National Park, Alaska (courtesy Julie Whittaker)

For example, the decision to create a regulated hunting season for black bears in New Jersey to reduce the incidence of bears breaking into homes was intended to have positive impacts for potential victims of these dangerous encounters. However, the action of hunting (not the intent of reducing house entries) mobilized stakeholders who wanted to keep bears protected from hunting.

Knowledge about stakeholders, how they value wildlife, and how to make decisions about wildlife to increase benefits and reduce costs (increase positive impacts and decrease negative impacts) all require systematic and rigorous science. Wildlife management has evolved in many ways and one of those ways is reliance on social science for reliable information on the human dimensions of wildlife management.

1.2. ESSENTIAL SOCIAL SCIENCES

A variety of social sciences provide useful information for wildlife management decisions, but the disciplines relied upon most frequently are social psychology, sociology, and economics. Knowledge of human values, beliefs, attitudes, preferences,

and expectations about interactions with wildlife or management actions improves the likelihood of achieving and sustaining outcomes or impacts desired by stakeholders. Managers are aided here by social psychology, which is the study of these human traits. Social psychology offers wildlife managers insight about the basis for stakeholders' perceptions of impacts because impacts typically are based on values but expressed in terms of attitudes and preferences. Understanding aspects of beliefs, attitudes, and value orientations described in Chapter 4 contributes to predicting human behavior and can be a basis for common management interventions such as communication or education.

Theories in social psychology, such as the theory of reasoned action–planned behavior, posit that behavior is a function of three influences: people's attitude toward a behavior, assumptions about the likelihood of a behavior happening, and social norms (established standards or patterns taken to be typical in the behavior of a social group). These theories help explain variations in stakeholder support for management actions. For example, they help explain why people may support lethal control of beavers if the animals are perceived to

Hunters and anti-hunters confront one another during a protest of New Jersey's first black bear hunting season in three decades (courtesy Joe Paulin)

carry disease but may not support lethal control if the reason is to prevent beavers from dropping trees in a woodlot or on a golf course. This situation-specific phenomenon has been described as impact dependency (Decker et al. 2006a).

Values, beliefs, and attitudes are not isolated from the influence of the societies in which people live and interact. As described in Chapter 5, sociology is concerned with what people do as members of a group or while interacting with one another. Sociology stresses understanding the social context in which people live and how that context influences their lives. Sociology reveals how individuals are influenced by society or social structure and how individuals, in turn, continually shape their society. Differences in social contexts could be expected between regions of the United States and even within a state (e.g., urban vs. rural communities). The complexity of wildlife management is largely attributable to the fact that the enterprise operates in a social context characterized by diverse values with respect to wildlife management outcomes and methods. Sociology can help managers predict and understand why those values emerge in the context of the society in which they exist and how values play out with respect to wildlife and management.

Economics is the social science that deals with the production, distribution, and consumption of goods and services among competing means to satisfy human wants (Gregory 1987). Economists study the flow of values—usually measured in monetary terms—through society. Chapter 6 describes economics as integral to wildlife management in several important ways. For example, positive and negative impacts from human–wildlife interactions can be conceptualized as benefits

and costs, respectively. Wildlife and management of wildlife have many direct, indirect, and induced economic effects on individuals, communities, states, regions, and even nations. Fortunately, even for attributes of wildlife that cannot be assigned a market value, economists have developed ways of non-monetary and non-consumptive valuation that are applicable to wildlife management.

Economic theory and methods are useful for evaluating proposed management actions such as habitat projects or communication campaigns. Benefit–cost analysis was developed to gauge efficiency of proposed projects by comparing estimated benefits of a project with the costs of carrying it out. Like any assessment method, benefit–cost analysis has limitations (e.g., it assumes that all of the benefits and costs can be predicted and enumerated). In addition, because wildlife management is a continual process of decision making in a world that is constantly in flux, achieving efficiency in the process is not always a socially preferred objective.

1.3. WILDLIFE MANAGEMENT PROCESSES AND DECISIONS

Wildlife management is not a discrete decision or event, nor is management a tidy, linear process that unfolds predictably over time. The process takes place in a management environment that has sociocultural, economic, political, and ecological components (Fig. 1.2). The process is cyclic, iterative, dynamic, and (as described in Chapter 7) it can be messy because uncertainty exists about the coupled sociocultural–ecological system. Unequivocal, single, or easy solutions to value-laden

Wildlife management in a nutshell

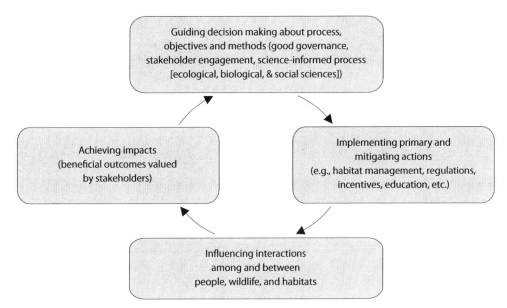

Figure 1.2. Wildlife management process demonstrating elements of definition used in this book (after Riley et al. 2002)

problems common in wildlife management seldom exist (Rittel and Webber 1973).

Wildlife management is a multi-faceted process (with several sub-processes nested within it) that includes developing broad goals and policies as well as setting specific objectives, selecting among actions, implementing those actions as management interventions, evaluating the interventions, and then revisiting the goals and policies. These facets of management call for several kinds of activities: collecting information (research and monitoring); analysis; planning and decision making (including articulating fundamental and enabling objectives and selecting actions); implementing various kinds of actions directed at wildlife, habitat, and people; evaluation; and readjustment. Involving partners (other agencies, NGOs, etc.) and stakeholders in the various facets of management adds to the complexity of wildlife management in practice.

The complexity of issues in wildlife management typically calls for a diversity of expertise. Although some wildlife managers necessarily operate in isolation because of constraints caused by budgets, geography, or other obstacles, much of their work is done in teams. In an ideal situation, teams of people from multiple disciplines (such as ecology, social psychology, sociology, economics, communication, and law enforcement) would be engaged in the work of management along with stakeholders.

This book is built around the premise that the core work of wildlife management is making and implementing decisions. Everything in wildlife management is related to making decisions about desirable conditions and outcomes and to obtaining information needed to take informed management actions. Although much of wildlife management is focused on actions, one should never lose sight that those actions are aimed at implementing decisions, evaluating the outcomes,

and refining knowledge to inform subsequent decisions. In wildlife management, or management for any purpose, decisions set the course of management, which may result in taking specific actions or purposefully not doing so. Effective wildlife managers are skilled at integrating the biological and human dimensions of management decisions. Essentially, to determine which interventions are likely to achieve fundamental objectives, decision processes integrate scientifically derived knowledge as well as experience-based insight about both human and biological factors and processes. If only the biological or the human dimensions are considered, the likelihood that the management decision will be effective (i.e., successful and sustainable) is diminished.

With few exceptions, wildlife management requires making decisions under conditions of uncertainty about both the sociocultural–ecological system in which management takes place and the likely effects of management on the system. In addition, one decision is typically linked to, and affects the need for, other decisions. Decision processes vary with context, necessitating careful selection of a suitable decision process, the specifics of which can differ depending on distinct problems or opportunities. In Chapter 8, a framework for making structured decisions is PrOACT, which requires rigorous description of Problems and opportunities, Objectives, Alternatives, Consequences, and Trade-offs.

Wildlife management has advanced to deal with complexity and uncertainties about various aspects of management systems (climate change, human demographic shifts, etc.) by devising decision processes that encourage learning while doing. Adaptive resource management (Lancia et al. 1996) and adaptive impact management (Riley et al. 2003) are such processes. Adaptive management relies on hypothesis testing by using management as an experimental endeavor, followed by measurement of effects and subsequent adaptation of management actions based on what is learned in the experiment. When the premium is placed on learning, such an approach is called active adaptive management (Meffe et al. 2002). Alternatively, in many situations where a greater premium is placed on outcomes relative to the resource rather than on learning about the system, a passive adaptive-management approach is more suitable (Meffe et al. 2002).

Human dimensions insight can inform decisions in several ways. First, determining the fundamental objectives of management—what management should strive to achieve with respect to desirable outcomes or impacts for people—is largely a human dimensions question. Second, when choosing among alternative management actions, human dimensions insight is valuable to understand what people will find acceptable. Third, if participation by people is necessary (e.g., hunters willing to harvest antlerless deer) to achieve an objective or carry out an action, human dimensions insight can help managers predict whether needed participants will do their part. Most management actions will have consequences that benefit some stakeholders and create losses for others. Such trade-offs are a necessary part of most decisions, and typically trade-offs are basically human dimensions considerations. Integration of adaptive resource management and adaptive impact management (Enck et al. 2006), where fundamental objectives are expressed in terms of impacts instead of the condition of a wildlife population or habitat, encourages stakeholders and decision makers to deal with impacts up front. These innovations in decision making take us a step closer to attaining good governance and are part of the continual effort to achieve more effective wildlife management.

1.4. CRITICAL SKILLS

Wildlife management relies on social science to inform decisions by improving knowledge of

1. how people value wildlife (impacts desired);
2. what benefits various stakeholders desire from wildlife management;
3. the social acceptability of management practices; and
4. how various stakeholders affect, or are affected by, wildlife and wildlife management decisions.

Wildlife managers have been trying to understand better what stakeholders think and do in regard to wildlife and the variety of interventions used in wildlife management since the profession was established. The demand for insights about the human dimensions of wildlife management, however, is increasing rapidly and systematic human dimensions inquiries are struggling to provide managers with the breadth and depth of information they desire. In today's world of increasing stakeholder numbers and complexity, managers are seeking insights into the types of impacts desired, acceptability of management interventions to achieve those impacts, and how well interventions work. Additional context-specific studies

are frequently required to gain reliable human dimensions information that informs decisions.

Some wildlife professionals (most of whom have formal education in ecological sciences rather than social sciences) are not comfortable planning human dimension inquiry or interpreting and putting into practice the results; yet human dimensions are much like ecological studies in many respects. The correct questions need to be asked, and the correct methods must be applied, to yield accurate findings. The same scientific rigor that is applied to study the effects of fire on Kirtland's warbler habitat should be applied to the study of stakeholder values derived from activities and uses associated with warblers. Ignoring accepted scientific procedures in social science will have the same pitfalls that result from ignoring conventions and protocols of biological study.

Planning and designing a study of the human dimensions of wildlife management involves many considerations for the wildlife manager. Chapter 9 reviews these considerations in detail and suggests steps that a manager can take to ensure that productive information is gathered to inform management deliberations. The strength of the conceptualization, planning, and implementation of any study is essentially shaped by pairing the appropriate social-science methodology with the study objectives, which should reflect the information needs of managers. Social scientists apply an extensive array of methods ranging from qualitative observations of wildlife recreationists' behavior (e.g., systematically observing birdwatchers to estimate disturbance to wintering waterfowl [Pease et al. 2005]) to quantitative methods, such as those used to estimate the net economic value of a wildlife recreation activity (e.g., economic value of birdwatching [Hvenegaard et al. 1989]). An increased reliance on scientifically obtained information about the human dimensions has stimulated improved application of social science methodologies in wildlife management. The material covered in Chapter 10 describes methodologies commonly used in human dimensions inquiry and encourages you to evaluate the strengths and weaknesses of various social-science methods applied in the context of wildlife management. Numerous valuable texts exist (Creswell 2009, Vaske 2008, Wholey et al. 2010, and others) that provide detailed, step-by-step instructions for conducting rigorous social science.

Human dimensions researchers commonly use three basic approaches: qualitative, quantitative, or mixed. These three approaches have a variety of connotations and applications in social sciences. For purposes of this book, a qualitative approach is inductive, interpretative, or open-ended. The normal purpose is to gain in-depth insights about the complexity of stakeholders' motivations, attitudes, and behaviors, rather than an ability to make general inferences across a population (e.g., *X*% of birdwatchers in the state spends more than *Y* days afield annually). The most common methods include behavioral observations, open-ended or unstructured interviews, focus groups, and content analysis of documents such as media or administrative records.

Quantitative methods produce data in the form of num-

A facilitator leads a citizen advisory group providing input to a wildlife management plan for Drummond Island, Michigan. Wildlife agencies have increased their efforts to engage stakeholders in management decisions over the past 20 years. (Courtesy Shawn Riley)

bers that are usually aimed at drawing inferences at broader scales than qualitative inquiry. Commonly applied methods include structured interviews, close-ended questionnaires (telephone, mail, and Internet), evaluations of demographic data, and quantitative measurements of human behavior (e.g., time budgets). Both qualitative and quantitative methods have strengths and weaknesses depending on the situation. Our experience in more than 100 human dimensions studies is that the complementary use of qualitative and quantitative methods—a mixed-methods approach—yields the strongest and most comprehensive inferences about stakeholders. The degree to which qualitative or quantitative methods are emphasized should be a function of the types of questions asked early in the planning stages of any inquiry. To achieve informed decision making, engagement of stakeholders relative to the decisions is often more important than scientific inquiry.

The trend during the past 20 years has been to actively seek input from stakeholders and involve them in more aspects of management (Decker and Chase 1997; Chase et al. 2000, 2002).

Stakeholder perceptions of transparency and fairness in decision making that are gained through participatory processes and reinforced by accountability increase satisfaction and trust in management (Lauber and Knuth 1997). Stakeholder engagement occasionally involves systematic inquiry or may be as informal as telephone conversations between stakeholders and managers. Chapter 11 provides a review of various forms of stakeholder engagement as a basis for promoting good governance and creating more sustainable decisions.

Stakeholder involvement may occur with individuals through NGOs representing member interests or through local grassroots groups who interact with local, state, and federal levels of government. Increasingly, wildlife management is a shared responsibility where good wildlife management is not simply exercising authority over, steadfastly retaining control of, or even taking sole responsibility for, wildlife resources. Good governance instead occurs through management that wisely shares responsibility for wildlife with partners and other stakeholders. Insight about stakeholders' beliefs and attitudes, patterns of behavior, and expectations of management with respect to wildlife are needed for success in all of the approaches identified. With some complex issues, stakeholder engagement approaches include formal social science inquiry. In almost every situation, however, skilled communication improves the quality of engagement.

Wildlife agencies continually identify communication skills or "people skills" as those most desired when hiring wildlife professionals. Agencies need wildlife professionals who can effectively develop and deliver information that improves stakeholder knowledge of an issue and therefore leads to their informed participation in management decision making. Effective communication skills, which are vital to establish public trust in wildlife professionals, are characterized by the use of language and channels of communication that stakeholders can access and understand. Communication skills also are required for information exchange among disciplines important to wildlife management.

Although communication—intentional or unintentional—is usually occurring, skills are most needed in purposeful communication directed to achieve increased understanding of, and by, stakeholders in wildlife management. The purpose may be to facilitate stakeholder involvement in decisions, inform or educate stakeholders, market an organization, or build public relations. The list of purposes for communication is wide-ranging and extensive; yet regardless of purpose, communication is affected by the use of language and channels of communication that receivers of the communication can recognize and access (see Chapter 12). Wildlife professionals today must be prepared for a different context and rate of communication than their predecessors; communication systems are now swifter, more highly connected, and rapidly changing.

1.5. APPLYING HUMAN DIMENSIONS INSIGHTS TO WILDLIFE ISSUES

The spectrum of issues facing wildlife managers around the world is seemingly endless; however, these issues can be classified into topic areas to make it more apparent how human dimensions insights can be applied. Wildlife management issues can be thought of as falling into three broad (although oversimplified) areas: management of abundant wildlife (Chapter 13), management of scarce wildlife (Chapter 14), and management of wildlife uses and users (Chapter 15). These are discussed briefly in the following paragraphs.

Management effort seems to have a bi-polar quality. Much attention is given to managing abundant wildlife on the one hand and scarce wildlife on the other. The profession grew because of concerns about wildlife scarcity, but success in restoring wildlife coupled with myriad land-use changes across North America has resulted in some species becoming abundant or even overabundant. Chapter 13 discusses differences between abundance and overabundance, emphasizing that this distinction is largely a human construct based on the social carrying capacity for human–wildlife interaction impacts experienced by stakeholders. Approaches to managing impacts of abundant wildlife range from specialized hunts intended to decrease population sizes to compensation programs that ease economic losses from wildlife. Determining thresholds of tolerance for negative impacts and identifying mitigation efforts acceptable to stakeholders are topics for human dimensions inquiry and stakeholder engagement. Abundant wildlife is one of the major challenges of modern wildlife management, especially in urban and suburban areas where human health and safety are of concern because of motor vehicle accidents (e.g., collisions with deer and elk), attacks on humans (e.g., by mountain lions, black bears, and coyotes) and zoonotic diseases (e.g., West Nile, Lyme, rabies).

Scarce wildlife is defined in Chapter 14 as including two different kinds of wildlife: species that naturally occur in low numbers as a consequence of their life history, and species that are rare because of anthropogenic effects on their populations (direct) or on their habitats (indirect). Restoration and protection of wildlife species adversely affected by humans is pursued in various ways, but human dimensions considerations abound. These include influencing public compliance with regulations that prohibit harming scarce wildlife and causing loss or degradation of their habitats. Understanding community and individual responses to proposals aimed at restoring a species within its historical range is a core component in assessing the feasibility of such efforts. Evaluation of restoration proposals is informed by human dimensions research at multiple levels (from community capacity to individual stakeholder reactions) and often includes some stakeholders who are crucial to restoration, such as private landowners (who may own the habitat required for restoration).

Wildlife uses and users have been important subjects of human dimensions inquiry for more than 40 years. Studies of wildlife-associated activity inventory and characterize wildlife users, monitor their use, and describe trends in wildlife recreation participation. Since 1955, the U.S. Fish and Wildlife Service (USFWS) has tracked national trends in wildlife recreation by surveying American households every 5 years. The trend data collected include hunter numbers, types of hunting, hunter days spent afield, and hunters' expenditures. Collection of additional information on non-consumptive activities (e.g., wildlife observation, photography) began in 1980. Wildlife agencies routinely survey hunters at state and sub-state levels to gather harvest data used to help estimate population abundance and trends of game species.

Chapter 15 describes how studies of hunters initially documented the extent and nature of hunting activity and later revealed important insights about the motivations of hunters and the dynamics of hunting participation, including the social factors that influence recruitment and retention of hunters. This body of inquiry became especially important as the numbers of hunters gradually declined while the need to maintain or increase harvest of species such as deer, elk, and Canada geese grew. In the 1980s, interest emerged in documenting non-consumptive wildlife-dependent or wildlife-associated activities such as birdwatching and other wildlife viewing. Human dimensions studies were conducted to inform wildlife education and viewing programs that grew across the United States since that time. Wildlife observation near people's homes or nearby green spaces accounts for much of the time spent watching wildlife, though trips for the purpose of wildlife viewing are common. The USFWS surveys reveal a demand for wildlife viewing that greatly exceeds that for hunting and trapping. Studies have provided insight about people who might value wildlife programs that facilitate wildlife viewing (e.g., Brown et al. 1979, Witter and Shaw 1979) and perhaps become potential donors to state and private wildlife programs (e.g., Brown et al. 1979, Witter et al. 1981, Manfredo 1988).

1.6. PROFESSIONAL AND PRACTICAL CONSIDERATIONS

One lesson that is imperative for wildlife professionals everywhere is that all the sociological and ecological science in the world does not convey to a wildlife manager what *should* be done in a given situation. Science informs managers of what can be done, what is desired by stakeholders, or what may happen with and without particular interventions. The question of what *should* be done, however, requires ethical considerations (Chapter 16). Ethics is a branch of philosophy dealing with morals of human conduct (the rightness and wrongness of certain actions). Human dimensions insights help wildlife professionals consider ethical dilemmas by clarifying pertinent values in a wildlife management issue, but it is a different process to determine whose values (e.g., values of all voters, certain stakeholders, or managers) should form the basis for management and who should ultimately judge the adequacy and acceptability of alternative management actions. That is, what the right thing is to do.

Russell Ackoff (2001:345), a preeminent scholar of management, offered the following thoughts on doing "the right thing":

> Paraphrasing Peter Drucker, there is a big difference between doing things right and doing the right thing. Efficiency is concerned with doing things right; effectiveness with doing the right thing.
>
> The righter one does the wrong thing, the wronger one becomes. If one corrects an error in the pursuit of the wrong thing one becomes wronger. Therefore, it is better to do the right thing wrong than the wrong thing right because it offers the possibility of improvement.

Much training and education in wildlife management is about how to do things "right" or "correctly" (e.g., how to estimate a wildlife population, evaluate condition of a habitat, or accurately reveal stakeholder attitudes). These are important skills to hone. In some cases, however, focusing on these skills may divert attention from other, more important management elements. Coming to a good decision about what should be done is challenging, but ethical considerations help provide a way of defining and deliberating about the issues.

Ethical discourse, however, is no cure-all in messy wildlife problems facing managers; that is, ethical discourse does not necessarily resolve value-laden problems for which multiple perspectives exist about what constitutes the right objectives and desired outcomes of management. Wildlife professionals are pulled in different directions by stakeholders with competing values. Aiding societal decisions about what to do and how best to do it in the public interest gives a vital purpose to the human dimensions of wildlife management.

A trait of a vital profession is the need for its members to engage in continual learning; this need is ubiquitous in the changing world of wildlife management. Vibrant research in all dimensions of wildlife management continually generates new knowledge, which requires career-long learning by people in the profession. This learning can occur in a variety of ways and contexts, many of which are described in Chapter 17. With respect to human dimensions, the education of future professionals is facilitated by pursuit of three fundamental objectives:

1. Understand factors affecting stakeholder values, beliefs, attitudes, decisions, and actions.
2. Understand organizations, communities, and institutions.
3. Improve skills needed to work effectively with people, be they co-workers, partners, or stakeholders.

Human dimensions educational needs simply cannot comprehensively be covered in the process of earning a typical undergraduate or graduate degree, so a motivated wildlife manager will look for extramural courses and in-service training opportunities that, over time, contribute important human-dimensions knowledge and skills to his or her professional capabilities. The chapters that follow in this book are an attempt to provide exposure to the three topics above and whet your desire for continual learning about the human dimensions of wildlife management.

SUMMARY

Wildlife management can be thought of as consisting of three specialized dimensions: wildlife, habitats, and humans. Everything that is not about wildlife and habitat is necessarily about the human dimensions of wildlife management. This first chapter introduced core topics, principles, concepts, and applications of human dimensions that are covered in greater depth throughout the book.

- Public wildlife management occurs throughout most of the world within a framework of governance, which is informed by human dimensions insights. Trends are toward collaborative governance to meet the challenges in wildlife management created from globalization and increased rates of change in the physical and sociocultural landscapes.
- Wildlife management is defined as the guidance of decision-making processes and implementation of practices to influence interactions among people, and between people, wildlife, and wildlife habitats, to achieve impacts valued by stakeholders.
- Impacts are the important effects on human values created by human–wildlife interactions. Impacts may be positive or negative and take a variety of forms. In addition to being produced by human–wildlife interactions, impacts arise from interactions among humans where the referent is wildlife.
- A stakeholder is any person who is significantly affected by, or significantly affects, wildlife or wildlife management decisions or actions. Stakeholders are people with various interests or stakes in wildlife, human–wildlife interactions, *and* management interventions. Sometimes stakeholders themselves may not even recognize their stakes, especially if stakes arise from management actions they never previously experienced.
- The social sciences, including social psychology, sociology, and economics, contribute theory and practical knowledge about how stakeholders value wildlife, the impacts stakeholders seek from wildlife, and how the trade-offs of management actions will be evaluated.
- Wildlife management is a system of processes rather than a discrete event. These processes are iterative and cyclical through time.
- A core task of management is to make decisions, which can be improved by following a structured process. Adaptive management is one such process developed to reduce the uncertainties associated with wildlife and management environments in which it occurs.
- The processes of wildlife management are improved through systematic inquiry and stakeholder engagement.
- Communication is one of the most important skills of a wildlife professional, and it is improved through consider-

ation of sources, messages, channels, and receivers of the communication.
- Science can inform decisions about what can be done, yet no amount of science can tell managers what *should* be done.
- Although not a cure-all for every dilemma a manager might face, ethical discourse can help provide a way of defining and deliberating about the messy issues in wildlife management.
- The only way for wildlife professionals to stay relevant in a fast-changing world is through continual learning.

Suggested Readings

Decker, D. J., T. L. Brown, and W. F. Siemer. 2001. Human dimensions of wildlife management in North America. The Wildlife Society, Bethesda, Maryland, USA.

Organ, J. F., D. J. Decker, L. H. Carpenter, W. F. Siemer, and S. J. Riley. 2006. Thinking like a manager: reflections on wildlife management. Wildlife Management Institute, Washington, D.C., USA.

Riley, S. J., D. J. Decker, L. H. Carpenter, J. F. Organ, W. F. Siemer, G. F. Mattfeld, and G. Parsons. 2002. The essence of wildlife management. Wildlife Society Bulletin 30:585–593.

GOVERNANCE OF WILDLIFE RESOURCES

BRENT A. RUDOLPH, MICHAEL G. SCHECHTER,
AND SHAWN J. RILEY

Wildlife management occurs in the context of the evolution of governance structures, which are affected by worldwide trends of democratization and globalization. Factors associated with these evolving structures are in many instances the same trends that change the nature of some of the most critical threats to natural resources. Insights needed to pursue more effective governance give purpose to the study and practice of concepts, principles, and skills conveyed in this book. This chapter provides a foundation for understanding governance by clarifying concepts and then using empirical cases to demonstrate the strengths and limitations of contemporary governance structures in which wildlife management is conducted. Despite the numerous challenges that exist when taking a collaborative approach, we suggest that collaborative governance holds promise to transform practices of wildlife management into a more effective and equitable process.

In the simplest sense, governance refers to "the totality of instruments and mechanisms available to collectively steer a society" (Bäckstrand and Jamil Khan 2010:8). That is, it is the means whereby governments, organizations, or individuals direct their programmatic activities, including the processes, laws, rules, and policies that collectively guide actions. Thus, as a way to distinguish governance from governments, governance can be thought of as what governments do. However, governance is increasingly something for which not only governments are responsible. Authority and resources are more commonly being shared through agreements among multiple governments (such as through treaties or formation of intergovernmental organizations; IGOs), coordination and collaboration with entities outside of government (such as non-governmental organizations [NGOs], universities, and industry groups), and greater interaction with and responsiveness to the public. As a consequence, a distinction may be drawn between *top-down* governance as the exercise of formal sovereign authority in an effort to regulate actions and more recent approaches to collaborative governance, which increasingly include networks of public–private partnerships, collaboration with community organizations, and other forms

of interaction with civil society (Box 2.1). On the international level, *regional* (e.g., European Union [EU]) and *global* governance has grown, involving multiple state and non-state actors, especially IGOs and NGOs.

Privatization of the former functions of the state through formation of various public–private partnerships at the national level and relinquishing some traditionally state functions to the international level (as with the EU) is seen as a way to overcome problems of "institutional arthritis resulting from progressive bureaucratization, capture of public agencies by special interests, [and] the impact of repressive measures carried out in the name of the state" (Young 1994:13). The World Bank, for example, called for "good governance," which many took to be another term for democracy (or at least pluralism) and thus an embrace of management processes in which civil society was increasingly important. Such advocates, perhaps most notably the prestigious Commission on Global Governance (Commission on Global Governance 1995), understand governance as a normative concept or one that values empowerment of a state's residents. They consider governance a better way to govern than state dominance of decision making. Others simply describe and explain decision-making trends, which increasingly involve civil society and interactions among a number of governments. This has been referred to as empirical governance as contrasted to normative governance (Bäckstrand and Jamil Khan 2010).

Adaptation to new processes of governance, and the time and effort demanded for coordination and collaboration among an expanded set of actors invested in these processes, creates some short-term costs; yet if this approach contributes to increased legitimacy and sustainability of decisions made to affect wildlife, then good governance advocates believe that management costs will be reduced and regulatory compliance will be increased over the long term. This new way of thinking about wildlife management—as an act of governance—is embraced by some managers and resisted by others. In some cases, new approaches are thrust upon agencies from outside forces or higher levels of government. In still other

Box 2.1 *DECENTRALIZED GOVERNANCE*

Co-management, a widely applied governance model, is defined as "a situation in which two or more social actors negotiate, define, and guarantee amongst themselves a fair sharing of the management functions, entitlements, and responsibilities for a given territory, area, or set of natural resources" (Borrini-Feyerabend et al. 2000:1). Co-management forms of governance cut across conventional state, market, and/or community boundaries and attempt to maximize the benefits and minimize the weaknesses of state-based, market-based, and rules-based approaches. Co-management has been implemented in many developing countries that do not have the institutional capacity to manage wildlife resources but where non-governmental organizations (often international non-governmental organizations) have interest in encouraging conservation and are willing to invest money and expertise. Although attractive for many reasons, co-management governance arrangements are difficult to design and even more difficult to implement given real-world complexities, particularly in developing

countries where rural people are dependent on natural resources and central government institutions are weak.

For example, the Ankeniheny–Zahamena Corridor (CAZ) is one of the largest blocks of rainforest in Madagascar that was recently granted protected-area status with a co-management governance arrangement involving multiple stakeholders at various levels. The governance structure of CAZ was designed and put in place after a series of meetings with stakeholders at multiple levels, including local communities, the Ministry of Environment and Forests (MEF), and various partners.

CAZ's co-management structure is divided into two main parts, a strategic orientation component and a management component (see figure in Box 2.1). The MEF delegates responsibility to the Protected Area Manager (an organization), which (along with the Orientation and Monitoring Committee) serves to define strategic priorities for management. The CAZ Manager and its staff, including six sector managers and the Local Management Unit

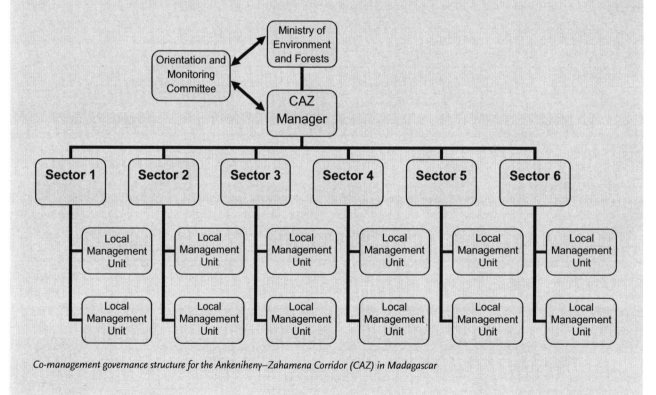

Co-management governance structure for the Ankeniheny–Zahamena Corridor (CAZ) in Madagascar

cases, changes simply may represent organizational evolution to survive the joint pressures of public scrutiny and expectations when economic recessions cause financial and personnel resources to dwindle. Improved governance provides opportunities for enhanced effectiveness of wildlife managers in achieving impacts desired by society and thereby enhancing relevance of wildlife and the wildlife profession to society.

2.1. INTENTIONAL COORDINATION AND COLLABORATION TO ADDRESS TRANSBOUNDARY AND TRANSAUTHORITY CONCERNS

A framework of governance that intentionally addresses transboundary and transauthority issues is emerging through the EU and its member countries. The proximity of numerous

Rainforest habitat in the Ankeniheny–Zahamena Corridor (CAZ), Madagascar (courtesy Conservation International)

(LMU) managers, ensure daily management functions and implementation.

Despite best efforts to communicate and share information widely and often, the geography and infrastructure at CAZ can be a barrier to information flow that co-management requires. Although actors come together regularly to make decisions, LMUs are widely dispersed over a vast landscape with poor roads, rough terrain, and no cellular telephone service. This significantly hampers communication, particularly about day-to-day events.

Efforts to enforce the rules are conditional on local realities as well. The Forest Service (in MEF) lacks adequate resources to patrol regularly or to respond rapidly. Locally led forest patrols are more effective, but LMU staff may not always have the level of authority needed to take action against infractions.

Although many responsibilities have been devolved to locally based components of the CAZ governance structure, ultimate decision-making authority rests with the government (i.e., MEF). Without more inquiry into the role of the consultative and deliberative groups, their influence, and the decision-making process that is used, it is difficult to assess the extent to which CAZ's structure responds to this element of good governance. In addition, the presence of an external third party, Conservation International (CI), necessarily affects the power dynamics among actors. CI has access to information and funding and therefore wields a certain amount of power while acting as an advisor to the protected-area management structure. The power dynamics among actors should be studied more rigorously before a conclusion can be drawn regarding how authority has been transferred at CAZ.

DANIELA B. RAIK

EU member countries and their extensive common borders result in sharing of many wildlife populations and environmental impacts in addition to the long history of movement of people across borders and interconnection of financial and economic interests. More than 300 directives, regulations, and other legal acts relating to the environment, many of which directly or indirectly address wildlife management, have consequently been enacted to coordinate approaches throughout the EU. Globalization trends toward substantial interconnectedness of the global economy, expansion of worldwide transportation networks, and increasing domination of ecosystems by humans have brought actions that were previously within the purview of individual national and local authorities under the shared responsibility of multiple sovereign states and agencies. Implementation of international conservation efforts and coordination among agencies with various jurisdictions and mandates has therefore increased throughout all regions of the world.

Comparatively larger sovereign territories and minimal sharing of borders within North America resulted in fewer recognized needs for international coordination to conserve wildlife. Some of the earliest efforts to coordinate international conservation efforts in North America included the 1916 Convention between the United States and Great Britain (on behalf of Canada) for the protection of migratory birds. This convention was followed by the 1936 Treaty for the Protection of Migratory Birds and Game Mammals between Mexico and the United States. These transboundary acts of governance were implemented to coordinate activities among states and provinces that collectively harbor waterfowl populations during various stages of migration. Canada, Mexico, and the United States later became parties to a number of treaties and conventions that guide multi-lateral cooperation related to the conservation of wildlife species and habitats. Examples of agreements to which all North American governments became parties include the 1940 Convention on Nature Protection and Wildlife Preservation in the Western Hemisphere, the 1971 Convention of Wetlands of International Importance Especially as Waterfowl Habitat, the 1973 Convention on International Trade of Endangered Species of Wild Fauna and Flora (CITES), the 1992 Convention on Biological Diversity, and the 1993 North American Agreement on Environmental Cooperation.

In addition to individually becoming party to each of these international agreements, the three governments established the Canada–Mexico–U.S. Trilateral Committee for Wildlife and Ecosystem Conservation and Management in 1995. The Trilateral Committee attempts to improve upon conservation efforts by establishing working groups that exchange information and coordinate activities among the countries rather than employing strict regulations or laws. This arrangement facilitates sharing of information and expertise that may aid countries individually as well as support stronger overall positions and participation in multi-lateral agreements. For example, these efforts may aid in preparation for CITES meetings, support development of jointly drafted positions, or support proposals. This sort of governance framework facilitates a continental approach to wildlife management and a coordinated approach to addressing increasing global threats and guiding actions under multi-lateral agreements.

Even at the national and sub-national levels, effectiveness of efforts to address challenges in wildlife management is hindered when there is insufficient coordination among agencies with different responsibilities, policy mandates, and geographic areas of authority. Coordination across states is increasingly recognized as a necessity given the unprecedented, landscape-scale threats facing our natural environment. The scale of contemporary conservation problems necessitates

that the governance approach must expand partnerships and adapt to rapidly changing conditions to achieve conservation outcomes. One innovative approach to strengthen scientific capacity and, therefore, conservation design, delivery, and evaluation is the Landscape Conservation Cooperatives (LCCs) effort. LCCs are a network of partnerships spanning the United States and portions of Canada and Mexico (Box 2.2). The intended purpose of LCCs is to increase efficiency and effectiveness among cooperating entities by sharing information, agreeing on conservation goals, identifying information gaps and research needs, and leveraging funding and other resources to improve coordination of conservation planning and delivery. The key to success of an LCC is overcoming institutional barriers (e.g., norms regarding geographic and other jurisdiction issues and data ownership) that have developed under previous governance regimes and to facilitate truly collaborative relationships.

Collaborative governance also is being used to address coordination across jurisdictions within states, as demonstrated in the way agencies are responding to wildlife diseases, which are a relatively new but significant management issue. For example, the first epidemic occurrence of bovine tuberculosis (TB) in free-ranging deer in North America was identified in 1994 within a population of white-tailed deer in northeastern Michigan. TB infections in free-ranging wildlife threaten public health and produce substantial economic impacts to the livestock industry through the testing requirements and elevated biosecurity measures mandated by state and federal agriculture departments. Assessment of, and response to, this disease involved efforts of scientists, managers, enforcement agents, and public information officers among local, state, and national natural resource, agriculture, and public health agencies. Laws, policies, and standard practices within each agency established the diverse authorities and obligations with respect to disease management. Achieving sufficient coordination among the agencies largely required establishment of new channels of communication and collaboration, such as formation of interagency working groups. Coordination was vital to ensure policies and practices were complementary in their efforts to benefit humans and wildlife while attempting to eradicate the disease. This mutual participation in combating a disease in wildlife demonstrates the types of coordination and collaboration occurring within a developing network of multi-level governance.

2.2. COLLABORATION TO POOL RESOURCES, EXPERTISE, AND INFLUENCE

Funding and other forms of support from private hunting and conservation organizations have played a role in supplementing agency resources for wildlife research and population management in both Europe and North America (Dwyer and Hodge 1996, Parkes and Thornley 2000, Williamson et al. 2001, Oldfield et al. 2003). These traditional partners and a number of other privately funded NGOs may enhance the capacity for conservation to occur where public resources are lacking or in

Box 2.2 *LANDSCAPE CONSERVATION COOPERATIVES*
An Example of Emerging Collaborative Governance Arrangements

In response to climate change, habitat fragmentation, and other large-scale changes, the Department of Interior initiated Landscape Conservation Cooperatives (LCC) and Climate Science Centers (CSC) to facilitate a collaborative and coordinated conservation approach. These new multi-disciplinary and multi-agency cooperatives represent emerging forms of governance to address complex environmental issues. A motivation for establishing such new governance arrangements was to knock down legal, cultural, and philosophical "walls" that stymied collaboration. Individual agencies, organizations, programs, funding, parks, refuges, sanctuaries, and interests in single species created a system where conservation seldom occurred with a broader and more comprehensive view.

LCCs are a nationwide network of landscape-scale conservation partnerships (there were 21 in 2011) that collaboratively develop conservation goals, identify information gaps, and fund research for use by decision makers in natural resources management. CSCs (eight planned) are partnership-based entities that will provide scientific tools and knowledge that assist managers in anticipating, monitoring, and adapting to impacts created by climate and other large-scale changes in the environment. Science needs identified by the LCCs may be addressed by the CSCs or directly through the LCC partnership. This innovative approach represents a paradigm shift in that it provides structure and incentives to improve effectiveness and efficiency in the way that the conservation community responds to a rapidly changing environment. With funding to support dedicated staff and provide seed-money for science projects, LCCs set a stage on which to reform a vision of conservation that meets shared conservation goals across large geographic areas.

The LCCs are designed to be self-directed partnerships governed by a Steering Committee, with full-time LCC Coordinators and Science Coordinators who also serve on National LCC Network teams. This double-layered approach provides opportunities for each LCC to identify the unique suite of priorities necessary to aid conservation activities in their respective regions and helps ensure relevance at national scales. The Steering Committee members represent federal, state, and tribal governments, and others who have an interest in conservation throughout the LCC. The Steering Committees decide the focus of the LCC (e.g., determine how the LCC can best serve the conservation community within the LCC) based on recommendations and input from a much broader suite of partners. These partners, in turn, may serve on technical teams or simply review and comment on proposed actions. Steering Committee members are decision makers within their respective agencies and organizations. This level of participation is critical to the success of the LCCs and should lead to greater collaboration between individual agencies and an increased focus of funds from multiple sources on high-priority issues and projects. The aim of this new institutional arrangement is identification of shared interests and conservation goals and greater synchronization of actions by individual conservation organizations. The hope is that this arrangement builds capacity for effective and responsive conservation.

CYNTHIA JACOBSON AND KAREN MURPHY

---

decline, such as where license fee revenue diminishes because numbers of hunters decrease (Wright et al. 2001, Jacobson and Decker 2006). In some instances, the enhanced role of NGOs in wildlife management has been due to outsourcing of government functions driven by intentional political-reform efforts to downsize the civil service (Peters 2000). Regardless of the primary cause, the growing number and complexity of threats to wildlife conservation have led to greater demand for the resources, management capacity, and influences NGOs may bring to conservation.

NGOs also may help overcome the lack of adequate training within government agencies, provide additional expertise, or contribute influence of their staff and membership to affect issues demanding elevated commitment or attention. A growing number of international agreements accordingly create roles for IGO and NGO assistance in collection, monitoring, and analysis of data. A significant role for many IGOs

established through international agreements is to provide the administrative functions necessary to coordinate communications and sharing of information among governments. Many of the international wildlife treaties were drafted by staff of IGOs or NGOs. The International Union for Conservation of Nature and Natural Resources, for example, played a major role in the formation of CITES, which was created to ensure that international trade in wild plants and animals does not threaten their survival. NGOs have played an increasingly prominent role in CITES with their attendance as observers to meetings and conferences. In some instances, personnel from NGOs have outnumbered participating governing organizations (Brown Weiss 2000).

Some comparative analyses suggest that among efforts addressing international environmental protection the "one key variable accounting for policy change . . . is the degree of domestic environmentalist pressure in major industrialized

Disease surveillance programs for tuberculosis or chronic wasting disease in deer require a coordinated, resource-intensive, interagency response. These photos show a cadre of professionals from multiple agencies in a daily debriefing during immediate response to discovery of CWD in deer in New York State in 2005. The emergency disease-surveillance program required lethal removal of hundreds of deer from private property in two counties. (photos courtesy NYSDEC/Jim Clayton)

democracies" (Keohane et al. 1993:14). IGOs and NGOs can set directions for, or help bring about, broad calls for conservation initiatives. They can also help increase public acceptance of conservation efforts. Their staff and other resources can be applied to campaigns to convince the public of the importance of conservation issues and build support for the means of intervening. These campaigns may influence the willingness of countries to become participating parties to agreements, shape the structure of agreements, or lead to crafting legislation to ensure meaningful participation of countries in implementing conservation initiatives. Such efforts are at least partially responsible for CITES being a far-reaching agreement among more than 170 countries that has been generally effective at preventing species extinction. New and elevated levels of threats to species' survival are emerging, however, as organized crime networks capitalize on economic globalization to gain easier access to illegal markets for trade in live wildlife and wildlife parts. The future conservation of wildlife vulnerable to trade will be even more dependent upon substantial efforts to coordinate activities among nations and agencies with diverse responsibilities, mandates, and cultural outlooks on the value of, and uses for, wildlife. That is, the future will depend on more effective collaborative governance.

2.3. ENGAGING PLURALISM AND EXPECTATIONS FOR RESPONSIVENESS

Another characteristic of collaborative governance is greater interaction with, and responsiveness to, less traditional stakeholders. Some of these adaptations are in response to overall increases in the general mistrust of government (Nye et al. 1997) and legal systems (Tyler 1998), which has led to greater public desire for direct involvement in decision making. In many settings, legal authority and responsibility for wildlife

management largely has been situated in government agencies (bureaucracies) and primarily regulatory in function (Jacobson and Decker 2006). Although the term "bureaucracy" has taken on a negative connotation in some political circles, bureaucracies originally were designed to be efficient, hierarchical organizations employing highly trained people who assumed responsibility for decision making. In the case of wildlife management, this meant that much of management authority came to reside with people trained primarily in wildlife ecology and the application of ecology in management.

Because the numbers and types of stakeholders active in wildlife management issues initially were few and shared similar backgrounds and values to those of managers, the empowerment of these agencies with decision-making authority was generally an acceptable and effective arrangement. Through time, however, stakeholders desiring engagement in wildlife management decisions increased in number and variety. Some of these stakeholders came to believe traditional approaches are a barrier to broad public engagement in wildlife

Multi-organizational collaborations, like the Saola Working Group shown here, are a form of governance that play a critical role in efforts to save endangered wildlife. The Saola Working Group is a coalition of conservation NGOs, universities, and government agencies in Laos and Vietnam. (courtesy Saola Working Group)

management. To fulfill their obligations within representative forms of government, state and federal agencies are moving toward new stakeholder approaches to ensure management attention is directed toward those issues relevant to the growing and diversified wildlife stakeholder base.

This shift may offset one consequence of the tension between previously dominant paradigms of bureaucratic management and growing public expectations for direct involvement—the increasing use of direct democracy (ballot initiatives and popular referenda) to shape public wildlife policy (Minnis 1998, Williamson 1998). The divisive nature of such campaigns and their preclusion of deliberation and compromise led to calls for changes to agency structures, decision-making processes, and funding models to further expand public engagement (Jacobson and Decker 2006). Carefully constructing opportunities for the public to be engaged in decision making may therefore aid in meeting societal expectations, as well as enhance effectiveness if stakeholders increasingly accept that rule-making and enforcement are being carried out by people and groups who can be trusted with the authority to do so. Detailed definitions and methods related to stakeholders and stakeholder engagement are provided in Chapters 3 and 11.

Regulations are pervasive throughout the world as an instrument to achieve objectives in wildlife management. For example, there are regulations on wetland development that aid in protecting wildlife habitat; regulations on hunting and take of game; and regulations aimed at changing human behavior, such as bans on transport of invasive species or feeding of wildlife. The list of regulations is long and comprehensive; the ability to achieve objectives in wildlife management is therefore dependent in great part on compliance with regulations. Regulatory compliance may be described by instrumental or normative models of behavior. Instrumental models focus on individual decisions that are made to balance the illegal gains of non-compliance with the perceived risk of being caught and punished, whereas normative models focus on an individual's

sense of internal duty as it arises from moral obligation, social norms, or an evaluation of the legitimacy of authorities (Meares 2000, Winter and May 2001, Blader and Tyler 2003b).

The influence of legitimacy is strongly determined by procedural justice perceived through views and opinions of fairness in decision-making processes and treatment of individuals by authorities. Instrumental approaches to protecting natural resources through a basic commitment to regulation and enforcement are important, but the effectiveness of this deterrence depends upon investment of resources to maintain or increase detection of violations. Such investments can be especially costly in wildlife management (for example, where hunters are engaged in activities in remote settings and often camouflaged or otherwise concealed to escape detection by their quarry). Factors contributing to procedural justice—participation, neutrality, trustworthiness of authorities, and treatment with dignity and respect (Tyler 2000)—are largely under the control of regulatory agencies. Furthermore, this perception of procedural justice also affects broad support for agencies and programs, which is needed to conduct management at any level (Blader and Tyler 2003b). Thus, meeting public expectations for increased engagement in decision making leads governance approaches toward a reduced emphasis on enforced compliance and increased attention on gaining voluntary compliance and broader cooperation (Pierre 2000). These changes come through increasing transparency of decisions via public participation, building trust among participants in decisions, and creating legitimacy for subsequent management interventions.

Returning to the case study on the self-sustained infection of TB in free-ranging deer in northeastern Michigan, the collective body of research and epidemiological evidence provided a substantial biological basis for pursuing eradication. The Michigan Department of Natural Resources implemented liberalized harvest regulations in an effort to substantially reduce deer densities and adopted feeding and baiting restrictions to reduce aggregation of deer. Management strategies apparently constrained the infection but did not eradicate the disease or totally avoid economic impacts on Michigan's livestock industry (O'Brien et al. 2006, Rudolph et al. 2006). Stakeholder opposition to strategies and regulations for reduction of deer populations and observations of continued illegal use of bait suggested that low stakeholder compliance with these strategies was limiting their effectiveness (Dorn and Mertig 2005). Human dimension studies during the past several decades provided insight suggesting a weak and decreasing influence of moral obligation and social norms against the use of bait by deer hunters in Michigan. Survey results indicated that only 22% of hunters surveyed just prior to implementation of a ban on baiting ascribed to a moral norm defining baiting as an unethical hunting practice, the proportion of hunters using bait increased from 29% to 48%, and the proportion of hunters indicating no opposition to use of bait by others increased from 33% to 61% (Langenau et al. 1985, Frawley 2000). Current management and research efforts are modifying public input and engagement processes and evaluating whether changing policies and practices through which regulations are adopted and enforced may enhance perceptions of agency legitimacy, compliance with regulations, and cooperation with management interventions (O'Brien et al. 2011).

2.4. OPPORTUNITIES AND CONCERNS

Collaborative governance creates opportunities to pool resources, expertise, and influence available to public agencies and private entities. The effect generally sought is a gain in public support and advocacy for conservation efforts required to sustain wildlife resources. This potential for enhanced effectiveness is needed in the face of increasingly numerous and complex threats to wildlife and habitats. Gaining an ability to control these conservation threats requires overcoming concerns that collaborative arrangements will lead to loss of control over resources and authority. This has been a point of debate regarding global governance. States are expected to resist entry into treaties and other international agreements that commit to a substantial departure from current practices because those agreements represent a threat to state sovereignty. More importantly, there is increasing recognition that the objectives and responsibilities for conserving wildlife demand actions from multiple states and agencies within the modern, globalized civil society. When faced with the decision to maintain complete control over a small portion of conservation threats or risk giving up some of that control in an effort to coordinate activities among all responsible parties through collaborative governance, the latter option is increasingly pursued as the preferred alternative.

Additional capacity may be provided by entities outside of government when financial or personnel resources are limited in public organizations. IGOs and NGOs also may provide an important link by facilitating communications between stakeholders and government authorities. Nevertheless, divergent views exist on the role of IGOs and NGOs in collaborative governance. For example, one frequently raised concern is that NGOs bear no specific obligation to meet the standards of transparency and accountability expected of government agencies or democratically elected officials. Public perceptions of the legitimacy of IGOs may be shaped by judgments of the governments who created the IGO and the IGOs' unique processes regarding issues in which they are engaged. These issues of accountability and legitimacy are critical given the influence associated with assessments of the legitimacy of management interventions. Both IGOs and NGOs, however, play important roles in the evolving approaches to wildlife governance by increasing accountability of decision makers. Such accountability often is a result of social, political, or economic influences placed on governments by the actions of NGOs or from within IGOs.

The tension between questioning whether collaborative governance approaches result in more effective conservation efforts or whether collaborative governance simply removes decisions from those people who can be held accountable is

illustrated by debates regarding the most prominent of CITES' efforts: conservation of elephants through limitations on the sale of ivory. Disagreement has persisted regarding the need for a total moratorium on ivory trade, as opposed to a limited ivory trade, based on the contested assumption that properly regulated trade can be sustainable and may provide an economic incentive to those poorer states where elephants live. Some of these countries resent what they believe were unfair influences of NGOs in lobbying for increased elephant protection against their interests as individual states (Brown Weiss 2000).

Some argue that broader NGO advocacy cannot be expected to be subject to the same demands for transparency faced by representative democratic governments but that these concerns over the expanding roles of NGOs within coordinated conservation efforts can be overcome when specific functions are delegated to NGOs along with identified obligations for accountability (Charnovitz 2006). Thus, when specific aspects of public trust obligations are delegated to NGOs, they should be held to the same standards of transparency and accountability as public agencies. If NGOs simply work through public information campaigns or other lobbying practices to

CITES efforts to conserve elephants through limitations on the sale of ivory are a highly visible example of collaborative governance approaches to wildlife conservation (top, courtesy Bret Muter; below, courtesy USFWS)

effect change that the public considers to be counter to local needs, however, then the basic issue of government accountability resurfaces.

Although recognition of the need to incorporate social science research to gain greater insight into stakeholder perceptions and motivations is occurring, support for altering decision-making processes based on social science insight is sometimes lacking. Among some managers this may be due to concerns about de-emphasizing the influence of expert opinion and analysis (Gill 1996, Mortenson and Kranninch 2001, McCleery et al. 2006). Despite these concerns, a growing number of case studies demonstrate that sharing or delegating management decisions can facilitate desirable wildlife-management outcomes where conflict has previously prevented implementation (Sandström et al. 2009). In some cases, however, agency obligations to management objectives set earlier in time (often through organic acts or legislation) may prevent shared responsibility for management. These cases usually preclude the sharing of decision making over certain management commitments. If the agency has a legal mandate to pursue specific objectives, the decision of whether or not to pursue them cannot be legally assigned to collaborators. In the case described previously of TB in white-tailed deer, eradication of TB from Michigan was a goal established through a 1998 Governor's Executive Directive (Michigan Executive Directive 1998-1), and decisions to maintain, discard, or modify the goal itself are therefore not under the purview of the Michigan Department of Natural Resources. Efforts to increase agency legitimacy and effectiveness by enhancing public engagement in this TB case focus on decision making regarding means of achieving this objective rather than changing the objective.

The rise in intergovernmental sharing of authority, collaboration with entities outside of government, and interaction with and responsiveness to stakeholders suggests future wildlife managers will be investing a growing amount of time and effort in these activities of collaborative governance. There likely will be costs to efficiency and resources, but benefits will also be gained in terms of legitimacy of decisions, attainment of desired impacts, and relevancy of wildlife management to society. Some of the costs can be mitigated by means other than engaging citizens individually. Consultation with or delegation of authority to advisory panels may provide a connection to local interests, or policies of subsidiarity (a governing principle that suggests management ought to be handled by the smallest, lowest, or least centralized competent authority) may be established to empower locally elected or appointed authorities (Box 2.3). The EU principle of subsidiarity, for example, delegates much autonomous authority to local agencies to establish local regulations and population objectives even though broader directives or policies intended to achieve larger scale conservation objectives are affected by such activities (Bromley 1997). Awareness of these potential issues of governance is beneficial to wildlife managers operating in a world of changing public expectations for governing organizations.

2.5. CONCLUSIONS

It is precisely when and where stakes related to wildlife are greatest that need for coordination and collaboration within a governance framework, informed by human dimensions insights, is greatest. Issues that affect, or are affected by, wildlife that also involve intersecting federal, state, and private interests center on complex matters with important implications for both humans and wildlife. The need for cooperation among agencies, NGOs and IGOs, universities, and the public to increase management capacity and identify durable interventions is readily apparent throughout the world. This is true at the scale of local communities wrestling with damage caused by deer and at the scale of multi-national IGOs addressing elephant conservation and the global trade of ivory.

What we learn from the literature and experiences discussed in this chapter is that governance is closely linked to normative ideals of democracy. A persistent move exists toward greater participation in governance by a growing set of stakeholders. The case of wildlife management is but one of many aspects of governance in which greater demand is occurring for *public* participation in *public* decisions about *public* resources. With globalization and increasing rates of change in ecosystems, greater and more rapid collaboration among all decision-making entities is needed for effective management of wildlife. Attributes such as trust, legitimacy, accountability, and representativeness—all important considerations in governance—likely will become even more important to practicing wildlife managers in the future.

Concerns will persist from some factions of the wildlife profession that activities such as increased public engagement, partnerships with non-government organizations, political bargaining, and coordination and collaboration among multiple agencies or governments will erode the technically sound basis for wildlife management. Others may believe these actions are an abandonment of responsibility or authority. Lack of collaborative governance, however, often leaves authoritative experts in charge of adopting unenforceable regulations that are restricted to activities and geographic scales insufficient to address expanding global conservation threats. One assumption underlying collaborative governance is that broader participation among stakeholders and agencies will result in more legitimate and effective governance. Although this outcome is not always achieved (Bäckstrand and Jamil Khan 2010), it is clear that resistance against collaborative governance can lead to challenges through referenda and other forms of decision making that involve less—not more—involvement by wildlife professionals (Jacobson and Decker 2008).

Approaching wildlife management through a framework of governance enhances relevance of wildlife management through attainment of impacts, as discussed in Chapter 1 and elaborated on throughout the remainder of this book. A collaborative governance approach thus conveys a progression toward the "inevitable fusion" of knowledge from biological and social sciences, as advocated by Aldo Leopold more

Box 2.3 *MANAGEMENT OF LARGE CARNIVORES IN FENNOSCANDIA UNDER EMERGING PATTERNS OF COLLABORATIVE GOVERNANCE*

Three nation states in Fennoscandia—Finland, Norway, and Sweden—are a good example of countries shifting responsibility for large carnivore management from central to regional, or even local, government authorities. In all three countries new approaches to large carnivore management recently emerged, including elements of decentralization that were intended to increase efficiency and improve the equity and transparency of the government toward, and participation by, the citizenry. Although these countries are similar in terms of their geophysical, ecological, and socioeconomic characteristics, they have chosen distinctly different decentralization strategies. In Norway, a representative model of decentralization has emerged, while the Finnish model has a corporatist character, and Sweden has chosen a mix of the two models.

The three governance models also differ when it comes to the decentralization of power. Although Finland (with its corporatist model) has an advisory role, the Norwegian large-carnivore committees have a decisive role (even though conditioned by responsible authorities). In Sweden, the recent institutional change from a corporatist system into a mix of a representative and corporatist implies a shift in authority both vertically, from the national level to the regional level, and horizontally, by the inclusion of stakeholders in the management process. In this new governance system, the regional level is further empowered and the mandate of stakeholder representatives is extended beyond information and deliberation. Game-management delegations, composed of political and organizational stakeholders and chaired by the county governor, have been established within each county since 2010. These delegations have decision-making power over regional management of large carnivores and herbivores. The three different governance models relating to large carnivores in the three countries demonstrate the effects of decentralization at the local or regional level.

At the time of this writing, the role of regional large-carnivore committees is continuing to change in Finland. The formal role of the regional Board of the Game Management Districts (representing mainly local hunters) is changing from a group acting as decision makers to regional collaborative forums. In 2011, new regional forums were nominated by the Ministry of Agriculture and Forestry.

In Fennoscandia responsibility for management of lynx and other large carnivores is shifting from central to regional, or even to local government authorities (© Friday-Fotolia.com)

The purpose of these forums is to support an open and transparent dialog between participants and to promote regional collaboration in large carnivore and ungulate issues. According to a governmental proposal, participants will represent local hunters, regional authorities, and elected politicians. A notable difference is that environmental NGOs are not members of the new regional forums and will play a reduced role in the large carnivore committees (Sandström et al. 2009).

CAMILLA SANDSTRÖM

than seven decades ago (Meine 1988:359), by considering the needs of human values—impacts—along with ecological and biological dimensions of wildlife management. Much of what you read in this book prepares you to practice wildlife *management* in a framework of *governance*. That is, to be effective and responsive to stakeholders, governance requires insights provided by the human dimensions of wildlife management, including knowledge of institutions. Theory, knowledge, and skill-building topics that follow in this book lay a foundation for wildlife managers to improve the practice of governance and thereby create lasting conservation for wildlife resources throughout the world.

SUMMARY

Wildlife management occurs within existing and evolving governance structures. Governance is affected by worldwide trends of increased democratization and globalization of transportation, trade, and information. Insights needed to pursue more effective governance give purpose to the study and practice of concepts, principles, and skills within the human dimensions of wildlife management.

- This chapter reviewed concepts and definitions of governance in general, and assessed strengths and limitations of contemporary governance structures in which wildlife management is conducted. In the United States, wildlife governance occurs within a "public trust resource" legal framework. Governance is closely linked to growing normative ideals of democracy, which is a widespread movement toward greater participation in governance by a growing set of stakeholders. Wildlife management is one of many aspects of governance in which greater demand is occurring for public participation in public decisions about public resources.

- Governance refers to the instruments and mechanisms available to collectively steer an organization or society— the means whereby governments, organizations, or individuals direct their programmatic activities, including the processes, laws, rules, and policies that collectively guide actions. Governance is distinguished from governments; governance is what governments do, yet governance is something for which not only governments are responsible.

- A further distinction exists between top-down governance and collaborative governance. The former is characterized by the exercise of formal sovereign authority in an effort to regulate actions. Collaborative forms of governance increasingly include networks of public–private partnerships, collaboration with community organizations, and other forms of interaction with civil society. The need for cooperation and collaboration among agencies, NGOs and IGOs, universities, and the public to increase management capacity and identify durable interventions is readily apparent throughout the world.

- Globalization, trends toward substantial interconnectedness of the global economy, expansion of worldwide transportation networks, and increasing domination of ecosystems by humans have brought actions previously within the purview of individual national and local authorities under the shared responsibility of multiple sovereign states and agencies. Issues that affect, or are affected by, wildlife that also involve intersecting federal, state, and private interests center on complex matters with important implications for both humans and wildlife.

- The need for collaborative governance is greatest when and where stakes related to wildlife are greatest. The motivation for this book is to prepare you to practice wildlife management in a framework of governance. That is, to be most effective and responsive to stakeholders, governance requires insights provided by the human dimensions of wildlife management.

Suggested Readings

Bäckstrand, K., and A. K. Jamil Khan. 2010. The promise of new modes of environmental governance. Pages 3–27 in K. Bäckstrand and A. K. Jamil Khan, editors. Environmental politics and deliberative democracy: examining the promise of new modes of governance. Edward Elgar, Northampton, Massachusetts, USA.

Hatcher, A., S. Jaffry, O. Thébaud, and E. Bennett. 2000. Normative and social influences affecting compliance with fishery regulations. Land Economics 76:448–461.

Jacobson, C. A., and D. J. Decker. 2008. Governance of state wildlife management: reform and revive or resist and retrench? Society and Natural Resources 21:441–448.

Nelson, F., editor. 2010. Community rights, conservation and contested land: the politics of natural resource governance in Africa. Earthscan, Washington, D.C., USA.

Nie, M. A. 2003. Beyond wolves: the politics of wolf recovery and management. University of Minnesota Press, Minneapolis, USA.

Peters, B. G., and D. J. Savoie, editors. 2000. Governance in the twenty-first century: revitalizing the public service. McGill-Queen's University Press, Montreal, Quebec, Canada.

STAKEHOLDERS AS BENEFICIARIES OF WILDLIFE MANAGEMENT

KIRSTEN M. LEONG, DANIEL. J. DECKER, AND T. BRUCE LAUBER

Following principles of the Public Trust Doctrine, wildlife in the United States and Canada belong to the people as common public natural resources; the government is entrusted to manage "the people's" wildlife resources in the public interest (Chapter 2). That mandate is the underlying reason for the existence of public wildlife management agencies to administer the public wildlife trust.

The management and decision-making challenges associated with governing the use and ensuring the sustainability of wildlife are similar to those found in the management of any common property. Although patterns of consumption and management traditions vary across cultures, regions, and resources, the pervasive human dimensions problem for professionals responsible for managing common natural resources is the same: recognizing the interests of many people and coordinating the use, distribution, abundance, and character of resources to provide benefits to people while sustaining the resources for future generations (Hardin 1968a, Dietz et al. 2003).

The management plan of the Bureau of Land Management (BLM) for Las Cienegas National Conservation Area (LCNCA) outside of Tucson, Arizona, exemplifies these challenges and points to one way to manage them. LCNCA is the centerpiece of a region known as the Sonoita Valley and encompasses 45,000 acres (18,200 ha) of relatively intact grassland and oak savanna woodland in a basin surrounded by high, isolated mountain ranges referred to as "sky islands." The riparian corridor supports a diverse plant and animal community, including six federally listed species and many other special-status species. Livestock grazing and a variety of recreational activities are ongoing. LCNCA is located only an hour's drive from Tucson so visitation is increasing and development has encroached on these protected lands, which has resulted in high-impact recreational activities, overuse of water, and the spread of invasive plant and animal species that threaten to transform the system. Examples include exotic bullfrogs, which imperil the threatened Chiricahua leopard frog, and heavy invasion of native mesquite into upland areas, which threatens habitat for open-grassland species, such as pronghorn.

BLM managed the lands before they were afforded the additional protection of National Conservation Area designation. Stakeholders ranging from ranchers to conservation organizations, off-road vehicle users, mountain bikers, birdwatchers, hunters, and eight government agencies all had competing interests in the region. After several unsuccessful attempts to develop a land-use plan based on traditional top-down approaches, BLM decided to try a collaborative approach. By listening to and learning from each other, stakeholders came to recognize that they shared a common core vision that fit within the BLM mission: to perpetuate naturally functioning ecosystems while preserving the rural, grassland character of the Sonoita Valley for future generations. By grounding management efforts in this common vision, the diversity of perspectives became a strength, which assures that no one voice dominated management decisions. The collaboration eventually grew into the ad hoc Sonoita Valley Planning Partnership (SVPP), which was instrumental in demonstrating to Congress the need for official designation as a National Conservation Area. The SVPP and other partners continue to work collaboratively with BLM on plan implementation and stewardship, assisting the agency in managing the LCNCA to support both a diversity of wildlife and a rich variety of related recreational opportunities and to protect it from ecosystem threats.

As demonstrated in the Sonoita Valley, common public resources such as wildlife can represent multiple values. People may value wildlife for aesthetic, economic, cultural, ecological, and other reasons. In the context of managing wildlife as a common property resource, several forces are at work: (1) the diverse and sometimes conflicting public interests, concerns, and uses of wildlife; (2) the impacts of wildlife on the public; (3) the public expectation for citizen participation in management decision making; and (4) many managers' inclusive view about who benefits from wildlife management and how beneficiaries can contribute to management.

In addition, different people can value the same wildlife resource for very different reasons, which leads to competing ideas about appropriate management objectives and actions. It

The Las Cienegas National Conservation Area, managed by the Bureau of Land Management, is the centerpiece of a collaborative effort to conserve a large, relatively intact grassland and oak savanna woodland in Arizona's Sonoita Valley. (courtesy Tana Kappel–The Nature Conservancy)

should not then be surprising that a major human dimensions challenge in wildlife management is to seek broadly acceptable fundamental management objectives, enabling objectives, and actions (means). How do wildlife managers approach that daunting challenge?

A significant development during the 1990s was the emergence of a management philosophy that actively seeks to incorporate diverse stakeholder perspectives in decision making. The manager adopting such a philosophy asks and answers these questions: Whose interests and concerns should be considered? Who should participate in wildlife management decisions? How should those people be involved in decision making?

This chapter explores the role of stakeholders in wildlife management. We begin by briefly reviewing the evolution of thinking about wildlife management beneficiaries over the past few decades. We then discuss the philosophy of engaging stakeholders in more detail, note some challenges of this approach, and describe some traits of successful managers. Chapter 11 describes the evolution of approaches to stakeholder engagement in wildlife management and provides an introduction to stakeholder engagement theory and practice.

3.1. FROM CLIENTS TO STAKEHOLDERS

During the 1990s, thinking about beneficiaries of wildlife management and who should be considered in decision making broadened the perspective of the wildlife profession. Unlike earlier eras in wildlife management, attention is now given not only to traditional clients (those who either pay for services of wildlife agencies or produce wildlife on their land) but to all significant stakeholders in each wildlife management situation. Demographic and socioeconomic changes such as aging and population movement (e.g., suburban sprawl, transportation and residential development), economic activity (e.g., resource extraction, commercial and industrial development),

and changing patterns of natural resource–based recreation activity have resulted in new, diverse, and interested stakeholders with expectations for input into wildlife management decisions (Jacobson and Decker 2008). Managers responding to these expectations developed both an expanded philosophy about stakeholder engagement and a variety of techniques for doing so effectively.

3.1.1. Stakeholder: A Critical Concept

A stakeholder is *any person who significantly affects or is significantly affected by wildlife management* (Susskind and Cruikshank 1987, Decker et al. 1996). Many agencies historically have used the term to refer to organized interest groups or "special interest groups," yet these are only one sub-set of people who may be interested in wildlife management. A stakeholder may be any citizen that has an interest (a stake) that could be affected significantly by a wildlife management decision or action. A stake may take the form of a recreational, cultural, social, economic, or health and safety impact from wildlife or the management of wildlife. Any of those kinds of impacts can be the focus of wildlife management. This perspective recognizes a larger set of potential beneficiaries of wildlife management than traditionally was considered (Decker et al. 1996).

The full range of stakeholders includes both those who benefit from positive outcomes of people–wildlife interaction and those who experience problems from such interactions (Box 3.1). They also include those who influence or make decisions about how a program is managed. Stakeholders typically include individuals and groups that have legal standing, political influence, power to block implementation of a decision, or sufficient moral claims in connection with the situation (Susskind and Cruikshank 1987).

People with many kinds of stakes may be stakeholders in a management decision, which is a notion that may be cumbersome to act on. A manager seeking to engage stakeholders must (1) identify important stakes, (2) consider using multiple methods for incorporating stakeholder input, (3) be strong enough to resist powerful special interests that might want decision making to exclude other stakeholder interests, (4) find ways to weigh the interests of different stakeholders in reaching management decisions, and (5) use effective strategies for communication to encourage constructive deliberation and understanding.

3.1.2. Top-Down Governance: Managing for Clients and Constituents

Before about 1970, the wildlife management profession focused almost exclusively on hunters, trappers, and landowners (farmers, ranchers, and forest owners). To use Aldo Leopold's (1933) terms, they were the people who produced or harvested the crops of wildlife that were of concern for management.

Hunters and trappers fit the traditional definition of a "constituency"—a group of people (constituents) who authorize or support the efforts of others (in this case, public wildlife professionals) to act on their behalf. They also were

Box 3.1 *DEFINING MANAGEMENT STAKEHOLDERS*

A stakeholder is *any person who significantly affects or is significantly affected by wildlife management.* For any given management issue the list of stakeholders may be long. Consider a suburban elk-management issue, for example. Stakeholders would include elk hunters and elk watchers; farmers, ranchers, and forest owners; businesses that cater to wildlife recreationists (gas stations, sporting goods stores, motels, restaurants, guides, and outfitters); homeowners, commercial property owners, and golf course owners; local government officials, highway officials, motorists, park users, and gardeners; and environmentalists, conservationists, and animal welfare advocates.

Some of those stakeholders, such as hunters and ranchers or environmental groups, may be well-organized. Others, such as homeowners stymied in their attempts to grow ornamental plantings, gardens, or flower beds in the presence of hungry elk, may not be organized at all.

Elected officials often become special stakeholders in local, state, and multi-state or national wildlife management issues. They must be considered in any management situation, because they can facilitate or block management, depending on their understanding of the impacts involved. Town supervisors, mayors, town council members, or village trustees have a stake in the elk management decisions that are implemented (or merely considered publicly) in their jurisdictions.

Even state and federal land managers, wildlife boards and commissions, and wildlife biologists and managers can be stakeholders under some conditions. For example, managers of a federal wildlife refuge might consider a state wildlife agency as a stakeholder during a process to develop operational plans for a refuge. Conversely, a state wildlife agency might consider federal wildlife refuges as stakeholders in decisions about deer management regulations. It all depends on the scale of the impacts being managed.

Suburban elk-management issues typically involve a diverse array of stakeholders, including homeowners and motorists (courtesy Bret Muter)

often thought of as "clients"—people who paid (e.g., through license fees or earmarked taxation) for professional or expert services. In this case, hunters and trappers bought licenses that paid for the services of public wildlife agencies.

Farmers, ranchers, and other landowners control much of the land base that provides wildlife habitat. Their land-use practices affect the productivity of their land for wildlife and the degree to which those lands are made accessible to other citizens, two vital considerations of the wildlife management profession (Brown and Thompson 1976). Yet wildlife are often seen as competitors or sources of problems for landowners as well (Brown et al. 1978). Wildlife can damage crops and landscaping, threaten livestock, retard forest regeneration, and so on, which negatively affects the activities that may be

the primary reasons for landownership. Helping achieve co-existence between wildlife and landowners and helping gain public access to wildlife dwelling on private land have been long-standing areas of management attention.

The terms "constituents" and "clients" reflected a special relationship between wildlife managers and the recognized beneficiaries of their work. The professionals were the experts who were paid to make decisions and carry out actions for the benefit of their clients. The professionals presumably had superior knowledge about all aspects of management, including being able to determine what the "right" goals and objectives were and the best ways to achieve them.

Because the managers served few constituencies, understanding the wants and needs of their constituents was relatively easy. Not surprisingly, experts could make many decisions about management objectives without special studies of users. In states that had them, wildlife commissions were surrogates for broader citizen participation. Managers would accept input voluntarily provided by stakeholders, but managers did not systematically seek stakeholder input.

The manager–client, or top-down, governance system encompasses the authoritative and passive–receptive approaches to management (see Chapter 11). This system functioned well for many years because either the clients and managers shared a narrow set of values regarding wildlife or managers were familiar with relevant stakeholders' values even if they did not share them. It was a virtually unchallenged model in most states until the early 1980s, when some managers began to consider groups with other interests in wildlife. With a growing diversity of stakeholder interests, managers found that the "constituent" concept no longer worked; newer interest groups did not support the status quo.

3.1.3. Public Input Addresses New Interest Groups

During the late 1960s and early 1970s, growing public interest in the environment—punctuated by Earth Day and evidenced in such legislation as the Multiple Use–Sustained Yield Act, the National Environmental Policy Act, and the Endangered Species Act—markedly influenced wildlife management. People other than traditional clientele of wildlife management were expressing keen interest in all wildlife (not just species consumed or causing nuisance). In addition, legal procedures requiring government agencies to consider public input were put in place. These non-traditional stakeholders increasingly sought to have wildlife resources protected for non-consumptive recreational, environmental, ecological, or existence value interests, which more fully reflected the spectrum of public values.

The wildlife management literature began to carry many references to "interest groups" or "publics." Those two terms helped focus managers' attention on groups of people who had interests in wildlife in addition to traditional consumptive use and had reason to be considered in management. The U.S. Fish and Wildlife Service's national surveys during the 1980s and 1990s confirmed that the number of non-consumptive wildlife users was increasing.

Partially in response to diversifying interests, laws requiring active solicitation of public participation in environmental decision making were introduced with the National Environmental Policy Act (NEPA § 102, 42 U.S.C. § 4321 *et seq.* [1969]) and state correlates (e.g., New York State Environmental Quality Review Act, 6 NYCRR Part 617 [1978]; California Environmental Quality Act, Cal. Pub. Res. Code § 21000 *et seq.* [1986]). The "top-down governance" model was no longer adequate and a new era of public input in government began.

While managers still acted as experts and accepted any unsolicited input about management, they became more inquisitive and began to seek stakeholder input through techniques to quantify the breadth of stakeholder interests and positions. Early human dimensions research revealed the broad range of beliefs and attitudes people held about wildlife and documented that many people valued wildlife for more than recreational use (Kellert 1980*b*). Expressions of people's beliefs, attitudes, and interests ranged from concern about problems wildlife caused for people (e.g., collisions with cars, Lyme disease, plant damage) to advocacy for the protection of rare and endangered species (e.g., northern spotted owl, piping plover) and animal welfare.

Managers also began to interact more with special interest groups to better understand their perspectives (for example, by regularly attending meetings of organized stakeholder associations). Although this approach helped managers better understand interests of individual groups, it did not provide stakeholders a forum to understand each other. Managers found themselves engaged in a form of shuttle diplomacy between stakeholder groups and desired a way to improve dialogue between stakeholders.

3.1.4. Stakeholder Engagement Expands the Capacity of the Wildlife Profession

By the beginning of the 1990s, expanded stakeholder involvement had become a common activity for some wildlife agencies and was a key ingredient in establishing their credibility with many stakeholders (Stout et al. 1992). The diversification of stakes and stakeholders led to a need for processes that more fully engaged stakeholders with managers and with each other and that facilitated mutual learning. In some cases, such as the SVPP described earlier, stakeholders became full partners in decision making and management implementation. Development of these dialogue-based processes broadened the scope of the wildlife profession to include a greater diversity of management options.

Managers now use a range of methods to incorporate stakeholders in management decision making, which depend on the specific context and stage of the management process. These methods encompass some of the same techniques from the top-down governance and unsolicited public input models but typically rely on dialogue and group interaction at key stages in management, such as setting fundamental and enabling objectives or developing management alternatives (see Chapter 11 for more detail on the evolution of stakeholder engagement in wildlife management). Managers embracing stakeholder engagement now are better able to include all people who may be affected by a decision (sometimes even people who do not recognize it themselves), not just those who feel empowered to make their views known to managers directly.

In addition, wildlife professionals have a trusteeship responsibility for the future and a responsibility for ensuring that tomorrow's citizens are considered stakeholders in today's management decisions. That is, they need to consider opportunity costs or options for future generations, sometimes referred to as existence and bequest values (Bishop 1987). Engagement of stakeholders to better understand their hopes and expectations for human–wildlife interactions in the future helps managers fulfill that responsibility.

People with traditional interests are not being ignored.

They now interact with other stakeholders in the decision-making process. The outcome is that management serves a broader cross-section of society. As managers gained experience with a diversity of interests, stakeholder engagement changed the fundamental focus of the management profession.

3.2. HOW DOES STAKEHOLDER ENGAGEMENT AFFECT WILDLIFE MANAGEMENT?

A management philosophy that embraces stakeholder engagement does more than simply respond to societal desires for increased voice in management decisions; it broadens the focus of the wildlife profession itself. First, a more inclusive view helps managers be adaptive. By routinely engaging a broad array of people holding wildlife interests, managers can better identify emerging needs and guide change to improve effectiveness. Second, it diminishes the probability of a few groups capturing the decision-making process. It distinguishes between reacting to pressures and anticipating or responding to needs and opportunities. Third, it allows professionals to address a spectrum of societal values, thereby assuring relevance of wildlife management for a greater segment of citizens.

Given the diverse range of stakeholders who desire involvement in decision making, wildlife management is no longer primarily a technical problem of producing or distributing a wildlife "crop." Instead, wildlife management has become a different kind of enterprise, one of understanding "wicked problems," where value differences must be elicited and understood in order to design and implement durable and sustainable management actions. Some stakeholders may have primary interests in the ends or conditions that managers seek to affect, while others may have stronger interests in the means used to achieve such conditions. All of these considerations require learning from and understanding stakeholders throughout all phases of management to improve management decisions.

3.2.1. Wicked Problems and Dimensions of the Management Environment

Wildlife management issues often are more than just "complex," they are also "wicked," or "messy" (Rittel and Webber 1973, Lachapelle et al. 2003). The term "wicked" was first used by Rittel and Webber (1973) to describe problems characterized by scientific uncertainty about cause–effect relationships and social conflicts over goals and management alternatives. Later work revealed a key distinction between complex, or technical, and wicked problems (Allen and Gould 1986, Lachapelle et al. 2003). Complex problems include interactions between many system components whose relationships are difficult to understand, but the natural laws that govern them (e.g., physical, biological, or market principles) allow managers a high degree of confidence in predicting outcomes of potential management interventions. Thus, although the structure of the problem may be highly complex, the problems being addressed are clear-cut; both the stakeholders involved and the desired effect on the environment are obvious and little contention exists about them. In addition, the outcome of any management intervention is easily classified as right or wrong, or as good or bad. Many engineering, computer science, or physics problems are examples of complex problems; with sufficient understanding of relationships between component parts, a clearly defined outcome is reliably predictable.

Wicked problems, while fundamentally complex, do not have a single "correct" formulation of the problem. Rather, the definition of the problem is articulated by the individual experiencing it, and how that person chooses to explain the problem determines the scope of possibilities for its resolution (Allen and Gould 1986). In other words, the problem may be understood differently by different individuals, depending on the values they associate with the resources and impacts of interest to them. As such, wicked problems often involve social conflict over goals and, therefore, appropriate solutions. With wicked problems, the process of problem articulation often is one of the most difficult aspects of the problem (Rittel and Webber 1973). Although it frequently is used to describe problems that are intractable or unsolvable, we use the term "wicked" to mean problems characterized by social conflict over ends desired and means of achieving them.

The complexity and wickedness of wildlife problems can be illustrated by considering the diverse factors that affect and are affected by wildlife management. These can be grouped into three major "spheres of influence" (Fig. 3.1):

1. Ecological sphere. Factors associated with the resource itself, such as resource condition descriptions, data, models, concepts, and working knowledge of the resource, typically grounded in biology and ecology of wildlife and habitat.
2. Institutional sphere. Laws, policies, and resource allocations that reflect societal values and priorities and guide management-agency responsibilities and activities; institutional dimensions also include managers' professional judgment.
3. Sociocultural sphere. Other factors considered in determining the goals of management, which include stakeholder values, attitudes, beliefs, and interests; societal norms; and other cultural, political, and economic considerations.

Wildlife management research often focuses on better understanding the complex relationships between the biotic and abiotic factors that comprise the resource dimensions. Simply understanding this complexity does not, however, tell a manager which resources or attributes of resources are most important to conserve. Sociocultural and institutional influences place value on resource factors and delineate the factors that are important to society and that fall within agency purview. Social science and political science can help managers understand the complexity within the sociocultural and institutional spheres, respectively.

Identifying the scope (factors included) and overlap of these

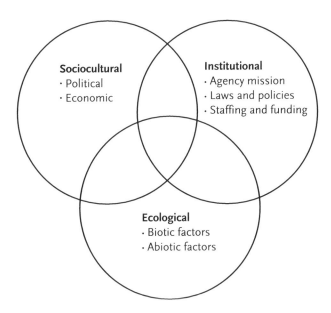

Figure 3.1. Spheres of influence on wildlife management

Management of the Florida panther, an imperiled species in Florida, is an example of a wicked problem characterized by staunch advocates of panther restoration and equally strident stakeholders who oppose expansion of this species. (courtesy USFWS; photo by Larry Richardson)

spheres can assist managers in understanding and addressing "wickedness" and determining which resource factors should be the focus of wildlife management. This type of analysis is inherent in a coupled-systems approach that recognizes the interaction between natural and social sciences in wildlife-management decision making. As demonstrated in section 3.1, stakeholders define the relevant human dimensions considerations, whether or not they are proactively engaged by managers (Box 3.2).

3.2.2. Stakeholders Have Interests in Different Types of Impacts

Stakeholder interests often are described in terms of concerns about the *primary impacts* of human–wildlife interactions; that is, concerns related to the desired outcomes of management interventions. Stakeholders with interests in primary impacts might have concerns about wildlife damage to crops, gardens, or landscaping; wildlife disease transmission; wildlife–vehicle collisions; wildlife viewing or hunting opportunities; biodiversity and species sustainability; or wildlife injury to people or pets. In other words, primary impacts often involve interests in reducing or increasing important types of human–wildlife interactions.

Management actions taken to achieve these outcomes may result in collateral and subsequent impacts (Decker et al. 2011; see Chapter 1). *Collateral impacts* are unintended impacts that occur simultaneously with the implementation of primary management actions, while *subsequent impacts* are unintended impacts that occur as a consequence of achieving primary management objectives.

Stakeholders who have interests in collateral or subsequent impacts may or may not have interests in primary impacts of management concern; that is, they may not be affected by, experiencing problems with, or deriving benefits from wildlife. Instead, these stakeholders are affected by management actions as they are implemented or by cascading effects from management meeting its enabling objectives. These stakeholders are *activated by* management interventions, not by *the original reason for* management interventions. In many cases, it would be impossible for a manager to identify these stakeholders until potential management actions are being considered and made known to the public.

In addition, these stakeholders may exist at different jurisdictional scales. A management unit, whether a state Wildlife Management Unit, federally or privately owned protected land, or even private residence, is embedded within other jurisdictional management frameworks. These extend from the individual unit to local, regional, national, and even international scales (Fig. 3.2). Stakeholders with interests in primary, collateral, and subsequent impacts can exist at all scales. Although these groups may overlap, some stakeholders may have interests only in primary impacts, while others may only have interests in collateral impacts or subsequent impacts.

This dynamic is especially prevalent for federal land-management agencies, where local communities may bear the brunt of land management decisions that are made for the benefit of the national public. For example, use of off-road vehicles (ORV) is often regulated at National Seashores along the eastern seaboard to protect endangered shorebirds (such as the piping plover). Environmentalists, from local to national, have interests in the condition of shorebird populations, which may be affected by unregulated ORV use (primary impacts) among other environmental stressors. ORV users at the seashore may or may not have environmental interests, but when regulation of ORV activity is considered as a potential management action, they would become stakeholders due to the collateral impacts of regulation. Similarly, local business owners may not have environmental interests but could experience collateral impacts if people reduced their visits to the seashore due to ORV regulations. National interest groups can become

Box 3.2 *LEARNING TO APPROACH CONDOR RECOVERY AS A WICKED PROBLEM*

The California condor was nearly driven to extinction by shooting, the pesticide DDT, egg collecting, and habitat loss. Serious population decline occurred throughout the twentieth century. The bird was designated as an endangered species in 1967. By 1985, only nine individuals remained in the wild. The U.S. Fish and Wildlife Service determined that the only way to save the species would be to move the remaining wild birds to a captive breeding program. Captive rearing was successful and officials began reintroducing condors to the wild in California in 1992 and Arizona in 1996.

The situation was largely viewed as a complex problem that could be fixed by understanding the necessary conditions for condor survival and breeding success in the wild. In the late 1990s, studies revealed that lead bullets fragment much more than previously recognized, which resulted in high potential for scavenger exposure to lead (Hunt et al. 2006). Wildlife officials concluded that lead poisoning was the leading cause of condor mortality and the largest obstacle to condor recovery. After much debate, in 2008 California banned the use of lead bullets for hunting big game and coyote in zones inhabited by California condors.

This action produced a backlash from hunting organizations and the National Rifle Association (NRA), who claimed that this action was evidence of an anti-hunting agenda. Opponents of the ban reframed it as "hunters versus condors." Also in 2008, additional studies about effects of lead on wildlife were published, which added fuel to the fire. As one hunter described in a blog, "the whole thing went out of control in CA as soon as it became a political issue and left the realm of wildlife management or conservation. On top of that, groups such as HSUS [Humane Society of the United States] got involved on the side of the condor advocates, while the 800-lb gorilla of the NRA stepped in on the other side . . . instantaneously polarizing the factions. The facts got buried in propaganda, distrust clouded reasonable argument, and when all was said and done, a decision was made that seemed to absolutely disregard the concerns of the hunting community" (The Hog Blog 2010). What initially appeared to be a complex problem of reducing condors' lead exposure became a "wicked" problem as soon as management actions that affected stakeholders were implemented.

The Arizona Game and Fish Department (AZGFD) took a different tack. The condor population in Arizona is designated a "non-essential and experimental" population under the Endangered Species Act, so regulation was not considered. Instead, they started by conducting surveys and focus groups of hunters to understand their knowledge of, and attitudes about, lead poisoning in condors and to test potential outreach messages. They used results of this research to design a variety of outreach efforts, such as including coupons for free non-lead ammunition with hunter licenses, increased media efforts, and incentives for bringing gut piles (where lead fragments tend to concentrate) to check stations. They continued to utilize surveys to evaluate voluntary lead-reduction actions and refine their outreach messages and techniques.

The approach taken by AZGFD implicitly acknowledged that condor restoration is a "wicked problem" (i.e., a problem characterized by scientific uncertainty and/or social conflicts over goals and management alternatives). It recognized that not only do stakeholders hold a wide range of views on restoration of endangered species generally, and condors specifically, but some stakeholders (in this case, hunters) are differentially impacted by management decisions. Rather than focusing on persuading hunters to change their attitudes about condors, AZGDF engaged these stakeholders to learn how to reduce barriers to hunters taking voluntary lead-reduction actions.

California condor (courtesy USFWS)

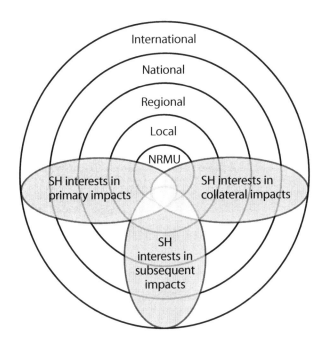

Figure 3.2. Jurisdictional spheres of influence applied to the wildlife management environment. Stakeholders (SH) with interests in primary, collateral, and subsequent impacts may exist and they may affect decision making in wildlife-management at all scales of influence. NRMU is Natural Resource Management Unit.

involved as well. Federal land managers must consider all of these interests within the context of their management mandates.

In addition to recognizing new groups of stakeholders activated at different stages of management, managers must decide which set of stakeholder interests is more important to address. A wildlife manager should recognize that the volume of noise created by a group does not necessarily correlate with its size or even with the importance of its stake. Loud minorities may appear to have more at stake in a decision than a careful analysis might find. Furthermore, the sheer size of a group does not necessarily correlate with the importance of its stake. In some situations, many people with very little at stake may not have as much bearing on a decision as a few people with much at stake. Learning to deal with the conflicts that may arise is a valuable skill (Box 3.3).

Detailed analysis is necessary to make the determination because choosing to ignore stakeholders has its perils. For example, in the 1990s, black bear hunting practices and regulations were publicly debated in Colorado. Despite the objections of several stakeholder groups and survey data revealing that the public was not supportive, the Colorado Wildlife Commission decided to continue to allow the hunting of bears during the spring (when females have dependent cubs) and with the aid of bait and hounds. Citizens initiated a grassroots effort to hold a statewide ballot to reverse the decision. The initiative passed and those bear-hunting practices were made illegal in Colorado. Although the Division of Wildlife did not expect the outcome to affect its ability to manage the black bear population, the Division believed that the outcome did reduce its authority.

Experience has shown that some stakeholders do not trust government agencies and will keep their views to themselves until sufficiently motivated to speak out. If that time comes well after a stakeholder involvement process is underway it may derail an effort that has already built some momentum among participating stakeholders. Conversely, proactive involvement of stakeholders can improve trust in the agency (Stout et al. 1992).

Sometimes concerns are expressed entirely outside the processes that managers have developed to involve identified stakeholders in decision making. For example, people may go directly to political representatives to make their concerns known. Progress hard-won through educational communication could quickly come to a halt or could even be reversed. Consideration of the range of interests involved (primary, collateral, and subsequent) can help identify potential stakeholders who may be likely to seek influence via other means.

3.3. STAKEHOLDER ENGAGEMENT

The various stages in wildlife management history demonstrate fundamentally different ways of thinking about the role of stakeholders in wildlife management. Both the public input (circa 1970 and on) and stakeholder engagement periods (beginning in the 1990s) reflect shifts in the wildlife management profession that indicate a new understanding of the role of stakeholders in management. Shifting to a new governance paradigm can be thought of as adopting an innovation (Rogers 1995), which typically takes time to diffuse broadly throughout society. A number of historical and institutional barriers (such as resisting the added responsibility of seeking broader stakeholder input, the perceived danger of an agency losing responsibility for management decisions, and difficulty of weighting stakes) once made it difficult for managers to proactively engage stakeholders, but today this is the norm for most agencies. Nevertheless, to engage stakeholders successfully, managers must find ways to address the challenges of doing so.

3.3.1. Accepting the Responsibilities

The contemporary manager's responsibility for assuring successful stakeholder engagement is greater than in the client-centered or top-down governance approach described earlier. Managers have to make sure that decision makers consider the interests and concerns of all significant stakeholders without falling victim to the special interest politics of one or more stakeholder groups. Consideration of the values of all significant stakeholders is essential in decision processes that will sustain wide public support.

Nevertheless, managers need to avoid adopting a customer-satisfaction approach. Given the contradictory goals of some groups, it would be chaotic to try to implement every action that every stakeholder group wanted. Following the wishes of a majority could be ill-advised if it leads to ecological disasters,

Box 3.3 *ANALYZING AND TRANSFORMING STAKEHOLDER CONFLICT TO CREATE DURABLE SOLUTIONS FOR PEOPLE AND WILDLIFE*

Wildlife professionals continually deal with the complexities of human conflict in their conservation efforts. Conflicts that are seemingly between people and wildlife are more often conflicts between people *about* wildlife. Unfortunately, issues at the center of conservation conflicts often serve as proxies for underlying stakeholder conflicts, such as struggles over group recognition, identity, and status. Sound conservation and wildlife-management programming often falters because it fails to address (or ignores) the multiple levels of conflict and the complex web of entrenched interests common to most conservation contexts. For instance, we may limit our understanding of, and capacity to address, the needs and concerns of various stakeholder groups by narrowly focusing our efforts on economic fixes, livelihood interests, and tactical solutions to wildlife depredation or other primary impacts, when more elusive factors such as empowerment, respect, and trust are among the unspoken and pervasive concerns that may ultimately undermine conservation's efforts (even those designed to offset the tangible costs of conservation to local people). Thus, it is critical that we train and promote wildlife professionals that are proficient at analyzing and addressing complex conservation conflicts.

The Human–Wildlife Conflict Collaboration's (HWCC) Conservation Conflict Resolution training (www.human wildlifeconflict.org) addresses the theory, principles, and practice of transforming complex conservation conflicts into positive, lasting change. Participants draw on best practices from both the wildlife conservation and manage-

ment and identity-based (deep-rooted) conflict-resolution fields. As such, they improve their capacity to analyze complex conflict dynamics, anticipate and address conflicts as they arise, and address long-standing conflicts that may impede new progress. For instance, one recent participant—working with Inuit in the Arctic to increase adoption of bear-proof measures around meat caches and therefore minimize human–polar bear conflict—realized that a lack of adoption of these practices had more to do with a perceived lack of respect by the wildlife authorities for traditional Inuit knowledge and practices and a need for better relationships between individuals involved, rather than a need solely for more education. Another participant working on wolf conservation in the western United States focused a significant portion of her efforts on building and disseminating scientific evidence to prove that predation by wolves did not harm elk populations, because this is the most common complaint made against wolves by hunters. However, after analyzing the conflict more deeply, she realized that for hunters the wolf symbolizes fears regarding a lack of control over a changing West and represents a threat to their way of life. Realizing the complexity of the conflict and what was truly at stake, her organization could redirect its resources to design processes that more effectively target the underlying causes for concern, rather than wasting money and time adding to scientific evidence that would only fuel further debate.

For many of us, designing conflict-resolving strategies and processes that target not only what people say but

---

extirpation or extinction of species, or failure to accomplish the long-range societal goals for wildlife conservation. Such decisions would fail to meet agency mandates and would effectively result in the agency abrogating its responsibility to manage in the public trust (Batcheller et al. 2010).

In considering the interests of diverse stakeholders, wildlife managers do not abrogate their responsibility for administration of the wildlife trust and do not become brokers who simply allocate resources to highly vocal or single-minded interests. Rather, managers are responsible for clearly communicating sideboards of the decision space (i.e., the legal, political, and scientific considerations that define the management question at hand). Some valid stakeholder interests may not be resolvable through agency management-planning efforts. For example, when providing input to National Park Service (NPS) ungulate-management plans, some local, regional, and national stakeholders often voice interest in establishing public hunting in parks. For most parks in question, however, hunting is prohibited in the enabling legislation that established

the park. A park's enabling legislation can only be changed by Congress, and NPS cannot lobby Congress to change the enabling legislation. Although these interests can be noted in an NPS wildlife-management planning process, the question of allowing hunting cannot be addressed in a park where that activity is expressly prohibited by law. In this case, interests at the national level (as reflected by legislation) may differ from and also constrain local interests.

The choice of whose values to consider and how to weigh them is one of the most difficult issues when engaging stakeholders. Ultimately, managers confront ethical questions in decision making (see Chapter 16). That responsibility is more evident as stakeholders become involved in more aspects of management. Values such as fairness, justice, and concern for the sustainability of resources are morally and professionally defensible in our society (Blader and Tyler 2003a), and decision makers should use them along with legislated mandates as guides. Similarly, managers must remind stakeholders of the overarching issues and help identify the consequences of

Participants interact during a training session organized by the Human–Wildlife Conflict Collaboration (HWCC) (courtesy Francine Madden)

also the meaning behind the words is critical if we are to be effective in addressing the complexity of conflict that we are presented with every day in wildlife management. By accurately and deeply analyzing conflicts and facilitating appropriate processes for addressing them, professionals can determine root causes, build a foundation for trust and respect among stakeholders, and unearth fertile ground for sowing and cultivating sustainable conservation solutions.

HWCC's professional training addresses the multiple levels of conflict characteristic of wildlife management and conservation efforts, with a focus on building wildlife professionals' capacity to design and implement stakeholder processes that support the development of sustainable solutions and supportive relationships.

FRANCINE MADDEN

alternative management decisions (the potential collateral and subsequent impacts) with an eye on fairness, justice, and sustainability concerns.

When invoking professional judgment about such values, however, managers must distinguish between what is scientifically defensible (the scientific insights they bring to the discussion) and what is morally or legally arguable and credible (the relevant values and laws). They should make clear whether they are talking about science, morals, or laws, and not confuse the three. Similarly, they must not mistake their own values and professional opinions for scientific "fact" (Decker et al. 1991). They also must consider how changing societal values affect the relevancy of the agency and whether and how it is appropriate for the agency to respond with institutional change (Jacobson and Decker 2006).

In summary, managers are responsible for ensuring that (1) decision-making processes consider the breadth of relevant stakeholder needs and interests, even those not advocated by special interest organizations, (2) those needs and interests

are addressed in the appropriate forum and given appropriate weight in decisions, and (3) decisions reflect the overall long-term public interest and important values of society. Application of these principles can result in effective partnerships of professional managers and diverse stakeholders working together to identify management goals and solutions to problems.

3.3.2. Addressing Challenges

Several challenges are inherent in a stakeholder philosophy of wildlife management. One challenge stems from lack of clarity about which stakes and stakeholders to consider. If interpreted too broadly, "stakeholder" becomes synonymous with "citizen" and thus is a useless term; not all citizens can, nor would want to, be meaningfully involved. If interpreted too narrowly, the larger public can rightly object that stakeholder involvement is simply a pretext for empowering certain interests. As a pragmatic matter, no decision can take every possible stake into consideration; judgments about which stakes and stakeholders to include or to omit must be made depending on the circumstances.

Having decided which stakes to consider, management professionals face another challenge: how to weight the many stakes that their decisions may affect (Carpenter et al. 2000). Under some alternatives one minority group of stakeholders may be subject to most of the negative consequences. Managers have to resist pressure from powerful majority interests and simultaneously avoid making decisions based solely on numbers from public opinion surveys because those may not best represent fairness, justice, and sustainability.

The important role of trust management for future generations is also a challenge for wildlife managers. If overdone, the trust manager role can be viewed by the public as a way for wildlife managers to veto other stakeholder interests. Moreover, wildlife professionals should be careful not to suggest that they alone (among all participating stakeholders) have a concern for future generations; understandably, that could be off-putting to others. In addition, decisions that reflect only the managers' perspective may not be implementable because stakeholders may then find alternate means to be heard, such as ballot initiatives, lawsuits, or end-runs to political appointees who can force agency decisions.

A related challenge is the use (and sometimes misuse) of science by stakeholders and managers to support their perspectives. Wildlife managers strive to incorporate relevant, credible scientific information when engaging stakeholders; however, stakeholders sometimes resist learning new information when it contradicts strongly held perspectives. Managers must be open to knowledge that stakeholders can bring, such as local ecological knowledge (Ballard et al. 2008). Open dialogue about the science-based knowledge that managers rely upon can help stakeholders accept it. For example, during the 2004 forest-planning process in the Green Mountain National Forest in Vermont, stakeholders had strong disagreements about the impacts of forest management on wildlife. Stakeholders shared and discussed published studies, but the credibility of the research was frequently questioned and stakeholders seemed reluctant to accept new information. To encourage open-minded learning, a panel of scientists was invited to present and discuss their research with stakeholders. Interaction with scientists is one method that has proven effective for helping stakeholders understand the process of scientific inquiry and validity of research results; however, use and misuse of science remains a source of conflict in wildlife management.

Aspects of wildlife management are likely to remain controversial well into the future. For example, much recent thinking about how we should treat animals challenges customary beliefs at a fundamental level. Advocates of animal rights and animal welfare want to enlarge the umbrella of protection that the states offer wildlife. Singer (1990) asserts, for example, that animals should receive the same moral considerations as humans. Depending on the extent to which citizens come to share that view, it may have profound implications for wildlife management.

Arguments about animal rights typically rest on strongly held beliefs that are usually part of an individual's worldview. Communicating and agreeing on wildlife management policies is difficult when competing worldviews exist. Stakeholder engagement, though challenging to implement under such circumstances, provides a framework for decision making.

3.3.3. Engaging Stakeholders Successfully

A management philosophy of stakeholder engagement is essential in modern wildlife management for at least three reasons. First, it considers the broad range of wildlife interests that exist in society and is open to considering others that may exist in the future. Second, because it is inclusive, more segments of the public understand and support management decisions. Third, the management of wildlife resources as a public trust resource calls for a philosophy that reflects the reasonable views of as many segments of society as possible.

The keys to successfully engaging stakeholders are (1) evaluating who is substantially affected by, or who can affect, wildlife management and who is therefore a significant stakeholder in decisions, (2) understanding stakeholder views, (3) seeking common goals underlying seemingly competing stakes, and (4) communicating with and learning from stakeholders.

Finding appropriate solutions to wildlife management issues requires clear-thinking, perceptive people. Those making decisions about management must consider a broad range of stakeholder interests and concerns, with the specific set depending on the situation. They are then responsible for determining the relative importance of those interests and concerns, which can be a daunting challenge. The major value of stakeholder engagement may be the help it provides wildlife managers by subjectively revealing relative weights people give to different stakes as they strive to find acceptable solutions (desired objectives for management and acceptable actions to achieve them).

Although there is no single definition of the "public interest," guidance is typically available in the form of laws that establish public agencies. The legal mandates are, in effect, interpretations of "public interest" goals for the agency. Even with well-executed processes for stakeholder engagement, managers who choose stakeholder input and involvement need to be willing to make value judgments and to label them as such.

As stakeholder engagement becomes more widespread, managers are gaining experience in designing practical strategies for involving stakeholders in decisions and for more explicitly articulating how decisions have been made. The range of approaches that is being applied is described in Chapter 11.

An ever-present challenge is accommodating stakeholders who refuse to recognize stakes other than their own. For some stakeholders dialogue with other stakeholders is all that is needed for productive participation. Other stakeholders are unwilling to listen to other interests that they believe will compromise their positions. The manager, having tried to include them, must move forward with the other participating stakeholders. The unyielding stakeholders have other methods of recourse, such as voter or legislative referenda, litigation, or seeking an audience with elected representatives.

Disgruntled stakeholders might turn to those methods despite the manager's best efforts, but those tactics are less often successful when well-conceived and well-executed stakeholder processes have been used and many stakeholder interests have been considered and incorporated. More and more, managers are designing stakeholder-engagement processes that utilize cooperative approaches rather than competitive approaches (Box 3.4). Although legislatures, courts, and populations of civic-minded citizens typically apply good judgment, stakeholder engagement will be a learning process for the profession and for stakeholders for some time to come.

Agencies proactively engaging stakeholders have realized success in a number of areas (see Chapter 11 for examples and discussion of advantages and approaches to stakeholder engagement). Stakeholder engagement also can have beneficial collateral and subsequent effects. In New York State, participation in a citizen task force that helped determine objectives for white-tailed deer populations led to greater trust in the agency as well as knowledge of the management system (Stout et al. 1992).

A management philosophy of proactive stakeholder engagement is essential to responsive, adaptive management.

The future of the profession will belong to the managers who adopt, refine, and practice this philosophy. The next section describes some traits we believe will help wildlife managers succeed.

3.4. BRINGING IN HUMAN DIMENSIONS INSIGHT

We describe in this section some key traits of wildlife managers who are successful at integrating stakeholder engagement and human dimensions scholarship into management as a regular way of doing business. Being receptive to, and inquisitive about, stakeholders' needs and interests is vital. Managers who are receptive and inquisitive do not assume that every problem has the same human dimensions characteristics. When faced with a management challenge, they ask what the nature of the problem, issue, or controversy is. They probe. They are analytic. They seek to understand both the biological and the human dimensions of the situation (i.e., they think about coupled social and ecological systems), and they are explicit in their assumptions and expectations for how the system works.

3.4.1. Be Receptive
What does being receptive mean? In any management situation, some people will seek out the wildlife manager simply to

Box 3.4 *FORGING COOPERATION IN MONTANA*

The Montana Partnership is an example of what an agency can achieve when it broadens its thinking about who its stakeholders are and works to engage these stakeholders in management. When Congress established the State Wildlife Grant Program in 2001, it required each state to develop a comprehensive assessment of its fish, wildlife, and habitat to ensure that federal funds would be used effectively. When Montana completed its Comprehensive Fish and Wildlife Conservation Strategy (CFWCS), it recognized that it did not by itself have the capacity to meet the conservation needs identified therein.

Consequently, Montana Fish, Wildlife and Parks (Montana FWP) collaborated with the Heart of the Rockies Initiative and the Five Valleys Land Trust to conduct a situation assessment as to whether the CFWCS could be used as an "umbrella" for all of the conservation work going on in Montana. A hired consultant interviewed more than 100 people throughout the state representing organizations with conservation interests. This situation assessment generated tremendous interest in the conservation community. It led to the recognition that considerable unrealized capacity for implementing the CFWCS existed, because numerous groups in addition to Montana FWP were doing work that advanced the CFWCS objectives. Realization of this potential could be facilitated by forming a broad-based

partnership to develop concrete proposals for building implementation capacity.

The creation of the Montana Partnership (originally called the Montana Conservation and Restoration Partnership) soon followed. This partnership is very diverse. As one might expect, it includes Montana FWP and other state and federal government agencies with conservation interests, as well as non-governmental conservation organizations. However, it also includes private landowners and representatives of ranching, petroleum, and timber industries—groups that are not always thought of as "conservation interests." Montana FWP recognized that for such a diverse partnership to be effective, it could not adopt a top-down approach to decision making, but needed to work side by side with other stakeholders; therefore, it focused on facilitating group formation, discussion, and decision making but avoided dominating the direction of the discussions and decisions.

The diverse interests represented in the Montana Partnership have been able to agree on a vision statement for Montana, which captures the core values that they recognized they share. They are working together on a series of initiatives to help achieve that vision, including improved communication and coordination among conservation interests and development of additional funding for community-based conservation.

share their views. Much more often, people want to convince the manager that their views are correct or most important. Managers should be receptive to their views.

Unsolicited input can take many forms—telephone calls, office visits, letters, columns in stakeholder newsletters, letters to the editor, editorials, news coverage of all types, posters, graffiti, demonstrations. Although such input is not as easily interpreted and generalizable as human dimensions data that are collected systematically in surveys, it should help the manager understand the landscape of public opinion surrounding an issue.

Do not overlook the usefulness of such information, but do keep it in proper perspective. Small-minority interests can sometimes have a large media presence, often as part of a strategy to influence decisions out of proportion to their actual stake in an issue. Being receptive (though it is necessary) is therefore not sufficient as an approach to learning about the human dimensions elements of a wildlife issue. What else is necessary?

3.4.2. Be Inquisitive

To avoid the limited perspective that can develop from passively being receptive to stakeholder views, go out and inquire about stakeholder needs and interests. You will quickly learn that multiple perspectives exist about most issues and, therefore, about their resolution.

The first step in being inquisitive is to determine the range of relevant stakes in an issue. Identify the stakeholders and learn about them. Who are the people with this stake in the management decision? What is the size of their group? What are their relevant beliefs, attitudes, and behaviors? How do they articulate their stake? Those and other questions come naturally to the inquisitive wildlife manager.

Sometimes answers to such questions are easily obtained or approximated. In other situations, obtaining answers may require special research. You may use scientific surveys, content analyses, and other social-science inquiry methods to reveal answers to your questions and illuminate the human dimensions components of the situation that can help you guide people toward socially acceptable and politically durable decisions. Useful techniques are discussed in Chapter 13.

3.4.3. Be Analytic

Effective managers need to be problem solvers, not just process coordinators. The entire management process is intriguing because of the intellectual challenges it presents. Applying your knowledge, intuition, and skills to a problem will make wildlife management consistently stimulating.

To be effective in problem solving, you need to carefully define the problem and scope out its important and sometimes subtle elements (e.g., the ecological, social, and political scales involved). You also need to develop a process or plan for approaching the problem. What kinds of decisions are inevitable? Who should be involved and to what degree? Is it enough to gather the views of stakeholders for this situation, or will they need to actively participate in discussions? Is seeking input suf-

ficient, or will stakeholders expect to be involved in making decisions? Is involvement in decision making adequate, or is it essential for stakeholders to be involved in implementing and evaluating the management efforts?

Counting and measuring stakeholder groups. The complexity of the human dimensions elements of a wildlife management issue is related to the number of stakeholder groups and the diversity of their beliefs and attitudes about the issue. An important step in comprehending an issue fully is to identify the stakes involved. Next, determine the approximate size of the stakeholder groups. Do not forget that managers have a responsibility to consider not only those stakeholders who make their presence known but also those who do not even realize they have a significant stake in the issue. The merely receptive manager could easily miss important stakeholder groups. The inquisitive manager seeks them out.

Describing stakeholder groups. Knowing the number and size of stakeholder groups is only a start toward scoping out the human dimensions context for wildlife management. Managers routinely seek information about demographic and communication characteristics of stakeholders. Familiar kinds of characteristics that are used to describe groups in society are also helpful to managers; these include age, gender, educational accomplishment, income level, geographic residency (urban, rural), race and ethnicity, sources of information (newspaper, television, radio), and trusted opinion sources (friends and neighbors, agency professionals, interest group leaders).

Basic information about stakeholders' characteristics can help managers understand how wildlife programs are received and how they can be improved. Although demographic characteristics are straightforward to measure, they may require purposeful, careful inquiry to obtain.

Attitudes, beliefs, and behaviors are other kinds of valuable descriptive information about stakeholders. They are difficult to measure, however, so the manager needs the expert assistance of a specialist in applied social sciences.

Recognizing agency stakeholders. Many wildlife management agencies regard staff within their agency as a special group of management stakeholders. Wildlife-agency staff members can be considered management stakeholders when they are influenced by, or when they help to design, implement, or evaluate, management programs. Natural resource agencies can also think of other agencies as stakeholders. For example, within the context of a specific land-management project, such as the restoration of a wetland complex across federal and state lands, several agencies may be stakeholders (or even partners, a particular role for some stakeholders).

A developing body of research is yielding understanding about "internal" wildlife-management stakeholders. The information can be used to help identify problems and opportunities within wildlife management organizations and to improve working relationships among professionals and between professionals and external stakeholders. Such improvement should contribute to greater job satisfaction for professional staff and ultimately translate into improved wildlife-management decisions, actions, and policies.

The most common studies involving internal stakeholders are those that compare the attitudes and beliefs of wildlife management professionals with those of external stakeholders. For example, studies have compared wildlife professionals and external stakeholder groups with regard to their attitudes toward animals (Peyton and Langenau 1985), perceptions of the influence of wildlife interest groups on state agency policy (Brown and Decker 1982), beliefs about the appropriate relationship of people to the environment (Shanks 1992), perceptions of damage to agricultural crops (Salatiel and Irby 1998), and beliefs about wildlife rehabilitation (Bright et al. 1997). In some studies of co-orientation, stakeholders and managers are asked to predict each other's attitudes and beliefs to inform the design of appropriate communication strategies (Leong et al. 2007, 2008).

Less common are studies of organizational culture in the natural resource professions. Some research has included consideration of wildlife professionals (Kennedy 1985) and wildlife management agencies (Brunson and Kennedy 1995), including organizational norms regarding employee engagement in professional activities (Lauber et al. 2009, 2010).

3.5. SERVING MANAGEMENT DECISIONS

The principal value of human dimensions insight (and biological insight) is to serve management decisions, big and small. Such insight can aid in planning and improve timely response to changing conditions. Incorporating knowledge about stakeholders—such as attitudes and values underlying wicked problems or interest in primary, collateral, and subsequent impacts—can guide wildlife managers in the daily performance of their duties. It can help them anticipate problems and even avoid them.

In addition to the day-to-day application of insight about stakeholders, another major role is informing specific policy or management decisions. Those decisions come in various forms and often are not clearly framed. Many managers believe that simply having more human dimensions data will help solve a problem, but collecting more data about an ill-defined, wicked problem that lacks decision criteria can lead to greater confusion. For that reason many researchers insist on knowing what decisions the findings from a proposed study are to serve. Managers' models (Decker et al. 2011) and other techniques can help identify and appropriately frame the problem. If management questions are not clearly articulated, it is difficult to provide useful biological or sociological data.

Inadequate articulation of decisions to be made is a major impediment to the integration of human dimensions findings. Studies are too often developed ostensibly to support decisions in situations with poorly defined criteria for guiding these decisions; predictably, the resulting data are not very useful. Had the manager and researcher worked together to scope out the problem, define the decisions, and ensure that the study would serve those decisions, they could have avoided wasteful and frustrating efforts. In transactional or co-managerial

Transactional forms of stakeholder engagement allow for multi-way communication between managers and stakeholders, and between stakeholders (courtesy William Siemer)

approaches (see Chapter 11), stakeholders often are brought into management decision making earlier in the process to help frame the problem and also to come to agreement on the necessary scientific base for decision making.

To frame and focus a human dimensions inquiry, it is useful to ask a few questions:

- What decisions need to be made to address the issue?
- Who should be involved in the decisions?
- Who will have to accept the decisions?
- How precise should the information be?
- From whom should information be obtained?
- When does the decision have to be made and the action taken?

The manager also needs to consider which characteristics of the situation to monitor and be sure to collect baseline information on those characteristics. People's beliefs and attitudes change and situations change, sometimes because of the inquiry itself and certainly because of the eventual management actions. Monitoring the changes is an elemental part of program evaluation and adaptive management. Measuring change and progress is difficult without baseline data.

The dynamic nature of both the biological dimensions and the human dimensions of wildlife management requires an adaptive approach. Solving a problem today does not mean that it will not resurface later or that a related one will not emerge. That is why evaluation and monitoring are important. By evaluating the influence of experience and program activity on stakeholders' beliefs, attitudes, and behaviors, you begin to better understand truly your stakeholders.

Cooperation, collaboration, and coalition building to overcome jurisdictional impediments and bridge values chasms will be critical to future successes in wildlife management. These coalitions and collaborative ventures can be challenging from the standpoint of human dimensions scholarship. Systematic assessment of the values and motivations of collaborators, cooperators, and partners may be a key to developing successful working relationships.

SUMMARY

This chapter explored the role of stakeholders in wildlife management. Stakeholder engagement helps managers fully realize their role in managing public trust resources. By incorporating the full range of stakeholders throughout all stages of management, agencies improve the durability of decisions and the sustainability of management activities. Lasting relationships gained in one decision or issue serve future decisions. As societal values develop and change, a manager practicing stakeholder engagement will be well-poised to identify and incorporate new interests in wildlife management.

- Several forces accelerated the emergence of public participation in decisions about public wildlife resources: (1) diverse and sometimes conflicting public interests, concerns, and uses of wildlife; (2) impacts of wildlife on the public; (3) public expectation for citizen participation in democratic decision making; and (4) an inclusive view among many managers about who are beneficiaries of wildlife management.
- A stakeholder in wildlife management is defined as *any person who significantly affects, or is affected by, wildlife management*. A full range of stakeholders includes those who benefit, as well as those who experience problems, from human–wildlife interactions. Wildlife professionals also have a responsibility for ensuring that tomorrow's citizens are considered stakeholders in today's management decisions.
- Stakeholders are distinguished from constituents and clients. A constituent (e.g., a hunter) is someone who authorizes or supports efforts of others (in this case, a state wildlife agency) to act on their behalf. Clients are people who pay (e.g., through license fees or earmarked taxation) for professional services. All constituents and clients are

stakeholders, yet not all stakeholders are constituents or clients.
- One of the most difficult issues when engaging stakeholders is the choice of whose values to consider and how to weigh them. Ultimately, managers confront ethical questions (see Chapter 16). When invoking professional judgment about whose values to consider, thoughtful managers distinguish between what is scientifically defensible and what is morally or legally arguable and credible.
- Stakeholder engagement does more than simply respond to societal desires for increased voice in management decisions; the practice of involving stakeholders in management broadens the focus of the wildlife profession itself.
- Wildlife professionals use a range of methods to incorporate stakeholders in management decision making, which depend on the specific context and stage of the management process (the subject of Chapter 11).

Suggested Readings

Chess, C., T. Dietz, and M. Shannon. 1998. Who should deliberate when? Human Ecology Review 5:45–48.

Chess, C., and K. Purcell. 1999. Public participation and the environment: do we know what works? Environmental Science and Technology 33:2685–2692.

Forester, J. 1999. The deliberative practitioner: encouraging participatory planning processes. MIT Press, Cambridge, Massachusetts, USA.

Pearce, W. B., and S. W. Littlejohn. 1997. Moral conflict: when social worlds collide. Sage, Thousand Oaks, California, USA.

Walker, G. B., and S. E. Daniels. 2001. Natural resource policy and the paradox of public involvement: bringing scientists and citizens together. Journal of Sustainable Forestry 13:253–269.

Wondolleck, J. M., and S. L. Yaffee. 2000. Making collaboration work: lessons from innovation in natural resource management. Island Press, Washington, D.C., USA.

PART II · SOCIAL SCIENCE CONSIDERATIONS

In Chapter 1, we indicated that everything that is not about wildlife organisms or their habitat is about human dimensions. Put simply, the field is immense! Unfortunately, we cannot cover the full breadth of relevant topics in this book and have left out contributing fields of study such as anthropology and law. Chapters 4–6 introduce you to concepts from social psychology, sociology, and economics that lay the theoretical foundation for the remaining portions of this book. We believe these concepts prepare wildlife professionals well for most human dimensions issues they encounter in their careers.

The basis for human values, an understanding of which is central to a focus on impacts in wildlife management, are reviewed in Chapter 4. Social psychology is the scientific study of the way in which people are influenced by their environment and how behaviors develop. Two theoretical approaches to understanding human behavior, cognitive and motivation–satisfaction, are emphasized in the chapter. Cognitive approaches help explain how values, attitudes, and norms influence human behaviors. Motivation theory seeks to explain why humans do what they do, and satisfaction theories attempt to explain how and why people evaluate their experiences in a given way. These topics are important in anticipating how wildlife management will affect the values and behaviors of stakeholders.

Continual changes in society, whether from economic globalization or human migration patterns such as the urbanization, suburbanization, and exurbanization occurring in much of the world, necessitate at least rudimentary knowledge of sociology to understand the current context for wildlife management and to anticipate how social changes will affect stakeholders in the future. Chapter 5 points out that sociology is concerned with what people do as members of a group or when interacting with one another—the patterns and dynamics of a society. Three theoretical approaches underlie specific sociological analyses that apply to wildlife management: functionalist, conflict, and interactionist. As you will read in the chapter, each offers perspectives on the process by which society is organized, emphasizes different outcomes, and implies different stakeholder-engagement strategies for wildlife managers.

Relevant economic concepts for applications in wildlife management are reviewed in Chapter 6. Economics is a social science that studies how society meets competing demands for goods and services in the face of limited resources. Economists measure (usually in monetary terms) the flow of values through society. The values of wildlife and human–wildlife interactions often are difficult to calculate solely on a market basis. Economists use non-market valuation methods, such as contingent valuation, to estimate the value of ecological goods and services produced by wildlife. Insights derived from non-market valuation provide a basis for estimating the costs and benefits, efficiencies, and cost-effectiveness of management interventions to produce the wildlife goods and services.

Social psychology, sociology, and economics are like any other science field—they reveal only what is, not what should be. Although these sciences inform the vital process of making decisions, wildlife managers and stakeholders engage in the often complex juggling of values to determine management objectives and choose management actions.

SOCIAL PSYCHOLOGICAL CONSIDERATIONS IN WILDLIFE MANAGEMENT

JERRY J. VASKE AND MICHAEL J. MANFREDO

Human dimensions research contributes to the knowledge about the public's thoughts and actions toward wildlife. Such knowledge is essential for accomplishing management goals such as encouraging participation in wildlife-related activities, reducing conflict among wildlife stakeholders, educating people about management practices, and predicting stakeholder positions on emerging issues. Management interest in this research stems from a desire to understand, predict, and influence the public's behavior in wildlife-related issues.

A scientific approach to the human dimensions of wildlife management has developed that knowledge during the past 30 years, especially by building on relevant theory and concepts in social psychology, such as values, attitudes, and norms (Manfredo et al. 2009b). Theory is important because it extends the generalizability of findings, improves the rigor and confidence in the research, provides a framework for integrating and building on previous findings, and contributes more to knowledge than purely descriptive research.

Social psychology is the scientific study of the way in which people's thoughts, feelings, and behaviors are influenced by their environment. *Cognitive approaches* have traditionally examined concepts such as values, attitudes, and norms that underlie the process that leads from human thought to action and the relationships among those concepts, especially to predict behavior. More recently, attention has been given to the importance of *emotions* in human–wildlife relationships. *Motivation approaches* seek to explain why we do what we do, and *satisfaction theories* attempt to explain why people evaluate their experiences in a given way.

This chapter reviews some of the basics of cognitive, emotion, and motivation–satisfaction approaches to understanding the social aspects of wildlife management, discusses important concepts for wildlife management, and presents examples of the concepts.

4.1. A COGNITIVE APPROACH

Cognitions refer to the collection of mental processes used in perceiving, remembering, thinking, and understanding, as well as the act of using these processes. The cognitive approach explores the relationships between values, value orientations, attitudes, and norms to understand how these concepts influence behavior.

Popular media commonly assert that values influence behaviors toward wildlife, but empirical evidence showing direct predictive validity is sparse. For example, research suggests that values have limited effects on predicting specific wildlife-related behaviors or support for management actions (Manfredo 2008). Cognitive theories offer explanations for these disparities, suggesting that attitudes and norms mediate the relationships between values and behavior. These theories distinguish stable but general values from more specific cognitions (e.g., attitudes and norms) that people use to evaluate objects or situations encountered in daily life. In social psychology, an *object* can be any entity that is being evaluated (e.g., a person, situation, wildlife, management action, or policy). Social psychologists differentiate concepts (e.g., values, value orientations, attitudes) based in part on the specificity of objects being measured. Specific attitudinal or normative variables predict behaviors better than more general cognitions such as values.

Such cognitions are best understood as part of a "hierarchy" from general to specific (Fig. 4.1). These elements build upon one another in what has been described as a "cognitive hierarchy," which has been applied to evaluations and behavior associated with wildlife (see Manfredo 2008 and Vaske 2008 for reviews).

4.1.1. Values

Values are commonly defined as desirable individual end states, modes of conduct, or qualities of life that we individually or collectively hold dear, such as freedom, equality, and honesty (Rokeach 1973). Values are general mental constructs that are not linked to specific objects or situations. Thus, a person who holds "honesty" as a value is expected to be honest when completing Internal Revenue Service tax forms, conducting business deals, or interacting with friends. Values reflect our most basic desires and goals and define what is important to us, such as honor and fairness. Values are often formed early in

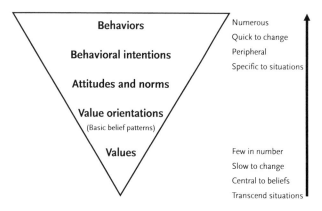

Figure 4.1. The cognitive hierarchy (adapted from Manfredo 2008)

life, are culturally constructed, and are tied with one's identity; therefore, they are extremely resistant to change.

People tend to have a small set of core values (Manfredo 2008). Such values can direct a large number of attitudes that express those values. For example, a person's attitude toward lethal control of a particular wildlife species may originate from values regarding respect for life. Much of the early applied work on values was not explained in the context of theory, so students (and even professionals) sometimes confuse the concepts of values and attitudes. Kellert (1980*a*), for example, created an item bank intended to measure nine different domains of thought about wildlife. Somewhat interchangeably, he suggested that the items measure both wildlife values and wildlife attitudes. More recent work has built on this foundation and advanced methodological procedures that are directly tied to cognitive theories (see Manfredo 2008 and Vaske 2008 for reviews).

4.1.2. Value Orientations

Because values tend to be widely shared by all members of a culture, values are unlikely to account for much of the variability in specific behaviors (Box 4.1). The notion of *wildlife value orientations* was introduced as a concept that describes the way that a value attains meaning for an individual. The value orientation concept initially described a pattern of direction and intensity among basic beliefs (indicative of a value) about wildlife.

Manfredo et al. (2009*a*) later proposed that, relative to wildlife, values are oriented by one of two cultural ideologies: domination (spawning utilitarian views of wildlife) and egalitarianism (giving rise to notions of equality and mutualism with wildlife). Utilitarian and mutualism wildlife-value orientations are measured using statements that depict different ideal worlds and the acceptable modes of conduct toward wildlife expected from people holding these different orientations. For example, an ideal world for a person with a strong mutualist orientation might be "animals and humans live side by side without fear," while an ideal world for an individual with a strong utilitarian orientation might be "fish and wildlife exist to benefit humans" (see Manfredo et al. 2009*a* for additional examples).

The Western Association of Fish and Wildlife Agencies commissioned a 19-state study of wildlife value orientations in the western United States. A map from that study shows that some states have a more utilitarian culture than other states (Fig. 4.2). States where there are higher percentages of utilitarians have higher percentages of hunters and anglers and, given conflict situations, are more likely to support lethal control of wildlife (Teel and Manfredo 2010). The study authors suggest that modernization (indicated by higher income, urbanization, and education) leads to a shift from utilitarian to mutualism value orientations (Manfredo et al. 2009*a*).

4.1.3. Attitudes

Attitudes are one of the most frequently studied concepts in the social sciences (Manfredo et al. 2004). *Attitudes* are defined as the favorable or unfavorable evaluation of a person, object, or action. Contemporary thinking divides attitudes into explicit attitudes and implicit attitudes. Explicit attitudes are formed from deliberate thought and processing, while implicit attitudes occur more automatically and often do not enter a level of conscious processing. Virtually all the research in human dimensions of wildlife has focused on explicit attitudes, although implicit attitudes are a critical area for future research.

Attitudes are a particularly important concept because they precede and direct behavior. While value orientations are believed to direct attitudes, attitudes are believed to directly influence behavior. Short-term behavior change typically will not become permanent unless one changes the accompanying constructs causing the behavior, such as underlying attitudes. Knowing what influences behavior helps us predict it more accurately.

Attitude questions used in surveys are typically framed in terms of like–dislike, good–bad, or positive–negative. Much of the human dimensions research dealing with opinions, preferences, and perceptions is actually an examination of attitudes (Manfredo et al. 2004). Perceived crowding, for example, is defined as a negative evaluation of a certain number of people in a given situation (e.g., birdwatchers at a National Wildlife Refuge). Defined in this manner, crowding is an attitude that people feel about seeing others while engaged in an activity. Similarly, determining whether a person likes new regulations that limit the number of people using a refuge is an indication of that person's attitude toward the regulation.

Attitudes have both an evaluative and a cognitive dimension. The evaluative component refers to whether the individual views the attitude object as positive or negative. The cognitive aspect refers to the beliefs associated with the attitude object. Beliefs are what we think are true, but are not necessarily objective facts. Both the cognitive and the evaluative characteristics of an attitude must be understood to predict behavior (Box 4.2). For example, one person may have a cognitive belief that wolves are dangerous to humans and will therefore evaluate wolves negatively because of fear. Another person may also believe wolves are dangerous but feel positively toward them because s/he is excited by the potential

Box 4.1 *WHY VALUES DO NOT DIRECTLY PREDICT WILDLIFE-RELATED BEHAVIORS*
An Illustration Based on the Cognitive Hierarchy

Based on the cognitive hierarchy, the figure below illustrates why values do not directly predict wildlife-related behaviors. Both person 1 and person 2 hold "respect for life" as a value. Person 1, however, attends anti-hunting rallies, while person 2 hunts. The reasons for this apparent discrepancy are the variables that mediate between the value and the behaviors. Person 1, for example, (1) believes that animals should have rights similar to humans (a basic belief), (2) has a personal norm against eating meat, and (3) feels that hunting is a negative activity. Person 2, on the other hand, (1) has a more utilitarian view of wildlife, (2) believes that wild game should not be wasted (a norm), and (3) has a positive attitude toward hunting. The specific attitudes and norms outweigh the influence of the more general basic beliefs and value in predicting the behaviors. Although the two people share the same value (respect for life), person 1 applies the value to both humans and wildlife; person 2 applies the value to humans but not to animals.

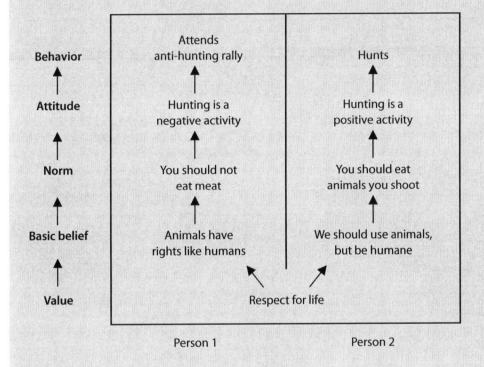

The cognitive hierarchy illustrates the reasons that values do not directly predict wildlife-related behaviors

danger or can avoid negative encounters with wolves. Both individuals share the belief that wolves are dangerous, but their evaluations of this belief are different.

Beliefs, attitudes, and behaviors are most strongly related when measured at "corresponding levels of specificity" (Whittaker et al. 2006). For example, to determine how people will vote to support wolf reintroduction, we should determine their specific attitudes toward *wolf reintroduction,* not just their beliefs about or attitudes toward *wolves* in general. Research has shown that while general attitudes can predict general behaviors (e.g., voting in general), specific attitudes are better for predicting specific behaviors (e.g., voting for a specific wildlife initiative). By framing items that measure attitudes so that they are context-specific, we can improve predictions of behavior. Ajzen and Fishbein (1980) identify four specificity variables across which measurements of attitude and behavior

should correspond: target (e.g., deer); context (e.g., deer are causing Lyme disease); action (e.g., conduct a special hunt); and time (e.g., next month).

Prediction of behavior is also enhanced when attitude salience is considered. *Salience* refers to how easily and quickly thoughts come to mind when an attitude object is introduced. For example, to determine the salience of a trapping ban in Colorado, Manfredo et al. (1997) asked voters why they voted to support or oppose it. The list of thoughts respondents provided indicated what was salient to them. Objects or ideas with more salience to a person are easier for the person to think about; they have higher *accessibility* and can be retrieved from memory or thoughts more easily.

Salience can help explain the reason that a person holds an attitude. The salient points are the types of things a person has thought about in forming an attitude. Salience also indicates

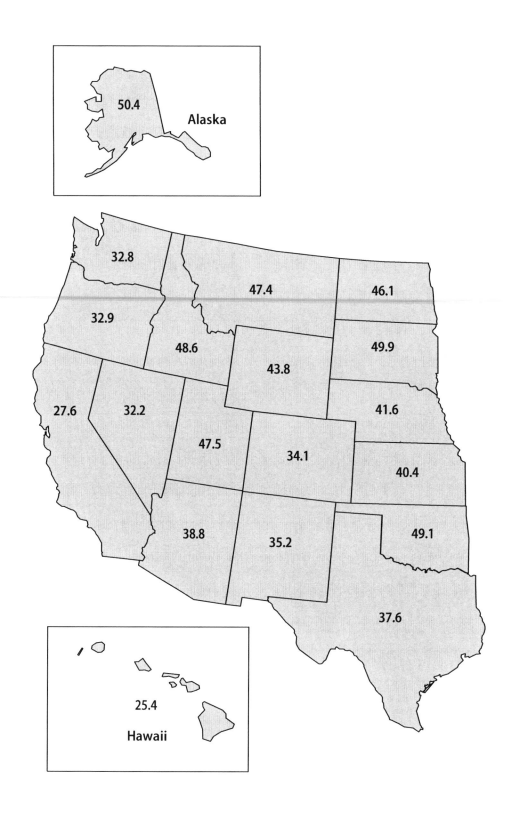

Figure 4.2. Percent (by U.S. state) of the population that exhibits utilitarian (domination-oriented) values toward wildlife (Manfredo et al. 2009*a*)

how much a person has thought about the attitude object. If an issue is very salient to a person, s/he is likely to have spent more time considering the issue and have more thoughts associated with it.

Strongly held attitudes are difficult to change. Attitude *strength,* however, should not be confused with how extreme a person is about something. Some people may have extreme viewpoints about a wildlife issue, such as trapping of animals, but not hold the attitude strongly because it is not a central issue to them. In such a case, the attitude has the potential to change given further information or persuasive attempts.

4.1.4. Norms

Norms can refer to what most people are doing (a descriptive norm) or to what people *should* or *ought to* do (an injunctive norm) in a given situation. *Social norms* are defined as "standards shared by the members of a social group" and *personal norms* are defined as an individual's own expectations, learned from experience, and modified through interaction.

In many definitions, norms are also intimately tied to the concept of sanctions—punishment for people who break norms or rewards for compliance with norms. Norms that are widely shared by most members of society (e.g., not littering, not poaching) often become legal mandates complete with formal sanctions (e.g., fines) for non-compliance. Such norms are also likely to be internalized, viewed as being right, legitimate, and hence obligatory. When there is less agreement or the norms are emerging, informal sanctions may be used to encourage acceptable behavior. Waterfowl hunters, for example, are not obligated to set up their blinds a certain distance from others in a marsh, but some degree of separation from other hunters is expected. Those who fail to comply with this personal distance norm are not formally sanctioned. Rather, informal sanctions such as "dirty looks" or shouts of disapproval serve to communicate and enforce the norm. If individuals internalize a norm, external sanctions are less necessary.

Norms help to explain why people (either individually or collectively) often act in regular ways, as well as aid our understanding of *irregular* human behavior. Anti-litter norms, for example, are strong and widely held; yet litter is often present, even in wilderness. Norms are interesting precisely because they vary by the proportion of people who hold them, their strength in an individual or group, the level of agreement about them, their influence on behavior, and their wider enforcement of social regularities. However, norms (like attitudes) are not static within or across people or situations. In a given social context, some people may have a well-formed norm that dominates their behavior or evaluation, while others may have only an emerging norm that barely influences what individuals do or think. Still others may be unaware of a norm and become bewildered when sanctions are brought against them for breaking it. Even well-formed norms may fail to influence behavior because of competing norms, attitudes, or motivations.

Different social psychologists define and use the concept of norms differently (see Vaske and Whittaker 2004 for a review). Some concentrate on the variables that serve to focus or activate a norm, while others address how social pressure can influence behavior or aid in the diffusion of ideas. Still others emphasize the structural characteristics of norms to help evaluate appropriate behavior or conditions. Knowing how different researchers use the same concept (e.g., norms) clarifies what theoretical approach is most appropriate for examining a given situation or problem. For example, if the issue involves promoting responsible environmental behavior (e.g., not littering, not poaching), norm focus–activation models are more appropriate. If the issue involves determining standards for acceptable human impacts, structural approaches are a better choice.

Norm theories also differ in how they measure the concept of norms. Norm focus–activation theories and the structural-norm approach measure norms at the individual level (i.e., personal norms) and then aggregate the data to derive social norms. The theory of reasoned action (Fishbein and Ajzen 1975), in contrast, focuses primarily on perceived social norms (i.e., subjective norms) and does not directly address the concept of a personal norm.

Norms can be linked to attitudes and are often construed as parallel constructs. Like attitudes, norms have both cognitive and affective components (the strength of obligation can be tied to emotions such as guilt) as well as the ability to influence behavior. Some attitudes and norms are more global than others, and the specificity of each is critical for determining whether the attitude or norm will accurately predict behavior. Norms are different from attitudes, however, because of the added dimension of obligation. Attitude measures focus on positive or negative evaluations, while norm variables examine acceptability evaluations (what a person, group, or institution should do). Beliefs about internal or external sanctions are additional components without parallels in attitude models.

A fundamental issue in understanding norms is the idea of *norm strength*. The ability of a norm to predict individual or group behavior is influenced by how strongly a norm is held by an individual or group. A norm does not just exist or not exist; there is a matter of degree. As a construct represented in a person's mind, a norm may be weakly held, difficult to access, without much sense of obligation, with no connection to moral values, and may be associated with low expectations of trivial external sanctions. As such, the norm is not expected to affect behavior. However, if the norm is strong, has a sense of obligation attached, and brings expectations of serious sanctions, it is likely to affect behavior. The research challenge is to measure the varieties of information that can be collected about normative concepts in people or across groups and then relate that to their behavior.

Norms have been important to wildlife management in three ways. First, norm research has been used to establish standards that specify the conditions that are acceptable to society or specific stakeholders. For example, wildlife mana-

Box 4.2 *THE BELIEF AND EVALUATIVE COMPONENTS OF AN ATTITUDE TOWARD A TRAPPING AMENDMENT*

In November 1996, Colorado voters passed a ballot initiative to amend the state's constitution by restricting wildlife trapping on public and private lands (Amendment 14). A telephone survey (*n* = 408) of registered voters was conducted immediately after the election (see Vaske 2008). Using the theory of reasoned action as the theoretical foundation, the study's objective was to explain the voting behavior by using measures of the respondents' attitude toward the amendment.

Prior to finalizing the telephone survey, an elicitation study (*n* = 50) identified four salient beliefs associated with the trapping ban. These four issues were asked twice in the final survey. In the first set of questions, respondents were asked whether each outcome was likely or unlikely (a belief). The second set of questions asked respondents to indicate whether each outcome was bad or good (an evaluation). The mean scores for those who voted for and those who voted against the trapping ban for both the belief and the evaluation components of the attitude are shown in the figure below. Each respondent's belief was

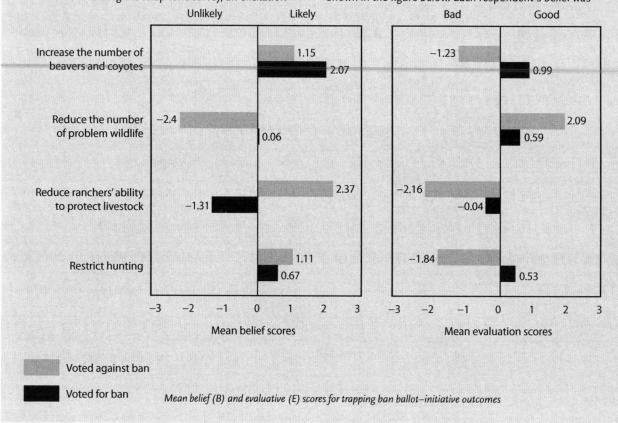

Mean belief (B) and evaluative (E) scores for trapping ban ballot–initiative outcomes

gers often seek to understand human tolerances for the social encounters, physical conditions, and services they offer hunters and wildlife watchers. The structural characteristics of the norms approach has helped in the development of indicators and standards that define quality experiences (Box 4.3).

Second, norms can predict behavior. The theory of reasoned action (TRA) and theory of planned behavior (TPB) hypothesize that people partially base their behaviors on subjective norms—what they think other people think they should or should not do in a given situation (Ajzen and Fishbein 1980). Natural resource applications of TRA and TPB models have shown that these subjective norms can predict behavior (e.g., predict hunting intentions and voting for a trapping amendment [Vaske 2008]).

Third, norms can influence behavior. The influence of others may be particularly important for people who are less informed about an issue. For example, one of the main sources of information for Colorado residents voting on a proposed trapping ban (Manfredo et al. 1997) was talking to others. In another example, drawing people's attention to the presence or absence of litter can affect their own littering behavior (Cialdini et al. 1991).

4.1.5. Emotions

The term *affect* in attitude and norm research refers to a general class of *feeling states* experienced by humans; *emotions* are subsumed under this category (Manfredo 2008). Because emotions are complex, a variety of definitions exists and a com-

multiplied by his or her corresponding evaluation of that outcome. This process resulted in four belief × evaluation (BE) product scores. Summing across the four BE products yielded the individual's attitude toward the trapping ban amendment, as shown in the figure below. Respondents who were against the ban consistently held a strong negative attitude toward Amendment 14. The averages of these BE products ranged from −2.5 to −5.65. Those who voted for the ban consistently held a positive attitude, but the BE scores were held with less conviction. These BE scores ranged from 0.16 to 1.06. This visual display facilitates an understanding of how the attitudes of the two groups differed.

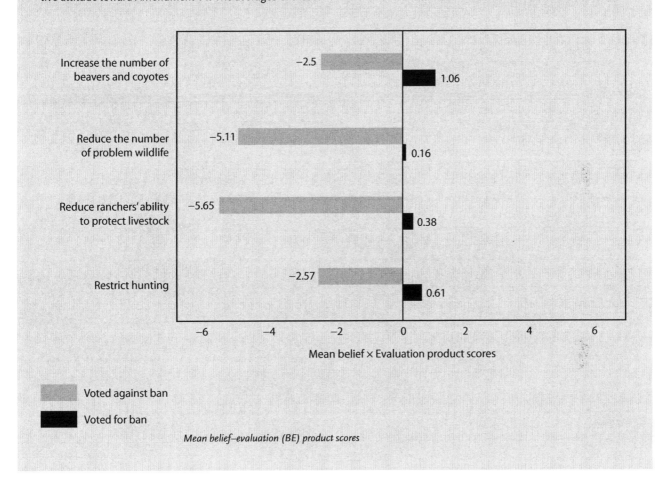

Mean belief–evaluation (BE) product scores

monly shared definition has not emerged (Barrett et al. 2007, Izard 2007). Many scholars, however, agree that emotional responses consist of (a) physiological reactions (e.g., increased heartbeat), (b) expressive reactions (e.g., smiling), (c) behavioral tendencies (e.g., approaching), and (d) emotional experiences (e.g., interpreting the situation, feeling happy; Cornelius 1996). The experience of emotion brings together, at a specific point in time, the affect, perceptions of meaning, and existing knowledge about the situation.

Emotional responses are at the heart of human attraction to, and conflict over, wildlife. The surprise and / or fear hikers experience when they encounter a wolf in the wild, or the anger that a rancher might express to a wildlife manager over a decision to reintroduce wolves, are emotion-laden events.

Such situations provide compelling justification to increase research on the emotional responses to wildlife. (For more information on emotions, see the special issue of *Human Dimensions of Wildlife* [Vol. 17, No. 1, 2012], Emotional Responses to Wildlife).

Compared with the cognitive approaches, empirical research on emotions is relatively scarce in human dimensions of wildlife (Jacobs et al. 2012). The studies that do exist have not explored the concept systematically and the findings have been fragmentary. This lack of attention given to emotional responses can be attributed to two reasons. First, the wildlife professional has traditionally emphasized *science* and sought to exclude emotional considerations from the decision-making process (Manfredo et al. 2009b). This emphasis has not encour-

Box 4.3 *INDICATORS AND STANDARDS FOR THREE ELK-VIEWING EXPERIENCE OPPORTUNITIES*

Indicators are the biological, ecological, social, managerial, or other conditions that managers and visitors care about for a given experience. *Standards* restate management objectives in quantitative terms and specify the appropriate levels or acceptable limits for the impact indicators (i.e., how much impact is too much for a given indicator).

Standards identify conditions that are desirable (e.g., availability of and proximity to animals) as well as the conditions that managers do not want to exceed (e.g., encounters with other people, wildlife flight reactions, incidents of wildlife–human conflict).

Elk viewing is among the most popular activities for visitors to Rocky Mountain National Park. Visitor expectations about crowding and other contextual factors influence their experience satisfaction. (courtesy Bret Muter)

aged researchers to embark upon an exploration of the role of emotion in human–wildlife relationships. Second, emotions research often employs techniques that use physiological measures, which necessitate laboratory-based, experimentally designed studies. Findings from these types of studies often have limited implications for an applied field such as human dimensions of wildlife.

The exploration of emotional responses to wildlife, however, may be one of the most intriguing and fruitful areas for future investigation (Manfredo 2008). First, emotions reflect our most basic reactions to animals. Research suggests that the rudiments of emotion are inherited and interact closely with cognitive functions to affect human behavior (Izard 2007). Second, although emotions may produce uncontrolled reactions (e.g., fear, rage, anger) they are critical to sound decision making (Cacioppo and Gardner 1999). Enhancing our understanding of human behavior will ultimately occur by exploring the interrelationship of cognitive concepts such as value

orientations, attitudes, and norms with affective concepts such as emotion.

When examining this interaction between cognitions and emotions, it must be emphasized that emotions and cognitions are theorized to be part of separate systems (i.e., emotions have an effect on behavior that is independent of thoughtful processing). Zajonic (2000) emphasizes the differences between emotions and cognition. For example, there are a limited number of emotions that are universal across cultures. By contrast, there are an infinite number of cognitions that vary greatly by cultures.

Different disciplines have examined emotions using expressive reactions, physiological responses, brain imaging, importance to appraisal, and subjective experience. Given that concepts employed in human dimensions cognitive models denote mental dispositions (e.g., value orientations), one potential starting point for studying emotions felt toward wildlife is to focus on emotional dispositions. Existing literature has

Impact indicator	Standards for different elk-viewing experience opportunities		
	Highly specialized back-country elk-viewing	Specialized front-country elk-viewing	General-interest roadside elk-viewing
Development			
Physical barriers separating humans and wildlife at prime viewing areas	0 barriers	≤10% areas with barriers	Barriers installed wherever needed to increase safety
Viewing blinds or hides			
Type	Temporary blinds	Temporary or permanent	Permanent blinds
Material	Natural colored fabric	Wood-clad, indigenous structure	Wood-clad, equipped with binoculars
Size	≤50 sq. feet (4.6 m^2)	≤200 sq. feet (18.6 m^2)	≤600 sq. feet (55.7 m^2)
Design and/or shape	Does not apply	Multilevel; irregular design to enhance privacy between viewers	Single-level; square or rectangular design; multiple access points
Trail surface	Indigenous mulch	Unpaved	Paved
Trail tread-way width	≤1 foot (0.3 m)	≤4 feet (1.2 m)	≤8 feet (2.4 m)
Crowding and/or norm tolerances			
Encounters with other groups on trails	≤4 groups/day	≤4 groups/hr	≤25 people/hr
No. of people in sight at one time	0	≤10 people	≤50 people
Percent of viewers feeling crowded	≤35%	≤35%	≤50%
Viewing distances to concentrations of elk	80% probability of viewing distance within 50 feet (15 m)	50% probability of viewing distance within 50 feet (15 m)	80% probability of viewing distance within 200 feet (61 m)
Regimentation			
Group size limits	≤4 people/party	≤8 people/party	≤25 people/party
Ranger escort	No	Yes	Yes
Freedom to roam beyond trails	Yes	No	No
Human–wildlife interaction			
No. of human–wildlife incidents involving:			
Injuries to humans	0	0	0
Disturbance to elk	0	0	0
Wildlife flight distance	≤200 feet (61 m)	≤50 feet (15 m)	≤50 feet (15 m)

Source: Adapted from Vaske et al. (2002)

defined "emotional dispositions" using two fundamentally different concepts. First, the term can reflect emotionally laden "personality traits" (Digman 1990). Using this definition, emotional dispositions can refer to a general tendency to be happy or sad (Shiota et al. 2006). Second, the term can denote criteria against which the emotional relevance of stimuli is appraised (Frijda 1986, Lerner and Keltner 2000). We use emotional dispositions in the second sense.

People do not exhibit emotional reactions randomly but rather in response to specific objects, events, or situations. The objective nature of a stimulus does not directly determine the emotional response (Scherer 1999). Rather, a process of emotional appraisal occurs. The evaluation of the stimulus (the appraisal) leads to an emotional response. Appraisal implies criteria that exist prior to appraisal, and those criteria are emotional dispositions. Only by virtue of having emotional dispositions is emotional appraisal possible (Frijda 1986).

Emotional dispositions, like all mental dispositions, are traits. While states reflect *how* you are, traits reflect *who* you are (Hamaker et al. 2007). As opposed to states, traits are always present even if they are not active. Knowledge that black bears are mammals is a property of an individual even if it is not part of current conscious thinking. As traits, emotional dispositions are relatively stable compared with states. Being scared by a bear is a temporary state that can switch on and off and differ in intensity depending on the situation; a disposition to fear bears is usually stable. The fact that many phobias are persistent illustrates the stability of emotional dispositions.

Research has revealed different types of emotional dispositions. Scholars of emotional appraisal list general criteria (labeled "appraisal dimensions" or "appraisal criteria") that are employed to evaluate the emotional relevance of situations and guide the unfolding emotional response (see, for example, Ellsworth and Scherer 2003, Sander et al. 2005). Although these lists differ in the number of and kinds of appraisal dimensions, considerable consensus exists about a limited set of primary dimensions. For example, theory and research suggest that humans evaluate the emotional relevance of stimuli in terms of (a) novelty (has anything changed?), (b) valence (is it good or bad?), (c) goals (is it obstructive or conducive to current goals?), and (d) agency (what is the cause and can it be controlled or predicted?). These appraisals evaluate situations as follows: Is there anything new (novelty), is it relevant (valence), are there consequences (goals), and can I cope (agency)?

To illustrate this general appraisal process, imagine that a person sees a moose. The appearance of the moose is appraised as novel and draws attention and interest. If the individual generally likes moose, then the appearance is rated as positive or pleasant. The appearance is then evaluated against current goals. If the moose is blocking the road, for instance, the appraisal might depend on whether the person is wildlife-viewing or driving to work. For the latter situation, the emotional response will vary depending on the person's perceived control and prediction regarding events in the near future (e.g., easily driving past the moose vs. thinking the moose will cause an accident).

People also have emotional dispositions toward specific objects (Ellsworth and Scherer 2003). Most humans have emotional dispositions toward wildlife. Snake and spider phobias, for instance, are ubiquitous (Cook and Mineka 1989). Appraisal theorists have not focused on these object-related emotional dispositions, perhaps due to their focus on generic principles that apply to every situation. For the study of emotional responses to wildlife, however, both specific object dispositions (e.g., the disposition to fear snakes) and general situation dispositions (e.g., appraising a situation in light of a current goal) are relevant.

Emotional dispositions have different qualities (Sander et al. 2005). The object versus situation disposition distinction is one source of variation within the total set of emotional dispositions. For example, emotional dispositions may also vary between coarse-grained criteria that foster fast and automatic appraisal that does not need consciousness and finer grained criteria that are slower to process and may require conscious thought. Evidence indicates the existence of different systems to process stimuli: (a) a primarily unconscious affective system; and (b) a primarily cognitive system that includes conscious thought (Ruys and Stapel 2008, Tamietto and de Gelder 2010). This distinction lies at the basis of dual process theories (Smith and DeCoster 2000) and between implicit (unconscious) and explicit (conscious) attitudes (Gawronski and Bodenhausen 2006).

Emotional dispositions can be innate (i.e., a consequence of biological evolution) or learned (Jacobs 2009). Wildlife was crucial to early hominids' survival, so humans inherited emotional responses to wildlife (Manfredo 2008). Due to these biologically evolved fear dispositions, people tend to fear those objects that were threats to our ancestors (e.g., large predators). Other dispositions are learned. The delight of a dedicated birdwatcher that sees a rare bird after a long search is a learned disposition in which the knowledge that the bird is seldom seen plays a role.

In summary, emotional dispositions vary with respect to specific situations or specific objects, level of consciousness in operation (from completely unconscious to fully conscious), and genesis (continuum from fully innate to fully learned). An emotional reaction results from an activation of different emotional dispositions. We believe that an examination of emotional disposition offers a starting point for integrating emotional and cognitive frameworks. The conceptual and empirical challenge lies in understanding the relationships between emotional dispositions and cognitive dispositions toward wildlife. For example, emotions can enforce and reinforce the values and norms important to a social group and can communicate social acceptance or rejection. A display of disgust or pleasure about a given wildlife recreational pursuit such as big game hunting, for instance, conveys the person's orientation. This revelation invites response and provides the basis of acceptance or rejection, commonality or difference, and approach or withdrawal from the individual. The display helps define social group boundaries.

The practical challenge lies in understanding the relationship between emotions and an agency's communication campaigns. The prevailing emotional state affects wildlife managers' ability to communicate with others, to achieve stakeholder consensus, and to reach conflict resolution. Negative affect inhibits these outcomes. As managers structure their interactions with stakeholders, an important first step is to establish a positive affective state prior to negotiations. This might be accomplished by focusing on areas of agreement, facilitating social engagement to make a person feel accepted, and eliminating physical barriers that separate managers from stakeholders. Wildlife professionals often communicate with the public in a highly factual, cognitive fashion, but people relate strongly to wildlife at an emotional level. Communication can be improved by developing strategies that evoke emotional reactions.

4.1.6. Organizing Social-Psychology Concepts via the Cognitive Approach

The cognitive approach examines how values, attitudes, norms, and emotions influence behavior. We have described some concepts involved in the cognitive approach and suggested relationships among them. Of interest to wildlife managers is that the cognitive approach suggests that people's values determine their attitudes and norms, and that attitudes and norms, in turn, affect behaviors.

Using the cognitive approach can benefit wildlife managers in several ways. First, by understanding the cognitive structure, from values to behavior, each concept can be examined to determine its influence on people's actions. The approach helps us understand, for example, how people's attitudes toward wildlife use predict their likelihood of supporting legal hunting.

Second, by understanding how the concepts work beyond a specific issue, we enhance the generalizability of the research. For example, attitudes may influence behavior in a specific way regardless of the particular issue. People who distrust an agency or organization may oppose its proposals regardless of the specifics of any single proposal.

Third, the cognitive hierarchy helps us understand regional differences in wildlife values, attitudes, and behaviors. Within a culture, we can assume those concepts operate in the same manner, although to different degrees, across individuals. That allows us to focus on the differences in attitudes from region to region. This approach may also help us discern why conflict among individuals is occurring and whether common ground can be found. For example, although two people may have drastically different behaviors while small-game hunting, they may have similar value orientations and norms regarding wildlife recreation, and management can emphasize that common ground.

Finally, the cognitive approach has the potential to help managers understand how the concepts work in different social groups. Managers can determine how specific attitudes differ and whether values and attitudes influence behavior similarly across cultures.

4.2. MOTIVATIONS AND SATISFACTION APPROACHES

While cognitive approaches improve our understanding of behavior, other approaches also contribute to that end. Substantial research has been directed toward understanding hunters' motivations for, and satisfactions associated with, participation in wildlife-related activities. Motivations drive individuals' interest in activities prior to participation. Satisfaction refers to individuals' evaluations after the experience.

4.2.1. Motivation

A *motivational approach* suggests that people are driven (motivated) to take actions to achieve particular goals (i.e., they seek certain outcomes from their experiences). Two enduring approaches to investigating motivations have emerged in the literature. One, introduced by Hendee (1974), emphasized multiple satisfactions. This approach suggested that recreationists seek a variety of benefits and outcomes. Although Hendee applied his arguments to demonstrate that hunters define satisfaction beyond merely harvest, this multiple-satisfaction idea is appropriate for all types of experiences. Wildlife watchers, for example, may seek outcomes such as solitude, being outdoors, or socializing with friends and family. The kinds of outcomes a person strives for can also change depending on the particular experience. A hunter pursuing squirrels or cottontails with his or her grandchildren has expectations and desires that may be very different from those s / he has when hunting alone.

Second, Driver and associates (Driver et al. 1991) emphasized the importance of understanding the bundle of "desired psychological outcomes" derived from recreation participation. Recreation was proposed as a way for achieving certain outcomes (e.g., achievement, stress release, family togetherness). The Recreation Experience Preference scales used to measure these outcomes were selected based on a review of the personality trait and motivation literature. In more than 30 studies, these concepts and variables have demonstrated their usefulness in helping to understand the nature of outdoor recreation experiences and recreationists themselves (Manfredo et al. 1996; Box 4.4).

Recognizing the diversity of experiences desired by participants in recreation activities, researchers have noted the importance of differentiating users into homogeneous and meaningful sub-groups. Bryan (1977:175), for example, proposed the concept of recreation *specialization,* which he defined as a "continuum of behavior from the general to the specific, reflected by equipment and skills used in the sport." Within the continuum, individuals may range from the novice to the specialist. Variations between user classes reflect differences in motivations, and the extent of prior experience with and commitment to an activity. As people become more specialized, they become more particular in their setting preferences and equipment. More specialized users are also more likely to have specific managerial requirements and are more likely to communicate with managers. Research has applied the concept of specialization to angling, hunting, and wildlife viewing (see Vaske 2008 for a review).

Other researchers have segmented hunters specifically on motivations for participation. Decker et al. (1987), for example, proposed three motivational orientations for wildlife recreation: affiliative, achievement, and appreciative. Hunters with an *affiliative orientation* participate in wildlife recreation for the enjoyment of being with others and to strengthen or affirm relationships through shared experiences. Those with an *achievement orientation* have specific goals; for example, to bag an animal that possesses certain traits (e.g., number of antler points on a deer). Hunters with an *appreciative motivation* seek peace in the outdoors and want to become acquainted with wildlife and the natural environment.

4.2.2. Satisfaction

Motivation research focuses on what initiates behavior, while satisfaction studies focus on the outcomes received from recreation experiences (Decker et al. 2004a). A number of different types of satisfactions may be associated with a recreation experience (e.g., time with family, enjoyment of the outdoors, exercise). Satisfaction is similar to Driver's notion of experience benefits (Driver et al. 1991). Satisfaction, however, can also refer to a feeling of pleasure or enjoyment derived from experiences. Using this latter definition, the concept of satisfaction becomes an attitude. In wildlife related contexts, we often are interested in satisfaction with a particular event or action.

Satisfaction is one of the most common topics of social inquiry in human dimensions because it appears simple to ask. The use of overall measures of satisfaction, however, is questionable because they tend to only measure major changes in the quality of service delivery (Decker et al. 2004a). An individual's satisfaction is complex and dependent upon a variety of aspects related to the experience, including one's expectations.

This recognition of the complex nature of satisfaction is part of the multi-faceted discrepancy model for satisfaction.

Box 4.4 *TYPES OF MOTIVATIONS IN OUTDOOR RECREATION*

Motivation theory has been associated with a goal hierarchy; motivations are proposed to be the impetus for achieving particular goal states.

Manfredo et al. (1996) conducted a meta-analysis of the Recreation Experience Preference scales to examine how certain motivations are associated with desired goals for recreation. They proposed that recreation is a way to achieve certain psychological outcomes, such as stress release. The items they used to measure those outcomes were based on a review of personality trait and motivation research. Some of the psychological outcomes or motivations studied in natural resource contexts include the following:

- *Achievement or stimulation:* Reinforcing self-image, gaining social recognition, developing skills, testing competence, having excitement, testing endurance, telling others
- *Autonomy or leadership:* Gaining independence, autonomy, control, power
- *Risk taking:* Taking risks, experiencing the risks associated with dangerous situations
- *Equipment:* Using and talking about equipment
- *Family togetherness:* Doing things with family, bringing the family closer together
- *Similar people:* Being with friends, being with similar people

- *New people:* Meeting new people, observing other people
- *Learning:* General learning, exploring, studying geography, learning more about nature
- *Enjoying nature:* Appreciating scenery, having a general nature experience
- *Introspection:* Examining spirituality, examining thoughts and feelings
- *Creativity:* Doing things that are creative, gaining new perspectives on life
- *Nostalgia:* Recollecting good times, reflecting on the past, bringing back pleasant memories
- *Physical fitness:* Getting exercise, keeping fit, feeling good after physical activity
- *Physical rest:* Feeling physically rested and relaxed
- *Escape from personal or social pressures:* Releasing tension, slowing down mentally, escaping role overloads
- *Escape from physical pressures:* Achieving tranquility, gaining privacy, escaping crowds, escaping physical stressors
- *Social security:* Being with considerate and respectful people
- *Teaching or leading others:* Teaching and sharing skills, leading others
- *Risk reduction:* Moderating risk, avoiding risk

In more than 30 studies, those scale items have helped researchers understand recreationists and their outdoor recreation experiences.

This model proposes that overall satisfaction is a function of more specific satisfaction with individual components of an experience. For example, overall satisfaction with a birdwatching trip may be a function of how satisfied one was with the weather, numbers and species of birds seen, encounters with other people, accessibility to the site, and the facilities there. Satisfaction with one of these particular components of a recreation experience is a function of the discrepancy between one's expectation for that component and what actually occurred. Therefore, if a birdwatcher expected to see a certain species of bird and did not, his or her satisfaction level may be low for that facet.

Despite its widespread application, there is still a need to understand further what influences satisfaction (the motivations and expectations that determine a person's evaluation of an experience). Managers are interested in the relationship between satisfaction and participation, which is not as direct as one might expect. A person can have a dissatisfying experience but continue to participate in an activity and vice versa. Certain satisfactions may be more important and outweigh others. It is important to determine the relative importance of different facets of satisfaction and the factors that motivate behavior.

Motivation and satisfaction approaches have made contributions in several areas. First, identifying the types of motivations provided in different environments and activities helps improve service delivery. This is particularly the case in market segmentation research and experience-based approaches to wildlife management (see Manfredo [2002] for a review).

Second, identifying the types of motivations that can be accommodated in different recreational environments and activities can help wildlife programs improve service delivery and provide more benefits. This is particularly important when hunter participation in various harvest schemes is essential for meeting management objectives.

Third, knowing the motivations can help identify the causes for conflict among stakeholders. Goal-interference models suggest that conflict occurs when the behaviors of one group are perceived to inhibit motivation fulfillment by another group (Vaske 2008). To examine a recreation conflict, a manager must first determine the motivations of the groups involved.

Fourth, an understanding of user motivations can help managers identify substitute activities—important when assessing the impact of allocation decisions. For example, restricting a wildlife-watching opportunity for which there

The concept of recreational specialization (now widely used to understand participation in wildlife-dependent recreation) was first developed to understand specialization among anglers (courtesy Jerry Vaske)

are abundant substitutes would have a smaller impact than restricting one that had few substitutes. A basic definition of *substitutability* holds that activities are substitutable only if they fulfill the same motivations.

Finally, understanding motivations can help a manager understand crowding. Perceptions of crowding are believed to stem in part from the types of motivations associated with an experience.

4.3. FUTURE NEEDS

Early human dimensions research was descriptive and useful only in the situation examined—essentially, case studies. Although more recent research has enhanced the generalizability and comparability across studies, more work is necessary.

There is a need to examine a broad range of human dimensions as they relate to wildlife issues. Because hunting generates most wildlife agency funding, research has concentrated on hunting issues. Wildlife managers are increasingly interested in applying social science concepts to other areas, such as wildlife viewing, trends in attitudes, reactions to techniques used in wildlife management, and habitat and non-game programs, where there is less research-based understanding of public desires and expectations.

There are several questions for which social psychological research is needed. First, what factors (such as attitudes, norms, and motivations) dictate the flow and nature of human–wildlife interactions? Second, what are the short- and long-term effects of human–wildlife interactions? For example, how do interactions affect knowledge about wildlife, wildlife value orientations, and attitudes toward wildlife uses? Third, how much can we influence and control human–wildlife interactions, and how should we communicate about wildlife and management?

Wildlife professionals should reexamine the widely held view that emotional response issues are trivial, unimportant, or non-informative. Emotional responses are a barometer of ideals that are deeply important to people and an important form of communication when management agencies deal with publics. Emotional displays frequently signify that something important is at stake to participants. More specifically, emotions reveal implications regarding threats to (or reinforcement of) a person's identity, values, and norms. Emotions merit careful consideration and thoughtful response.

The link between emotions and value orientations, attitudes, and norms has interesting implications for research on human–wildlife relationships. Research should explore whether there are predictable relationships among specific situations, value orientations, and emotional responses. An important question becomes the extent to which emotional intensity evokes certain forms of behavior over and above traditional attitude and normative measures.

Wildlife managers' need for social information in wildlife management will increase as conflicts in attitudes and interests become more contentious. That is evidenced by the increasing numbers of wildlife ballot initiatives and stakeholders' interests in becoming involved in wildlife decision making. If we understand what people do, why they do it, and what they think, we have a foundation upon which wildlife managers can work with their stakeholders.

SUMMARY

Human dimensions research that applies the discipline of social psychology contributes to knowledge about stakeholder thoughts and actions toward wildlife. This chapter introduced key principles and theory related to social psychology, which is the study of the way in which people's thoughts, feelings, and behaviors are influenced by their environment.

- *Cognitive approaches* examine concepts underlying the process that leads from human thought to action. The theories that underlie these approaches suggest that people's wildlife-value orientations influence their attitudes and norms toward wildlife and that attitudes–norms, in turn, affect wildlife-related behaviors.
- *Values* are commonly defined as desirable end states, modes of conduct, or what we hold dear. Values are general mental constructs that reflect our most basic desires and goals. They are few, are formed early in life, and are resistant to change. Compare this use of the term to that of economic value as used in Chapter 6.
- *Basic beliefs* reflect our thoughts about general objects or issues; they essentially operationalize values. *Value orientations* are patterns of direction and intensity among basic beliefs. Basic beliefs and value orientations help explain how positions toward specific issues evolve from broad values. Because value orientations directly influence attitudes and indirectly impact behaviors, understanding them can help wildlife managers predict support for management actions.
- People who share a value (e.g., respect for life) do not always share value orientations. Differences in value orientations can lead people with shared values to have different opinions about a specific wildlife-management issue.
- An *attitude* is a person's evaluation, either favorable or unfavorable, of something. Attitudes can predict and influence behavior. Knowing what attitudes and beliefs influence a behavior helps predict the behavior more accurately.
- In addition to their evaluative dimension, attitudes have a cognitive dimension—the beliefs associated with the attitude object. Beliefs are what we think are true and are not necessarily objective facts. To predict behavior, it is important to understand both the cognitive and the evaluative characteristics of an attitude.

Hunters and hunting were the focus of pioneering social-science applications to wildlife management. Managers are increasingly interested in applying social science concepts to other areas, such as wildlife viewing, trends in public attitudes about wildlife, and reactions to techniques used in wildlife management. (courtesy Jerry Vaske)

- To use an attitude as a predictor of behavior, researchers need to understand at least three of its characteristics: specificity, salience, and strength. Specific attitudes are better at predicting a specific behavior than are more general cognitions. The more salient an attitude, the more likely it is to influence behavior. More strongly held attitudes are more difficult to change.
- *Norms* are standards of behavior that specify what people should do or what most people are doing. *Social norms* are standards shared by the members of a social group. Both personal and social norms influence behavior.
- *Emotion* is part of affect, or the feeling states, of individuals. Emotion is examined from many perspectives: expressive reactions, physiological responses, importance to appraisal, and subjective experience. All of these perspectives illuminate a complex human process.
- Some situations evoke more emotionally based processes than others. Persuasive appeals that evoke emotion can be highly effective, though somewhat contextually dependent.
- People hold various motivations for participating in wildlife recreation. *Motivation theory* helps wildlife professionals identify the reasons people participate and understand the outcomes and benefits they are seeking.
- *Satisfaction* refers to benefits received from experiences. Satisfaction can also be a feeling of pleasure or enjoyment derived from experiences. In wildlife contexts, we are often interested in the satisfaction with a particular event or action.

- A person's satisfaction is influenced by the extent to which individual and situational factors associated with an experience are fulfilled. Different experiences provide different types of satisfactions.

Suggested Readings

Decker, D. J., T. B. Brown, J. J. Vaske, and M. J. Manfredo. 2004. Human dimensions of wildlife management. Pages 187–198 *in* M. J. Manfredo, J. J. Vaske, B. L. Bruyere, D. R. Field, and P. Brown, editors. Society and natural resources: a summary of knowledge. Modern Litho, Jefferson City, Missouri, USA.

Manfredo, M. J. editor. 2002. Wildlife viewing in North America: a management planning handbook. Oregon State University Press, Corvallis, USA.

Manfredo, M. J. 2008. Who cares about wildlife? Springer, New York, New York, USA.

Manfredo, M. J., J. J. Vaske, P. J. Brown, D. J. Decker, and E. A. Duke. 2009. Wildlife and society: the science of human dimensions. Island Press, Washington, D.C., USA.

SOCIOLOGICAL CONSIDERATIONS IN WILDLIFE MANAGEMENT

RICHARD C. STEDMAN

Successful wildlife managers tend to have at least a rudimentary understanding of sociology. They apply knowledge of how society works, the role of institutions, and social trends to inform long-term planning and day-to-day work. This understanding may come from formal study or professional experience. However their sociological analysis skills are developed, wildlife managers have ample opportunities to use them as they seek to manage multiple species for a diverse public across a wide range of social contexts. Rapid population growth and diversification, especially on the urban fringe, affects both wildlife habitat and rural people. Our society is also becoming truly globally connected. All of these social changes have increased the usefulness of sociological approaches for helping wildlife managers understand and respond to social change.

Sociology is concerned with understanding what people do as members of a group or when interacting with one another. It stresses the social contexts in which people live and how these contexts influence people's lives, and it emphasizes how people are influenced by society–social structure and how they, in turn, reshape their society. The interplay of these social influences can cause wildlife conservation and management dilemmas and provide potential solutions as well. To assist the manager in thinking about such influences, this chapter presents some central tenets of sociology, describes broad theoretical approaches within sociology, briefly introduces more specific sociological perspectives pertinent to wildlife management, and closes with a discussion of important social trends that may affect wildlife management.

Compared with social-psychological and economic approaches, sociological concepts and research have been used relatively little in wildlife management. Until recently, sociology as a discipline has not focused on the relationship between social behaviors and the natural world. Environmental and natural resource sociology has emerged to address this gap, but has not focused on issues related to wildlife. We can expect this to change because, as society continues to change and wildlife management becomes more complex, under-

standing sociological principles of power, class, and culture can help managers negotiate the shift from "management to governance" (see Chapter 2). This shift emphasizes the diversification of stakeholders, increased partnerships, an emphasis on norms rather than laws, and diverse criteria for assessing management effectiveness.

This chapter offers a "primer" on sociological theory relevant to wildlife management application. For example, basic sociological theory can be applied to a real-life situation of great interest to wildlife managers: explaining changes in hunting participation. Participation in licensed recreational hunting has been declining in much of the United States. This decline is of concern to wildlife managers who seek to sustain political support of programs, maintain revenues, and control populations of some overabundant species of wildlife. The decline in hunters may be explained through social psychology (e.g., changing attitudes toward hunting), economics (trade-offs and cost / benefit ratios), or through sociological explanations. Each of these explanations suggests very different avenues through which managers might respond. A sociological perspective on hunting trends is interjected at various points in this chapter.

5.1. FUNDAMENTALS OF SOCIOLOGY

As social *science,* sociology is concerned with systematic prediction of behavior—the logical and persistent patterns of regularity in social life (Babbie 2007). Sociology, like other science, reveals what *is,* not what should be. Social science cannot settle debates on value; it cannot tell wildlife managers what the optimal solution is ("should" and "ought" questions are the realm of ethics; see Chapter 16). Further, prediction of human behavior is difficult, whether you are interested in individual behavior or aggregate social behavior.

The question, "How much freedom do people have to act?" is an ongoing topic of discussion in the social sciences. This question involves the relationship between structure and "agency," defined as the capacity of individuals to act independently and to make their own free choices. Social structure, which can be defined as the enduring, orderly patterns of relationships among elements of society (roles, institutions, or expectations), is emphasized in sociology as a strong influence on human behavior. Psychology tends to argue that individuals have a great deal of freedom to behave as they choose. An intermediate perspective is that social structure is created through the interaction of people and is constantly in the process of being renegotiated. This is referred to as "structuration" (Giddens 1990) or the mutual dependency of structure and agency. Much as a river's banks constrain (or structure) its flow, the flow itself is constantly in the process of reshaping the banks. So, too, with structure and agency; society shapes human behavior, while people are constantly in the process of reshaping society as a whole. Both individual-level and societal-level factors are at work in most issues of concern to wildlife managers.

Considering hunting participation again (Box 5.1), social-psychological explanations might focus on hunter attitudes toward hunting: have they changed over time? Have hunter cognitions changed? In contrast, sociologists would place more emphasis on changes to the social structural *system* that may have affected these attitudes or the relationship between attitudes and behavior. For example, has the landscape become more urbanized so that it is more difficult to gain access for hunting? Have the rural communities that have been an important source of social support for hunting participation declined in population relative to urban areas? Have rural areas themselves changed in character?

Sociology studies social facts rather than individual-level explanations of behavior. Social facts are the values, cultural norms, and social structures *external to the individual.* These are expectations that come from the social community, which socializes each of its members. Durkheim's analysis of suicide (1951) is a classic example of the analysis of social facts. Although the decision to commit suicide is made by the individual, individual-level explanations (e.g., someone is depressed) are unsatisfactory and cannot explain differential *rates* of suicide (across places, groups, and/or times). For insight into phenomena such as this, the social structural conditions that drive this individual behavior need to be revealed. For example, Durkheim's analysis reveals that social support strongly predicts aggregate suicide rates: suicide is less common among people who are married and who are members of more tightly knit religious groups (e.g., Jews and Catholics).

Hunting participation can also be analyzed using "social facts." Hunting is ultimately a decision made by an individual person, but individual explanations ("I don't have enough time, or a place to hunt, or people to hunt with") are not sufficient to understand the phenomenon. Taking a sociological perspective, the manager would ask, "What are the times, places, and contexts that are associated with these phenomena being more common?" The sociological perspective enables us to develop an awareness of the relationship between our own experience, or someone else's, and what is going on in the wider society. Mills (1959) distinguishes between "troubles" and "issues,"

where troubles are the experience of an individual and issues refer to our understanding of how these troubles may come to be shared or become widespread. The decision to continue hunting can be interpreted using the "troubles versus issues" framework. Surveys of current and past hunters commonly ask them to describe reasons they have ceased participation and often include response options such as "not enough time," "nowhere to go," and "friends stopped hunting." These reflect individual-level "troubles"; a particular hunter has less time, has lost access to a favorite spot, or has no hunting companions. The *distribution* of these troubles invites a sociological analysis: has there been a widespread decline in leisure time? If so, is it so for everyone, or just for some segments of the population? Has posting of land or land conversion increased? Has it been more common in some geographic areas than others? These questions apply social change analysis to the study of effects on an individual-level decision such as hunting participation, which offers greater possibility of explaining a behavior that has multiple influences.

5.2. THEORETICAL APPROACHES IN SOCIOLOGY

Three major theoretical approaches underlie more specific sociological analyses that apply to wildlife management: functionalist, conflict, and interactionist. Each of these offers very different perspectives on the process by which society is organized, emphasizes different outcomes, and implies different engagement strategies for wildlife managers.

5.2.1. The Functionalist Perspective
The effects of Charles Darwin's theories on evolution had impacts beyond just how we think about nature. Early social theorists, influenced by Darwin's thoughts about evolution and the apparent interdependence of various elements of society, used evolutionary or "organic" analogies to describe the co-emergence of relationships among elements of society. This perspective emphasizes that the presence of any particular institution or activity has consequences for other "parts" of society. Each component contributes to the maintenance of the whole system; thus, systems are seen as trending toward stability—equilibrium, with perturbations to this basic working system requiring correction. This orientation has been manifested strongly in cultural perspectives. For functionalists, culture is seen as the basic causal force for social organization where collective consciousness or shared values hold society together. Individuals are taught these basic values; the individual is rewarded for "learning the rules," but as these rules are learned, they are reinforced and the social order is further supported. From a functionalist view, power is relatively benign; the role of management is simply to define into law public goals that emerge from collective agreement. Therefore, management reflects the will of the people (broadly construed) rather than particular powerful interest groups.

Much of the human dimensions of wildlife work that emphasizes values, socialization, and norms are (at least implicitly) based in the functionalist perspective. For example, tradi-

Box 5.1 *SOCIOCULTURAL AND DEMOGRAPHIC INFLUENCES ON HUNTING PARTICIPATION TRENDS*

Among the recreation activities that social scientists have studied, hunting is among the most multi-faceted in terms of effects of social and demographic trends on participation in the activity. This is in part because hunting is as much a cultural activity involving people not engaged in hunting per se (social world of hunting) as a recreational activity in the sense of a sport or leisure diversion.

Many factors affect hunting participation trends; some have reinforcing effects and some have dampening effects. Trends in social and demographic characteristics of the population in the United States influence trends in hunting participation—the number of people participating, their traits, and the extent of their activity. A few research-based observations about demographic and social trends are presented here.

Demographic factors. Most demographic trends that influence hunting tend to have a dampening effect. The single greatest demographic trend correlating with constraining or reducing hunting participation has been urbanization of the population. This directly affects hunting recruitment (because fewer people are originating from rural families) and indirectly affects participation because of less land available for hunting, and because ownership of rural land shifts from people with traditional rural backgrounds to people with urban backgrounds (who generally lack understanding of hunting culture). In addition, hunting participation is highest among white males, but the fastest growing portions of the U.S. population are in other ethnic segments that are typically concentrated in urban environments.

Sociocultural factors. Hunting has strong cultural importance for several million Americans who have been introduced to the activity as part of their cultural heritage, but cultural norms change over time. Leisure-time use options have grown markedly with new technology and mobility, especially those options that require less preparation and travel time (which are barriers to hunting for urban dwellers). This is exacerbated by the decline in discretionary time among working parents. Furthermore, the fastest growing population segments in the United States are not, in general, those where hunting has a strong cultural or socially supported role.

Knowledge of how the sociocultural and demographic trends correlate with trends in participation in hunting and other wildlife-related activities can be a powerful tool for planning future program demands and income (in a fee-based system for funding wildlife management). Further research into the mechanisms that connect the broader trends to wildlife uses can improve both prediction of use and feasibility of interventions that might be designed to influence participation in the future. Sociological insight will grow in importance to wildlife management in the future.

---

tional hunter-recruitment thinking suggests that core hunters are socialized into widely accepted patterns of values, attitudes, and behavior (social structure shaping the individual) around hunting and rural life. As people "learn the rules" for how to behave appropriately, these rules are further reinforced and strengthened. More recently, functionalist theories have somewhat fallen out of favor because they are less able to account for instability, conflict, and social change. In periods of widespread, rapid, or otherwise acute social change there are more likely to be challenges to rules that were formerly taken for granted. For example, the accepted rules of behavior in rural areas (norms) may not be as strong in instances where there is rapid growth and land conversion. Similarly, the notion of what it means to be a hunter may also change rapidly. For example, witness the emergence of the "locavore" hunter phenomenon (see, for example, Pollan 2006), where hunting may be less tied to traditional rural ways of life and may instead be tied to concerns about where one's food comes from and the effect of one's eating practices on environmental, economic, and social well-being.

5.2.2. The Conflict Perspective

The conflict-based perspective, which suggests social organization emerges not from shared cultural values but from on-going battles over resources, is quite different from the functionalist perspective, which focuses on the maintenance and creation of social order. From a conflict-based view, society has winners and losers, and the questions of interest are: Who are the winners and who are the losers? How do they emerge? And what social structures influence this? These questions address inequalities in power and resources that are not adequately explained by cultural concepts of norms and values.

Two specific manifestations from the conflict perspective have had strong influence on sociology: Max Weber's (1958) notion of power and Karl Marx's treatment of class (Marx and Engels 1986). For Marx, class was defined via the relationship people have to the production of goods in society. He divided society into two primary classes: those who own property (capital) crucial to the production of some good or service (the capitalists) and those who only have their labor to sell (the proletariat). This original meaning has been largely supplanted in the popular consciousness with strictly economic terms (i.e., those with more wealth are seen to be of a "higher class" than those with less wealth); therefore, wealth is one crucial way by which society is stratified (ranked hierarchically along some dimension of inequality).

The idea of power is especially relevant to wildlife governance and management. According to Weber, power is the

Charles Darwin, circa 1874.

force driving social organization. Although there are multiple definitions of power, we can condense these definitions by defining power simply as the ability of individuals, groups, or institutions to achieve their goals or impose their will over the wishes of another. Social organization is created as society becomes dominated by those who gain and maintain control against the will of others (note the difference from the functionalist perspective, which sees power only reflecting the "public" interest rather than self-interest). Formal organizations (government) *use* culture to legitimate their power by convincing others that the organization's actions are in their best interests.

This perspective is readily applied to our hunting example, where wildlife management agencies can be seen as influenced by powerful interests (e.g., consumptive recreation interests) that provide much of the funding and other resources upon which the agency depends. Wildlife conservation and management policy comes to favor powerful organized interests. This puts stakeholder groups and management authorities in a very different light; public wildlife management would not necessarily be expected to reflect the broader interests and will of society but the interests of a particular group (for example, consumptive wildlife interests). To the extent that a power perspective is played out by wildlife agencies in their interactions with a narrow set of stakeholder groups, they might be criticized, but their use of power might go unnoticed as long as there is widespread social agreement about the importance of the wildlife-management stakeholder group in question.

Social change plays an important role here as well. Weber believed that the use of power would become more transparent in contexts of rapid social change, resulting in a decline in the power that is based on consensus. Weber suggested that in times of social change organizations would either lose power or have to revert to more obvious forms of control. For example, many people may be critical of the way that wildlife is managed in the United States (perhaps they are concerned about an overemphasis on consumptive interests and game species). They might suggest that management agencies that are supposed to manage for a broad range of public values have been captured by special interests with particular forms of power (in this case, funding for wildlife management through the sale of licenses and taxes on sporting arms and ammunition). These kinds of questions are more likely to surface as social change continues and perspectives on wildlife management diversify (Jacobson and Decker 2008). As Coser (1956) emphasizes, conflict is not necessarily something to be avoided at all costs. Conflict is pervasive if positions held by stakeholders are in some ways fundamentally incompatible. Acknowledging social conflict can thus provide important functions, such as clarifying the perspectives of multiple stakeholder groups, as well as clarifying the positions held by managers or the processes by which the interests of some groups are advanced over those of others.

5.2.3. Symbolic Interactionism

In contrast to functionalist and conflict perspectives that see the creation of social order as unfolding more or less "automatically," symbolic interactionism emphasizes the active role that humans have in creating and assigning meaning to social life. Blumer (1969) emphasizes that humans behave primarily on the basis of the meanings they ascribe to phenomena (e.g., self-reliance achieved from hunting), and that these meanings are created through social interaction (e.g., with family, interest groups [birdwatching clubs]) and reflection by the individual. Further, our behavior depends on how we define ourselves and others and how we interpret our particular context or situation. These definitions and interpretations are shared by members of a society or cultural group, but they differ between groups. Research in symbolic interactionism tends to use a "micro" level of analysis, which seeks to understand how meaning is produced at the level of the social interaction rather than broader-scale social trends.

In keeping with our hunting example, Dizard (1999) offers an insightful symbolic interactionist-based account of the contested meanings attributed to deer, nature, hunting, and hunters at the Quabbin Reservoir in Massachusetts. He argues that conceptions of these phenomena (by competing groups) were at the heart of public response to a proposed managed deer hunt. In particular, non-hunter attributions of what deer and nature (including the forest) were "supposed" to be like (and the actions required to "bring things back to normal") contrasted dramatically with the meanings attributed to these phenomena by hunters and wildlife managers.

5.3. SPECIFIC SOCIOLOGICAL THEORIES

This section of the chapter introduces more specific theories within sociology that have emerged from the larger macro-theoretical approaches. These are organized loosely from those that put the most emphasis on the individual to those that put relatively greater emphasis on social structure as influencing behavior. Specific theories considered are (1) social roles and identities, (2) social group influence (conformity and innovation), (3) culture and socialization, (4) community, (5) social movements, and (6) social position as defined by inequality and stratification.

5.3.1. Social Roles and Identities

Drawing primarily from the symbolic-interaction approach described above, identities include the meanings someone holds for himself, or herself, based on self-reflection and feedback from others (Stryker 1980). They are the sum of roles or positions in social structure: parent, investment banker, birdwatcher, Republican. Each of these positions includes socially generated expectations for behavior ("How are birdwatchers supposed to behave in national parks?"). Some of these expectations are well-defined and consistent across multiple situations, and others more variable and negotiable. We all have multiple roles, and our identity is based on a hierarchical set of roles within the social structure.

Roles are socially constructed in multiple ways. First, the very existence of these social categories (i.e., that there is a category called "hunter" with well-recognized attributes) implies the process of social negotiation over time. Closely related is the relative prominence of each of the roles; although many of us may expect to play the role of "parent," fewer will play the role of investment banker. Society also influences the meanings and expected behaviors we associate with these roles, and each role carries a set of behavioral expectations; that is, we know how a parent should act. Context matters in several ways. How people "define the situation" (Goffman 1967, Blumer 1969) affects what aspects of identity people choose to emphasize and the particular way they choose to play a role. For example, the "student" role includes multiple behaviors, and acceptable behavior within the student role may be very different on a Saturday night than on the morning of a final exam! In the wildlife management realm, Kuentzel (1994) found that situational influences (observing other hunters shooting at birds out of range) resulted in behavior change among hunters (who also chose to engage in such behavior) even when their self-professed attitudes did not support such behavior.

5.3.2. Social Group Influences

A considerable amount of research examines the effect of the social group on the individual's attitudes and behavior. Although this area of inquiry could reveal a great deal about stakeholder positions and actions, little research on social group influences has been pursued in the context of human dimensions of wildlife. Research in this area has focused on conformity or the subjugation of the individual to the group.

Moscovici (1985) argues that social psychology in particular has had a "conformity bias" that emphasizes how individuals with deviant positions modify their views in accordance with group expectations. In most circumstances disagreement is undesirable and causes individuals to modify their position to reduce this conflict. The degree to which an individual will modify his or her position to conform to the group is tied to several factors: the level of discrepancy between the positions, the nature of alternative positions, one's level of commitment to one's own position (linked to one's identity as described above), behavior strategies of the negotiators (i.e., How extreme is their position, how consistent are they, or is there a hierarchy between the parties?), and the relative size of the majority compared with the minority. Despite the extensive use of groups in wildlife management (e.g., stakeholder groups, advisory committees, and task forces), there has been relatively little formal study of group dynamics and their effect on majority–minority influence and the convergence or divergence of majority or minority sub-groups (i.e., how do the group members' initial views or preferences differ from those that result from group discussion?).

5.3.3. Culture and Socialization

Given the reliance of wildlife management on citizens' voluntary compliance with regulations, managers have reason to want to understand factors leading to people's socialization into expected standards of behavior shared among a particular cultural group. The previous chapter (Chapter 4) introduced the norm construct, where norms are standards of behavior (social norms are those standards of behavior that are shared by a social group). Socialization is firmly rooted in the functionalist perspective described earlier. Here we address socialization as the process underpinning the social origin of culturally normative behavior. Culture includes learned, socially transmitted customs, knowledge, material objects and behavior, and meanings that are shared by a collection of people and that are expressed in symbols, rituals, stories–narratives, and values–worldviews. Normative behaviors are, thus, established standards of behavior maintained by society. Socialization is the process whereby people learn the attitudes, values, and behaviors appropriate for members of a particular cultural group. Building cultural "competency" helps an individual function well in society. From the standpoint of society, this process simultaneously reinforces and strengthens social agreement about expected standards of behavior. Socialization occurs through human interaction and includes three core sets of variables: agents (those doing the socializing), targets (those being socialized), and the context in which the socialization is occurring. Research in socialization emphasizes the importance of primary (parental) socialization during childhood, but important socialization also occurs through the influence of peers, teachers, extended family, and institutions.

Hunting is a more traditional aspect of rural than urban cultures (Fitchen 1991, Heberlein and Stedman 1997). Stedman and Heberlein (2001) demonstrate an interaction between agent, target, and rural context that plays into predicting hunt-

Box 5.2 *HUNTING AND RURAL SOCIALIZATION*

Hunting generally is viewed as a typically "rural" activity or a "way of life" in rural areas. Rural people are more likely to participate in hunting, more likely to be committed hunters, and more likely to initiate their participation through primary socialization channels (i.e., being taught by a family member, most often a father). Socialization involves learning expected standards of behavior appropriate to a particular situation: key variables include characteristics of the *target* (the person being socialized), the *agent* (the person or institution doing the socializing), and the *context* (or setting in which the socialization takes place).

Stedman and Heberlein (2001) used national-level data from the General Social Survey to explore the relationship between three key variables—a people's gender, whether they grew up in a rural area, and whether their fathers hunted while they were growing up—to more precisely examine how rural socialization works to foster hunting behavior. Each of these three variables is (in isolation) associated with increased propensity to participate in hunting, but no work had examined their joint effect in contributing to hunting behavior.

Their findings on each of the individual socialization variables reflected conventional wisdom: males were more likely than females to have tried hunting at some point in their lives (57% vs. 14%), as were people whose fathers had hunted while they were growing up (54% vs. 22% for those who grew up without a hunting father). About one-fourth (24%) of respondents who grew up in urban areas with populations greater than 500,000 had tried hunting, compared with 44% of respondents who grew up "in the country."

The more interesting results emerge from the interaction between these socialization variables. A strong majority (82%) of males who grew up in a household with a hunting father had tried hunting, and urban versus rural upbringings made no difference (81% vs. 83%). Here, Stedman and Heberlein found strong evidence for the impact of primary socialization (as defined by characteristics of the agent) that rendered the context (growing up in rural areas) virtually irrelevant.

When the effect of a hunting father is explored for females, we see a much more modest (and not statistically significant) effect of primary socialization: 27% of females with a hunting father had tried hunting and this varied only

slightly between rural and urban upbringings; however, only 6% of females who grew up *without* a hunting father had tried hunting and rural upbringings had no effect on these participation rates. Thus, we see very strong effects of *primary* socialization for both males and females: the presence of a hunting father "trumps" the rural versus urban context in which the socialization takes place.

When do rural upbringings matter? They matter only in one case: males that grew up in rural areas without hunting fathers were much more likely to have tried hunting (59%) than males who grew up in urban areas without hunting fathers (34%). This suggests that other socialization agents present in the rural setting (i.e., extended family, friends, or institutions) are relatively likely to step in and teach the normative nature of hunting to males. In contrast, and as described earlier, no such corresponding effect of rural socialization was observed for females. These findings, taken in combination, suggest that rural upbringings foster hunting participation primarily by standing in for absent primary socializing agents (i.e., hunting fathers) when the behavior is widely perceived as normative (hunting is more normative, even in rural areas, for males than for females).

Research suggests that the presence of a hunting father plays a greater role in propensity to hunt than does the rural or urban context in which youth socialization takes place (courtesy Liz Bihrle)

ing participation (Box 5.2). They use national-level survey data from the U.S. Fish and Wildlife Service to examine how rural socialization is differentially associated with hunting participation based on agent–target characteristics. They found that rural socialization was associated with increased probability of hunting only for male targets that did not have hunting fa-

thers in the household. They surmise that in rural settings, secondary agents of socialization (i.e., extended family, peers, or institutions) foster hunting participation in the absence of a primary socialization agent, but only in instances where characteristics of the target (i.e., gender) were consistent with dominant norms.

5.3.4. Community

We tend to view the well-being of the environment and of society through the lens of community, which serves as an intermediate realm between the individual and society. "Communities of interest" may reflect shared interests only (such as when people donate to a particular conservation cause; Cox and Mair 1988), but most definitions of community include shared interests coupled with particular geographic location and social interactions. Wilkinson's (1991) interactional approach to community emphasizes three basic elements: (1) local territory (interactional communities are based on a physical setting), (2) local society (set of structures and institutions that help meet local needs; geared toward local issues), and (3) social interaction (interaction with each other). In the manner of other social groups, communities faced with a novel or urgent situation may produce behavior that would not be predicted from knowledge of beliefs and attitudes of individual community members. For example, community response to a natural resource issue may be manifested in the passing of local preservation ordinances, which may represent a compromise but not reflect the sum of the attitudes and values of community members.

There has been a strong push recently toward community-based management of natural resources, including wildlife (e.g., Gibson and Marks 1995). Proponents argue that community control better serves local environmental and socioeconomic needs and is more in tune with local environmental realities. Many proponents believe that transferring control to local communities increases the odds of sustainable natural-resource use (Baker and Kusel 2003). Bradshaw (2003) challenges this thinking, suggesting that credibility and capacity of local communities to manage their local natural resources effectively needs to be assessed. Credibility refers to questions of whether communities are genuinely concerned about a broader stakeholder community as opposed to just their particular local and immediate interests, whether communities may seek short-term solutions (due to financial constraints), and whether community-based solutions represent the will of the entire local community as opposed to an elite faction. By implication, stronger "distant" centralized government may (ironically) have the effect of protecting some local citizens from their neighbors. Capacity issues relate to whether communities can effectively manage their own resources based on differences in resources and / or capacity that include technical expertise, management skills, financial resources, and political access. Under community-based resource management, strong intercommunity inequalities may emerge based on these differences in capacity.

5.3.5. Inequality–Stratification

Grounded squarely in the conflict perspective, stratification and inequality concepts recognize that the unequal distribution of resources (wealth, power, prestige) in a society affects individual and group-level behavior. Social stratification is defined as the processes by which resources and opportunities are distributed among various social actors according to hier-

archical "rankings" along some dimension or socially agreed-upon set of criteria. Most commonly invoked is economic inequality, which refers to disparities in the distribution of assets such as income or wealth. Although the term typically refers to inequality among individuals and groups within a society, it can also refer to cross-national inequality. Inequality is increasing over time in the United States and in most other developed countries. For many people, a lack of financial resources is a significant barrier to participation in wildlife-related recreation as costs of equipment, transportation, and land access become prohibitive. Wildlife managers therefore face the prospects of losing the interest and potential support for conservation of large segments of the population that may cease actively participating in wildlife recreation.

Inequality matters to wildlife management in other ways as well. Boyce (1998) suggests that greater inequality produces environmental degradation because environmental degradation entails winners and losers. Winners can impose costs on the losers for a number of possible reasons: (a) losers do not yet exist (i.e., future generations); (b) they do not know they are losers (they lack information); and / or (c) they lack the power to resist. Obviating inequality in wildlife management outcomes and unequal access to the policy and management decision processes are at the root of the United States public trust doctrine's emphasis on the premise that wildlife is not owned by a person or particular group of people, but by all citizens, and is held in trust by governments for the public benefit of current and future generations (Jacobson et al. 2010). The costs of degradation in traditional economics are "impersonal"; one need not ask *who* wins and *who* loses, but whether the total social benefit exceeds total social cost. If so, social efficiency exists (the winners could theoretically compensate the losers and still be better off). According to Boyce, the reality is that the winners ignore the losers if they can. Losers with more relative power make it more "expensive" to do business (through government sanctions, regulations, etc.); therefore, relative power matters a great deal and greater inequality implies greater ability of winners to impose power on losers and greater resulting degradation. This phenomenon underlies attempts at international conservation where vast inequalities exist between developed nations and the local systems—often in developing nations in the global south—in which they are attempting to practice conservation (e.g., Brechin et al. 2003).

5.3.6. Social Movements

The specific sociological perspectives examined thus far have emphasized how social structure acts to influence the behavior of the individual. It is also worth considering the emergence of *collective* behavior. Social movement scholars describe the emergent properties of group interaction around a common purpose. A social movement is a collective, organized attempt to act with some continuity to change the social order (Snow and Oliver 1995), or it is individuals coming together as a group to foster change (Box 5.3). Understanding how these phenomena develop will help the wildlife manager understand how wildlife management issues arise and are addressed, in-

Box 5.3 *ENVIRONMENTALISM*
An Enduring Social Movement

Few social movements fully achieve their goals and most fail to survive for more than a few years. The environmental movement that began to emerge at the end of the nineteenth century in the United States is an exception. Scholars disagree on the degree to which the environmental movement has slowed ecological deterioration, but its enduring nature and widespread influence are undeniable. The environmental movement is commonly regarded as one of the more successful social movements of the latter half of the twentieth century, in the United States as well as Europe. In addition to the impressive staying power and large organizational base of environmentalism, the movement has clearly had significant institutional and cultural effects within most industrialized nations and beyond. A key reason for the success of environmentalism, relative to that of most social movements, is that its goal of environmental protection is widely supported by the general public. Public support is a crucial resource for any social movement and the largely consensual nature of environmental protection has given the environmental movement an advantage over movements that pursue more divisive goals.

Today the environmental movement is diverse and multi-faceted. It encompasses traditional as well as newer, typically more radical elements. Analysts usually distinguish between at least two broad wings of contemporary environmentalism found in the United States and especially in Western Europe: conservationism and environmentalism. Conservationism, which has roots going back a century or more, is often depicted as being primarily interested in the preservation of wildlife and aesthetic environments and, particularly in the United States, the conservation and efficient use of resources.

Social movements are an organized attempt to change the social order through non-institutional means. Like any new social movement, conservationism began as an attempt to change the status quo. Over time it became institutionalized, and its challenge to the status quo is seen as limited today. Environmentalism, by contrast, is seen as encompassing the broader goal of environmental protection and entailing a more exacting critique of the status quo. New movements such as ecofeminism, deep ecology,

social ecology, and environmental justice challenge the traditional political values of industrialized societies. These newer streams of environmentalism embrace and encompass other contemporary movement goals in addition to environmental protection. This tendency is particularly evident among the "Greens," who have developed political parties and earned followers throughout Western Europe and to a lesser degree in the United States by advocating an amalgamation of concerns, including social equality, human rights, and world peace, as well as environmental protection (Mertig and Dunlap 2001).

ANGELA G. MERTIG

Theodore Roosevelt (left) and John Muir (right) were leading figures in two factions of an environmental movement that began in the United States in the late nineteenth century. The movement gave rise to institutions such as the National Wildlife Refuge System, the U.S. Forest Service, and the National Park Service.

cluding the roles of stakeholder engagement, communication, and coalition building in management.

Social movements can be studied from multiple theoretical perspectives: cultural change, changes in the distribution of resources, or changing symbols and/or meaning. Several theoretical approaches have addressed social movements. *The issue-attention cycle* (Downs 1972) suggests that most social movements have a natural "life cycle," beginning with an

initial excitement phase, which leads to solutions often embedded in institutional change (i.e., passing legislation). These victories are more "symbolic" than real and apparent "success" of this type can sometimes result in the movement losing momentum as people assume the problem has been solved. Wildlife managers encounter this phenomenon, for example, when marked progress is achieved during the first year of a multi-year community-based wildlife damage-abatement program. Too often community interest in the multi-year work required for a lasting solution wanes; for a brief period, community leaders declare success, only to have the problem recur in the following year or two.

The *resource mobilization* perspective (Snow and Oliver 1995) contrasts with traditional models that start with grievances as the rational response to "real problems" by suggesting that the nature of the grievance is secondary. In other words, this perspective holds that movements form because of changes in group resources and/or opportunity structure (i.e., the grievances are always present, but the capacity to act on them is not). Mobilization is the ability to secure collective control over the resources (human capital, financial, political, etc.) needed for collective action. Movement *framing,* which emerges from symbolic interactionism, emphasizes active negotiation about the nature of the problem. Meanings of phenomena do not automatically flow out of structure but are politically contested. Frames are tools for interpretation; they render events meaningful (i.e., they help answer the question, "What is this event an example of?"). The power of a movement, therefore, is not necessarily embedded solely in the event itself but in the resonance of its story, how it is portrayed, and to what other phenomena it is linked.

The animal rights movement offers an interesting case. Beginning with the writings of several philosophers (e.g., Singer 1990), arguments that were based in "rights" for animals were able to utilize the arguments (and in some cases mobilize the resources) of other "rights-based" movements (i.e., women's rights and minority group rights). This framing and mobilization of the animal rights movement has brought it into mainstream thinking with wide institutional support. Some aspects of the movement emphasize direct action (e.g., freeing captive animals from factory farms, research labs, and cosmetics factories; disrupting hunts). Others concentrate on the institutional level through education and media. A testament to the power of framing and the strength of the movement is a recent countermovement to frame direct action activities as "eco-terrorism" (manifested in the 2006 "Animal Enterprise Terrorism Act").

5.4. KEY DRIVERS AND CHALLENGES: THE CHANGING SOCIAL CONTEXT OF WILDLIFE MANAGEMENT

This chapter has offered a number of sociological perspectives for understanding the relationship between humans and wildlife, and among humans, as relates to wildlife presence and use. Sociological perspectives generally argue that "context matters" and that particular behaviors—rather than being based in quasi-universal "rules"—are likely to play out differently in different contexts. It thus becomes crucial to understand some of the key social changes occurring in the United States and beyond that comprise this "context" and, by association, the workings-out of the relationship between people and wildlife. A few such trends are described below.

A diversifying population. The population of the United States is becoming more diverse every year due to differential rates of immigration and reproduction for various groups in American society. Especially evident are the rate differences between non-Hispanic whites and other racial–ethnic groups. These diverse populations present managers with new and different understandings and/or traditions of interacting with nature, including wildlife. This almost certainly will pose interesting opportunities for managers to develop novel programs in coming years as the configuration of wildlife stakeholder groups changes substantially and traditional assumptions about (for example) socialization into expected patterns of behavior (e.g., hunting, wildlife viewing) are likely to change.

Population redistribution and urbanization. Although the population of the United States is growing only moderately, the growth is not evenly distributed across social class or place. With respect to class, population growth is fastest among lower income groups. With respect to place, migration patterns show a large-scale exodus from the Midwest and Northeast to the South and Southwest. In the United States, we also are seeing substantial outmigration from urban city centers and toward higher amenity rural places adjacent to these cities. Thus, despite a deconcentration of population density, the footprints of most urban areas continue to expand rapidly. Coupled with this is a decline of "traditional" rural places. Some—especially those that are located proximate to thriving urban centers or that have natural amenities such as water, favorable climate, scenery, and/or abundant public land—are transforming from landscapes characterized by traditional activities such as forestry and farming to landscapes characterized by tourism or second home development. Many rural areas that are not undergoing these transitions are being "hollowed out" (e.g., through depopulation). From the symbolic interactionist perspective, new urbanites are actively negotiating and creating new meanings about nature and wildlife. The power relations articulated in the conflict perspectives suggest that these newer meanings are likely to proliferate and challenge established rural norms in places undergoing rapid change.

Globalization and interconnectivity. Globalization and mobility of ideas, capital, and people makes management far more complex and potentially strains managerial capacity. Globalized trade ushers in the specter of cause–effect relationships that are spatially and temporally distant, hard to identify, and characterized by complex interconnections. It is far more difficult for a single managerial body to have regulatory authority over an entire production–consumption–disposal chain when wildlife resources and habitat in one part of the globe are sought by distant societies. Even the most rural areas of the United States are intricately connected to the rest of the world. This requires a shift from management to governance (see Chapter 2) based on "supply and demand" factors. As the demands of management become more complex, governmental wildlife management agencies in many states (and nations) find that they have inadequate capacity. Much of this capacity to manage wildlife resources and their uses is being taken up by NGOs and other private citizen groups and / or partnerships between government and the private sector.

SUMMARY

This chapter examined the application of sociological theory and research to wildlife management. Sociology is concerned with systematic prediction of behavior (i.e., the logical and persistent patterns of regularity in social life). Sociology stresses the social contexts in which people live and how these contexts influence people's lives; it focuses on how humans are influenced by society and associated social structure, and how people in turn reshape their society.

- Sociologists study values, cultural norms, and social structures external to the individual, rather than individual-level explanations of behavior.
- The shift from traditional management to alternative governance structures is partially based in the realization that the context of wildlife management is variable across geographical regions and cultures, even within one jurisdiction (e.g., a nation or a state). Sociological perspectives help clarify the myriad contextual factors affecting the behavior of individuals, groups, and institutions.
- Three major theoretical approaches underlie more specific sociological analyses that apply to wildlife management: functionalist, conflict, and interactionist. Each of these offers different perspectives on the process by which society is organized, emphasizes different outcomes, and implies different engagement strategies for wildlife managers.
- Many of the human dimensions of wildlife management that are about values, socialization, and norms are based in the functionalist perspective, which focuses on the maintenance and creation of social order. A conflict-based perspective proposes that social organization emerges from ongoing battles over resources as opposed to developing from shared cultural values. In contrast to functionalist and conflict perspectives that see the creation of social order as unfolding more or less "automatically," interactionism emphasizes the active role that humans have in creating and assigning meaning to social life.

- Given the reliance of wildlife management on citizens' voluntary compliance with regulations, managers seek to understand factors leading to people's socialization into expected standards of behavior. Socialization, firmly rooted in the functionalist perspective, is the process by which people learn the attitudes, values, and behaviors appropriate for members of a particular cultural group.
- Human populations around the world are rapidly changing due to differential rates of migration and reproduction in an increasingly globalized society. Combined with increased rates of transportation and communication, these demographic changes pose opportunities for wildlife professionals to create novel programs and partnerships to reach ever-changing stakeholders.
- Sociology, like any other science, reveals what is, not what should be. Sociology cannot settle debates about value; it cannot tell managers what the optimal solution is (i.e., "should" and "ought" questions are in the realm of ethics). Sociology, however, does provide valuable insights into social drivers that influence perception of impacts and acceptability of management interventions to achieve wildlife management objectives.

Suggesting Reading

Field, D. R., and W. R. Burch. 1988. Rural sociology and the environment. Greenwood Press, New York, New York, USA.

Humphrey, C. R., T. L. Lewis, and F. H. Buttel. 2002. Environment, energy, and society: a new synthesis. Wadsworth, New York, New York, USA.

Stedman, R. C., D. Diefenbach, C. Swope, J. Finley, A. Luloff, H. Zinn, G. San Julian, and G. Wang. 2004. Integrating wildlife and human-dimensions research methods to understand hunters. Journal of Wildlife Management 68:762–773.

ECONOMIC CONSIDERATIONS IN WILDLIFE MANAGEMENT

RICHARD C. READY

Society has many expectations for management of wildlife, expectations that are felt by local, state, and federal governments and by non-governmental organizations engaged in wildlife or land management activities. In a world where needs always seem to exceed the resources available to meet them, agencies and non-governmental organizations are constantly making decisions about where to direct scarce resources for greatest impact. This is the situation faced by most state wildlife management agencies. For example, hunters who pay license fees that fund the agency want improved access and hunting quality. They want more opportunities to hunt, they want higher quality animals, and they want to be allowed to take more animals during each hunting season. Citizens (including many hunters) also want less browsing by large herbivores on native plants, crops, and ornamental plantings; less depredation by predators on livestock; more non-game wildlife for viewing; and more protection for endangered species. Finally, business owners and local governments in rural areas want the agency to manage wildlife in ways that encourage hunters and non-hunters to spend money in their communities. The agency has a limited budget to spread over land acquisition, habitat improvement for game, law enforcement, and programs for non-game species. How does the agency make decisions in the face of these competing, even contradictory demands? The field of economics can help.

6.1. WAYS ECONOMICS ASSISTS WILDLIFE MANAGEMENT

Economics is a social science that studies how society meets competing demands in the face of limited resources. It can help evaluate the trade-offs that a wildlife management agency or wildlife conservation organization must make in distributing resources among its competing objectives. Economics can be particularly useful to wildlife managers in three areas:

1. *Non-market valuation.* Even though native wildlife in North America is not typically bought or sold in markets, it has value to society. For example, if an action is taken to increase the number of eagles in a local population, that action will increase the value of the population to people who like to watch eagles, which generates benefits to the eagle watchers. Similarly, waterfowl hunters value the opportunity to shoot ducks. If additional ducks are produced through wetland management, then the value of the opportunity to hunt ducks in a flyway increases, which generates benefits to the hunters. To determine whether the cost of accomplishing these improvements is justified, we need a way to measure the value of wildlife in situations where the eagle watcher or the duck hunter does not actually pay some unit price for the additional eagles or ducks. Economists have developed non-market valuation methods to estimate the value of public goods (goods that people value but that are not sold in markets).

2. *Economic impact.* Although wildlife may be a public good that is not traded in markets, it often provides the foundation upon which other markets and economic activity depend. For example, according to the National Survey of Fishing, Hunting and Wildlife-Associated Recreation (NSFHWAR), hunters, anglers, and wildlife watchers in the United States spent $122.3 billion (all dollar values in U.S. currency) on these activities in 2006. These expenditures support jobs in the manufacturing and sales of equipment as well as in hotels, restaurants, and other service industries in the areas where wildlife-associated recreation occurs. All of this economic activity generates tax revenues for local, state, and federal governments. Economists have developed methods for projecting how expenditures related to wildlife-associated recreation will affect local and state economies, employment, income, and tax revenues.

3. *Modeling and simulation.* Ecological and economic systems are deeply interconnected. Human choices and subsequent behavior have profound influences on ecological systems; conversely, ecological systems create powerful incentives or disincentives that drive human use of wildlife

A family watches birds at Baskett Slew National Wildlife Refuge in Oregon. Many economic values are associated with non-consumptive uses such as wildlife viewing. (courtesy USFWS)

The money spent by wildlife watchers and big game hunters supports local businesses, jobs, and tax revenues in the regions surrounding Africa's wildlife parks and reserves. Economists have developed methods to estimate the impact that such expenditures have on a local economy. (courtesy Mark Needham)

resources. Economists and others use a family of quantitative approaches to model dynamic feedback between wildlife populations and economic inputs and outputs. Insights from these bioeconomic models can help managers apply knowledge of coupled ecological and economic systems to help resolve wildlife issues. Bioeconomic modeling provides one quantitative framework for considering the benefits and costs of alternative wildlife management policies.

This chapter defines the concept of value from an economic perspective and presents a brief introduction to concepts fundamental to economic valuation in the context of wildlife

management. Every wildlife professional should have a basic understanding of economic valuation, how economic data are gathered, and how insights from economic analyses may be used as part of an information base that can inform decisions. This understanding will help wildlife managers identify when collaboration with economists is needed for more effective wildlife management.

6.2. DEFINING AND MEASURING THE VALUE OF WILDLIFE: WHAT IS "VALUE"?

Economists mean something very specific by the term "value," which can differ from how this term is used in other fields that seek insight about the human dimensions of wildlife management (e.g., social psychology) or from how it is used in everyday speech. To develop a concept of economic value, start by considering a good that is bought and sold in a market, such as a custom-made hunting rifle. Such a rifle might sell at a price of $5,000. But what is its value? The answer depends on whose value we are interested in. A wealthy gun enthusiast might be willing to pay much more than $5,000 to get the rifle, perhaps as much as $8,000, while someone with limited wealth might be willing to pay only $1,000 to get the rifle. If we sold the rifle to the first person, we have sold him something he values at $8,000. If he chooses to buy the rifle for $5,000, then the value he gets from it exceeds his cost by $3,000. This difference is called consumer surplus. It is the value you get from having something minus the cost you paid to get it. If instead we sold the rifle to the second person (and prohibited him from reselling it) we have sold him something he values at only $1,000.

This is how economists define the value of something: it is the greatest amount of money that an individual would be willing to pay for an opportunity, experience, service, or object. This value is called the individual's willingness to pay (WTP). How would this definition of value extend to wildlife, which does not have a price? Note that the value of the rifle to each person did not depend on its price. So the fact that wildlife does not have a price does not mean it does not have a value. Still, there is an important difference between a market good such as a rifle and a non-market good such as wildlife. A rifle can be owned by only one person at a time. If I own the rifle, you cannot simultaneously own the same rifle. In contrast, if I get pleasure from watching an eagle, you can also get pleasure from watching the same eagle.

For goods such as wildlife, it usually does not make sense to try to value one unit of the good; instead, we value changes in the level of the good. So instead of trying to value one animal in isolation, we would value changes in the total number of animals in a population. To illustrate this point, let Q be some measure of the quality of a wildlife resource. Q could measure any aspect of the wildlife resource that people care about, such as the number of animals in a population, their health, wildlife diversity, and so forth. The value that an individual places on a change in Q from Q^0 to Q^1 is defined as the largest amount the individual would be willing to pay to have Q^1 instead of Q^0.

Having established a measure of the value to an *individual* of a change in a wildlife resource, how does that relate to the value to *society* of the change? Economists make the critical assumption that if a representative individual has been made better off by a certain amount, then society has been made better off by that same amount per individual. This assumption is an important foundation for conducting cost–benefit analysis. It asserts that we can add up all of the benefits (values) that are generated by a policy action irrespective of who receives those benefits. This assumption sidesteps the issue of who should receive the benefits from public actions and who should bear the costs. This is a common concern for wildlife managers that is not addressed by economic analysis. The convention among most economists is to assume that WTP measures value to the individual and value to society. You should recognize that in practice this definition raises some ethical issues.

Another important economic assumption is that wildlife has value only if someone cares about it. If some action would affect a wildlife resource but no human in society cares about the effect (i.e., there are no stakeholders), then the effect has no value. Our definition of economic value, therefore, does not allow for "intrinsic" values, which is the concept that wildlife has value independent of how it is viewed by humans. This does not, however, imply that wildlife has no value unless it is used by humans. As we will see, individuals may value wildlife even if they never personally interact with wildlife.

In some cases, people may experience or expect negative values for changes in wildlife abundance and distribution. For example, in several regions of the United States, many people believe there are too many deer or elk, which is leading to excessive browsing of native and ornamental plants and elevated risk of motor vehicle accidents. For people with such concerns, an increase in the number of deer or elk would be viewed as a bad thing. To be consistent, we define the (negative) value that such a person would experience from an increase in Q from Q^0 to Q^1 as the smallest amount that the individual would have to be paid to accept the increase in Q. This concept is called the individual's willingness to accept (WTA) compensation for the change from Q^0 to Q^1, and it represents a cost to the individual. WTA, in this case, would include compensation for both monetary impacts (e.g., damage to ornamental plants or to vehicles) and non-monetary impacts (e.g., lost enjoyment during hiking due to a lack of wildflowers in the forest).

For most wildlife species, whether an increase in the size of a wildlife population would be viewed as a good thing or a bad thing will depend on the initial size of the population. The top panel in Figure 6.1 shows how the value that an individual places on a wildlife population can vary with population size. The bottom panel in Figure 6.1 shows the amount the individual would be willing to pay for a small increase in Q, which is called the marginal value of Q. It measures the slope of the value curve in the top panel. At very small population sizes where species survival is jeopardized, an increase in Q might be highly valued. This can be seen by the steeply upward-sloping total-value curve and the large positive marginal value. At medium population sizes, an increase might not be worth

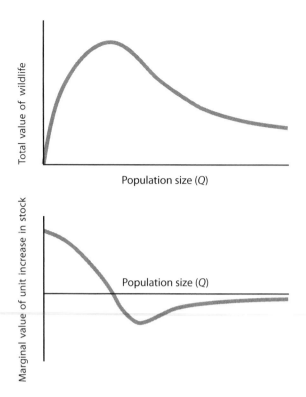

Figure 6.1. The total value an individual places on a wildlife population as it varies with population size (*top panel*), and the marginal value of each unit increase in the population size (*bottom panel*)

as much; here, the total value curve levels off and the marginal value curve is close to zero. At high population sizes where the individual feels that there are already too many animals to tolerate (i.e., the negative impacts exceed limits of acceptability), an increase in Q would have a negative marginal value; here, the total value curve falls and the marginal value curve is negative.

More than one person simultaneously receives value from a wildlife population, but different people will have differently shaped value curves for Q. The value to society of a specific change in Q from Q^0 to Q^1 would be the sum of the WTP values for people who hold positive values for the change minus the sum of the WTA values for people who hold negative values for the change.

The number of individuals in a wildlife population may not be the only trait that people care about. We can redefine Q to represent other dimensions of wildlife that people value. Examples could include the health of the wildlife population, the average size of individuals in the wildlife population, the probability of seeing wildlife, the probability of survival for an endangered species, the probability of hitting a deer with your car, or the level of damage to ornamental plantings from wildlife browsing.

6.3. TAXONOMY OF ECONOMIC VALUES OF WILDLIFE

It is useful to classify the different reasons for which people value wildlife into a taxonomy. This allows analysis without

Table 6.1. A taxonomy of wildlife values

Use values
   Consumptive use values
      Commercial harvest
      Recreational harvest
   Non-consumptive use values
Non-use values
   Option value
   Existence value

double-counting, and recognizes that the same person could value wildlife for more than one reason or could hold both positive and negative values simultaneously. Table 6.1 provides one taxonomy of different types of values that people in our society hold for wildlife.

6.3.1. Consumptive Use Values
When we think about how people relate to wildlife, most examples that come to mind involve direct contact with the wildlife. For example, a hunter shoots a deer or a birdwatcher sees a bird. In both cases, the human must have some interaction with the animal(s). In these examples, we say that the human *uses* the wildlife, which generates *use values.*

Use of wildlife that includes removal of animals from the population is called consumptive use. Consumptive wildlife use would include harvest for sale (e.g., furbearers), recreational harvest, and subsistence harvest. We think about the value of wildlife differently in these three contexts. Where the wildlife is being harvested commercially, the marginal value of additional harvest is fairly easy to define. If the productivity of a population of furbearers is increased so that harvest can be increased, then the value to the trapper is given by the market price of the additional animals taken minus the cost of taking those additional animals. A trapper would theoretically be willing to pay up to the difference between price received and cost to harvest an animal to be able to take the additional animal.

For recreational harvest, the value to the hunter is not given by a price per pound of the harvested animals. In fact, it is usually not appropriate to define a value per animal harvested. For many applications, we instead are interested in the value that the hunter places on the opportunity to hunt. It costs money to go hunting. The hunter must pay for things such as hunting licenses–tags and other fees, supplies, fuel, lodging, and food. If the hunter is willing to pay more than the costs, then he will decide to go hunting and will receive a *consumer surplus* from going hunting. That consumer surplus represents the value to him of the opportunity to participate in a recreational activity through which he expects to obtain any of a variety of satisfying outcomes (e.g., being with friends and family, appreciating nature, demonstrating personal competence, and obtaining meat).

The consumer surplus from a hunting trip will depend in part on the quality of the wildlife resource. Before a hunt, the hunter expects some probability of taking an animal. If we were to increase the wildlife population of interest, then the

probability of success would increase. A hunt with a high probability of harvest generally is more attractive than one with a lower probability of success, *all other factors being equal.* One value of interest is the hunter's WTP for an increase in the probability of success. We might alternatively measure success on a more continuous scale (e.g., number of ruffed grouse flushed per hour or average amount of time spent afield to harvest one deer). By redefining Q, we could also switch from valuing changes in hunting success to valuing changes in the quality of the wildlife population (e.g., the average size of the male deer or the proportion of mature male turkeys). WTP for a change in success or quality could be defined for a single hunting trip or for an entire hunting season.

6.3.2. Subsistence Values
Valuing subsistence use of wildlife can be challenging. In developing countries, wildlife harvest often is an important source of protein for local communities. A management action that restricts harvest, such as creation of a national park that precludes hunting, will impose a cost on the communities whose harvest must be reduced. The cost to those communities of the harvest reduction is defined as the amount that they would have to be paid in compensation for the reduction. Assuming no cultural or other values are relevant (seldom a realistic assumption), this could be approximated using the local market price of similar meats.

Where subsistence use is culturally important, valuing a change in a wildlife population or access to it is more difficult. Subsistence hunting can be an important activity connecting a community to its traditions and reinforcing community values. These values can result from cooperation during the hunt and sharing of meat after harvest. It has been argued that WTP to engage in subsistence hunting (or WTA compensation to forego it) does not adequately measure the value to communities of the cultural aspects of subsistence hunting. One difficulty in applying value concepts to this situation is that definition of value is measured at the individual level, while cultural values are held by groups of people. Economists disagree about whether cultural values associated with subsistence hunting can be quantified monetarily.

Even if we cannot assign an economic value to the cultural benefits from subsistence hunting, economic concepts can be useful when considering those benefits. For example, the concept of marginal versus total value can help guide us when considering the effect that a management action would have on subsistence use. The cultural benefits from subsistence use will not be sensitive to marginal changes in the wildlife resource in many situations. For example, while harvesting caribou may be of great cultural importance to some Native Alaskan communities, the cultural value to those communities may not be affected by a 10% increase or decrease in the size of the caribou population. However, a 50% decrease might have profound cultural impacts.

6.3.3. Non-consumptive Use Values
Non-consumptive uses of wildlife are more common than consumptive uses and involve more people. More than twice

as many people engage in non-consumptive wildlife viewing as engage in hunting in the United States, according to the NSFHWAR.

Many values are associated with non-consumptive uses, such as various kinds of wildlife viewing. Similar to the examples for hunting, we can determine the consumer surplus from a day spent birdwatching, or the increase in consumer surplus per day from an increase in the quality of birdwatching. Considering that more people engage in wildlife viewing around the home than take trips to view wildlife, we can define the value of songbird diversity or abundance as the amount people would be willing to pay in return for an increase in, for example, the number of different songbird species living in their neighborhood. That value would not be defined per trip or per viewing occasion but might be defined per year.

Non-consumptive uses of wildlife are not restricted to wildlife viewing. Any way in which wildlife directly impacts the well-being of humans but that does not involve harvest of the animal(s) would be included in this category. Examples would include pollination or germination services provided by wildlife (a form of ecological services); damage by wildlife to crops, forests, or ornamental plantings; property damage and injury from motor vehicle collisions with wildlife; impacts of wildlife on water quality; and transmission of human disease through wildlife vectors.

The wide variety of ways in which wildlife affects people generates a wide variety of wildlife values in society, and the same person can hold several different values for a given wildlife species simultaneously. An individual who hunts might hold a positive value for a higher deer density because it could increase his probability of hunting success, and at the same time hold a negative value for higher deer density due to damage deer cause to his vegetable garden. The net value this individual places on an increase in deer density could be positive or negative, depending on which value dominates.

6.3.4. Non-use Values

If we define the value to a person of a change in a wildlife resource as the amount that person would be willing to pay to get the change, then it must be true that any change to a wildlife species that someone is willing to pay to get must have value. For example, people will donate money to organizations to preserve habitat for wildlife populations that they will never "use" in any way in places they never intend to visit.

A non-use value is a WTP for a change in a wildlife species that does not involve direct use of it. You might be willing to pay to support restoration of an endangered lemur species in Madagascar, even if you have no intention of ever visiting that country. Or you might be willing to pay to support protection of polar bears, even though you expect never to see one in the wild. In both cases, you value knowing that the endangered species still exists. For this reason, non-use values are often called "existence values." While it is easy to imagine someone willing to pay to preserve an endangered species, non-use values can exist for more common species as well. From an economist's perspective, the enjoyment you get from seeing a bird

(use value) and the enjoyment you get from simply knowing that the bird exists without seeing it (non-use value) are both legitimate values. If you will never personally see the bird, then any value you place on it must be motivated by altruism, which is the enjoyment you get from knowing that someone or something else is experiencing something positive. Note that existence value is not the same thing as intrinsic value (the worth of wildlife independent of whether anyone cares about it). It cannot be measured in dollar terms because it is not a value held by humans. In contrast, non-use or existence value is the amount of money that someone would be willing to pay to improve a wildlife resource even if no one would ever be able to use it. It only exists if someone cares about the wildlife stock (the number of animals or wildlife species being considered).

6.3.5. Option Values for Wildlife

Someone may not use a wildlife population now, but may value having the option to use it in the future. Such a value is called an "option value." For example, a person who has only hunted white-tailed deer in his home state of Pennsylvania may hold a value for maintaining elk-hunting quality in the western United States because he wants the option of hunting elk in those states in the future. Option values can apply to use values and non-use values. A value held for anticipated future use by other people would be a bequest value, which is another type of non-use value.

6.4. VALUING HARM TO WILDLIFE

The discussion of use and non-use values has so far focused on the value of an improvement or increase in a wildlife stock. We can also consider the cost (negative value) of harm to a wildlife stock, whether it be from a habitat loss, an accident, or a change in climate. We can define the cost of harm to a wildlife stock in two different ways. Consider an event that reduces quality of a stock from Q^1 to Q^0. First, we could define the cost as the amount that someone would have to be paid to accept the harm (WTA:$Q^1 \rightarrow Q^0$). Or we could define the cost as the amount that someone would be willing to pay to keep the harm from occurring (WTP:$Q^0 \rightarrow Q^1$).

Note that even though the levels of Q are the same, these two values will not generally be equivalent. WTA:$Q^1 \rightarrow Q^0$ typically will be larger than WTP:$Q^0 \rightarrow Q^1$. One reason for this is WTP to avoid harm to a wildlife stock is bounded by a person's wealth. Regardless of how strongly you feel about the wildlife stock, you cannot pay more than you own to protect it. Conversely, there is no upper bound on WTA:$Q^1 \rightarrow Q^0$. Which value definition should be used? Economics does not provide clear guidance. It is often easier to measure WTP:$Q^0 \rightarrow Q^1$ than to measure WTA:$Q^1 \rightarrow Q^0$, so WTP:$Q^0 \rightarrow Q^1$ is often the measure applied in practice.

6.5. MEASURING WILDLIFE VALUES

Having defined how the value of wildlife is conceptualized in economics, we turn to measuring those values. Typically the task is to estimate the amount that someone would be willing to pay to get a change in a wildlife resource. The difficulty is that, in contrast to goods bought and sold in stores, there is no observable price for Q. To overcome this difficulty, economists have developed several methods for valuing non-market goods, which can be divided into two classes: methods that rely on observable behavior (revealed-preference methods) and methods that rely on survey questionnaires (stated-preference methods).

6.5.1. Revealed-Preference Non-market Valuation Methods

Although it is not possible to buy wildlife in a store at a market price, people make choices that affect the level of Q that they personally experience. The most common revealed-preference technique for valuing wildlife uses information on recreational trips that are made to hunt or view wildlife. Consider a ruffed grouse hunter who is choosing where to go hunting on a particular day and has two options: Area A and Area B. Area A has a high density of grouse and a high expected harvest rate, Q^A. Area B has a lower grouse density and lower expected harvest rate, Q^B, with $Q^B < Q^A$. However, Area B is closer, so that it takes less time and money to get to Area B than it does to get to Area A. Let TC^B be the cost to travel (roundtrip) from the hunter's home to hunting Area B, and TC^A be the cost for Area A, with $TC^A > TC^B$. Assuming nothing else differs between the two areas, the hunter must then choose between the two pairs, (Q^A, TC^A) and (Q^B, TC^B). If we observe that the hunter chooses Area A, then we know that his WTP to get Q^A instead of Q^B is larger than $TC^A - TC^B$. We do not know how much larger, but we at least have a lower bound on his value for this quality difference. Conversely, if the hunter chooses Area B, then we know that his WTP to get Q^A instead of Q^B is smaller than $TC^A - TC^B$. This technique for inferring value from observed recreation choices is called the travel cost method, because it relies on differences in the cost of traveling to recreation sites.

In the above example, we obtain either an upper or lower bound on one quality difference for one hunter. In reality, hunters (or birdwatchers) will choose among many different recreation sites and provide more information about relative values from each observed trip. Further, different recreationists will face different choice sets. Finally, different recreation sites will differ in more than one way. For example, hunting sites might differ in terms of harvest rate, ease of access, and crowding by other hunters, so that our measure of Q should be multi-dimensional. With many observations of site choices by many different people, it is possible to develop a site choice model that estimates the utility (enjoyment) that the hunter will get from each site and assumes that the hunter chooses the site that gives the highest utility. From this estimated site-choice model we can also estimate the value of changes in individual dimensions of Q. For example, the model could be structured to tell us the amount that a hunter would be willing to pay for an increase in the probability of hunting success, or the amount that a birdwatcher would pay for an increase in bird species diversity.

A second type of travel cost model focuses on how often the recreationist visits a specific site. Data on visitation are collected from many different recreationists who visit the same site. Different recreationists will face different travel costs to get to the site. A trip frequency model is estimated by statistically regressing the number of trips to the site on the travel cost to the site. Additional explanatory variables, such as relevant characteristics of the recreationists, are also included in the regression.

Figure 6.2 shows what a trip frequency model might look like for a site. A given recreationist, i, who faces travel costs TC_i, would be expected to visit the site V_i times over a season. Recreationists who live close to the site and who face low travel costs are expected to visit the site frequently, while those who live farther away and face higher travel costs will visit less frequently. The curve drawn in Figure 6.2 is exactly analogous to a demand curve for a market good. It tells us how many units (visits) of the good a consumer will "buy" at different price (travel cost) levels. By interpreting the trip frequency model as a demand curve, we can use it to measure the value that a recreationist receives from being able to visit the site. The shaded area in Figure 6.2 measures the consumer surplus (CS_i) the recreationist receives from being able to visit the site at cost TC_i. It is the extra value the recreationist places on visiting the site V_i times over and above the cost the recreationist must pay ($TC_i * V_i$).

To calculate the total benefit to society being generated by the site, we would calculate V_i and CS_i for each visitor to the site, and sum all of the individual consumer-surplus values. This generates a measure of the value to society of the site. If the site changes due to a change in the quality of the wildlife at the site, then the benefit generated by the site will change. The value of the change in the quality of the wildlife at the site is measured by the change in the total consumer surplus.

A second technique for valuing non-market goods from observed behavior is the hedonic pricing method. This technique infers values for environmental goods by looking at differences in real estate prices. While the hedonic pricing method is most often used to value environmental quality measures such as air quality and noise pollution, it can in certain cases be

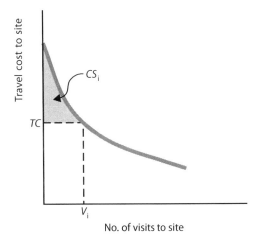

Figure 6.2. A trip frequency model for wildlife-associated recreation

used to value differences in wildlife quality. Take, for example, the value of land that can be used for a hunting cabin. Land parcels located in areas with good quality hunting (e.g., plenty of game, plenty of land with access, little crowding) will tend to cost more to purchase than parcels located in areas with poor hunting quality when other factors are held equal.

The hedonic pricing method infers the value of differences in wildlife quality from these differences in land values. In an application, information on the market value (as observed through the sale price) and on wildlife quality are collected for many different parcels, along with information on other factors that would affect the market value of the parcel (parcel size, topography, soils, proximity to populated areas, natural gas or timber values, etc.). A regression equation called an implicit price function is estimated, where parcel value is regressed on all of the other measures. The regression coefficient for wildlife quality from this estimation is called the marginal implicit price for wildlife quality. It tells us by how much a parcel's value would change from a change in wildlife quality near the parcel. The total value of a change in wildlife quality that would affect a larger area would be estimated by calculating the impact of the wildlife quality change on each parcel and summing over all affected parcels. A similar approach could also be used to look at the relationship between game density (or game quality) and lease rates for hunting leases on private lands.

6.5.2. Stated-Preference Non-market Valuation Methods

Although economists have developed clever ways to infer how much someone would be willing to pay for a change in wildlife quality from observing their behavior, in many situations values cannot be inferred from observed behavior. You may place a high value on knowing that wolves exist in Yellowstone National Park, but unless you visit the park you may never do anything that can be externally observed that reveals that value. In such cases, the only way that your value can be estimated is by asking you about it.

Techniques for asking people about the values they hold for non-market goods are called stated-preference valuation methods. These are administered through surveys, either in person, by telephone, by mail, or through the Internet. Reliably measuring the value that a respondent places on a change to a wildlife resource requires a highly structured and rigorously designed and tested survey instrument. It is not enough to ask, "Would you pay more for wildlife?"

The goal of a stated-preference survey is to identify the largest amount that the respondent would pay for a specific change in the wildlife resource. The difficulty is that real-world situations where individuals have an opportunity to "buy" a change in wildlife are rare. Instead, a hypothetical scenario is constructed where the respondent typically must make a decision that determines which of two options (A and B) will occur. In one option, Q is set to some low level, Q^A. In the other option, Q is set to some higher level, Q^B. In option B, however, the respondent's disposable income, I, is lower than in option

A. If the respondent states that he would make a choice that results in option B, then we know that his WTP for the quality change $Q^A \rightarrow Q^B$ is greater than $I^A - I^B$. If the respondent states that he would choose option A, then we know that his WTP for this quality change is less than $I^A - I^B$.

The hypothetical scenario constructed must be credible and understandable to the respondent. Several different aspects must be included in the hypothetical scenario. First, the good being provided must be described accurately in a way that the respondent can understand and can relate to. Second, the two levels of the good, Q^A and Q^B, must be described. Third, the respondent must understand the means by which Q will be affected. The provision mechanism must also be neutral (i.e., it must not have positive or negative side effects that the respondent will care about). Finally, the means by which the respondent's income would be affected, called the payment vehicle, must be described. This might be an increase in taxes, user fees, or a general increase in prices for goods and services. The payment vehicle must be credible and must be seen as a fair way to pay for the good. In addition, the payment vehicle must be designed so that the respondent enjoys Q^B only if he pays the money. For this reason, donations are not a valid payment vehicle in stated-preference valuation surveys. A person could value a wildlife resource very highly yet choose not to donate to improve it, relying instead on donations by other users. Such behavior is called free riding. When we observe someone donating money to improve wildlife, we know that person has a positive value for the improvement, but the amount of the donation does not provide a reliable measure of the size of that value.

Having described the hypothetical scenario to the respondent, he is then asked which option he would choose. Called the elicitation question, this question can take many forms. The referendum format is commonly used for a good such as wildlife. In this format, respondents are asked to imagine that there will be a public referendum on a program to improve wildlife habitat, populations, and so forth. The program would cost money that would be raised through taxes or through an increase in hunting license fees. After having heard a description of the program, the respondent would be asked a question of the form, "Suppose that if the referendum passes, the program will be implemented and you will have to pay $X more in taxes or fees than you do now. Would you vote for or against the referendum?" By asking the same question to many different people but with different values for X, it is possible to trace out the distribution of WTP for the change in Q.

Another approach to the elicitation question is the open-ended format, where the respondent would be asked a question of the form, "What is the largest tax increase that you would vote for?" Or respondents might be given a card listing several possible tax increases, and asked which increases they would vote for and which they would vote against.

Yet another elicitation alternative is called the stated-choice format. It presents each respondent with a series of choices among two or more options. In this method, each option is described by several different measures of quality, called the attributes of the option. For example, in a survey to value deer-hunting quality, hunters might be asked to choose from among several different hunting areas. The measures of quality of hunting in each area might include probability of bagging a deer, density of hunters in the area, and average size or condition of the deer. Each option would require some payment by the hunter. In this case, the payment vehicle might be the cost of getting to the area to hunt. The respondents would be presented with a set of hunting areas that differ in their attributes and their cost, and they would be asked which hunting area they would visit. With many respondents answering several different stated choice questions of this type, it is possible to estimate the value of a change in each of the attributes described. For example, we could calculate the average WTP for a specific increase in the probability of success, for a specific increase in deer quality, and for a specific decrease in hunter crowding.

A concern with stated-preference valuation, regardless of which alternative approach is taken, is that it is based on responses to hypothetical questions. In reality, when a hunter makes a decision about where to hunt, there are penalties if he makes a mistake, either through paying more than he should have to get to the hunting area or by hunting in an area that is of lower quality than the area he should have chosen. In contrast, there is no penalty to a survey respondent for providing a "wrong" answer. For this reason, we should be concerned about whether responses to stated-preference surveys are reliable. Methods for examining reliability have been developed, but discussion of these falls outside the scope of this chapter. Suffice it to say that because stated-preference methods rely on hypothetical behavior, economists tend to prefer revealed-preference estimates of value when they are available. In many situations involving wildlife, however, revealed preference approaches will not work, and stated-preference approaches are the only way to learn about wildlife values.

6.6. USING WILDLIFE VALUES IN COST–BENEFIT ANALYSIS

If we know the value of a change to a wildlife resource, we can compare that value to the cost of the change. A cost–benefit analysis is a formal accounting of all of the costs and benefits of a proposed action. In keeping with standard terminology, this action is called a project. The project being analyzed could be a traditional project that involves spending money (for example, removing a dam or purchasing land for endangered species habitat) or it could be a change in laws and regulations (for example, adjusting hunting-season lengths and bag limits or increasing hunting-license fees). In each case, the various values associated with the project are identified and estimated and then summed to determine whether the project generates positive net benefits (i.e., the sum of benefits is greater than the sum of the costs). While in principle this is straightforward, there are important issues regarding how the different values are summed.

6.6.1. With–Without Analysis versus Before–After Analysis

To measure the value of a change in a wildlife resource resulting from a project, the analyst must predict both Q^1 (the level of quality that will occur with the project) and Q^0 (the level that will occur without the project). It is important to remember that Q^0 is not necessarily the same as the level that occurs at the time of the analysis. For example, for a project aimed at preventing the decline of a wildlife stock, Q^0 is likely lower than the current level of the wildlife stock. Climate change and land-use change will affect wildlife populations independent of any actions taken by wildlife management agencies. For long-lived projects, it is particularly important to generate projections of Q both with and without the project, and to compare those, rather than comparing the level of Q after the project to the level that exists before the project.

6.6.2. Accounting Frame

Having identified the levels of Q that will occur with and without the project, the analyst must decide whose benefits and costs will be counted in the analysis. For example, over what geographic unit will benefits and costs be measured? The geographic scope of the analysis typically will be chosen based on the agency making the decision. A state agency will usually count benefits and costs that accrue to its residents, even if its decisions affect people living in neighboring states. A federal agency will usually count benefits and costs that accrue nationally, and so on.

The choice of geographic scope can make a critical difference to the analysis, particularly when non-use values are important. We expect that for most wildlife resources, non-use values will decline as distance from the resource increases, but some wildlife resources (such as polar bears or California condors) might generate non-use values that are independent of geography. The choice of the geographic scope can be decisive for an analysis. A project that generates $20 per person in non-use values may fail a cost–benefit test if those benefits are summed over only the 200,000 people who live in the immediate county, but it may pass easily if they are summed over 300 million U.S. citizens. Geographic scope issues are even more complex for resources of international importance (rhinos in Africa, tigers in India).

After the geographic scope of the analysis is determined, the analyst must decide whether the preferences of all people living within the geographic area will be counted. Should benefits and costs to non-hunters be counted in an analysis of a project that will be paid for with hunting license fees? Should benefits and costs to people who own vacation homes in an area be counted, or only those that affect permanent residents? These are questions that must be decided, often on practical or political grounds. Stakeholders can be engaged to aid with such judgments.

6.6.3. Choosing a Time Horizon

Another consideration is the choice of the time horizon for the analysis. Some projects are very long-lived. Land purchased for habitat or access will likely be used for those purposes for a very long time. Values generated by the project are typically estimated per season or per year. Over how long a period should the analyst sum those values? There are no rigid rules. Rather, the analyst must make a trade-off. If the analyst uses a time horizon that is too short, he will exclude relevant values. A too-short time horizon will tend to undervalue a capital-intensive project such as a land purchase or construction of a facility. In these types of projects, costs tend to occur up front while benefits are generated over time. In contrast, a project where benefits are front-loaded and costs are spread out over time would tend to be overvalued with a short time horizon. An example would be a mining permit where extraction occurs over tens of years while damage to habitat lasts much longer.

Using a very long time horizon, however, introduces another type of bias. A cost–benefit analysis of a project that uses a very long time horizon assumes that choices made today will not be changed during that time horizon. In reality, wildlife management decisions are adjusted over time in response to changing conditions or new information. An analysis of a proposed change in bag limits should not assume that those new limits will necessarily be in place for decades. They likely will be changed again at some point. The choice of the time horizon must then be made on a case-by-case basis, taking into account the expected duration of the impacts from the decision, the likely duration of the decision, and the ability of managers to adjust over time.

6.6.4. Discounting Future Benefits and Costs

When a project's impacts last for several years, the analyst must compare impacts that occur earlier in time to impacts later in time. Are impacts that occur at different times directly comparable? The general answer is no. To see why, consider an impact that will generate benefits to hunters 5 years from now. We know that 5 years from now, when the benefit appears, each hunter will be willing to pay $B for the predicted change in Q. How much would those hunters be willing to pay now for that promised future benefit? That is the value that should be counted in the cost–benefit analysis.

The value calculated is called the present value (PV) of the future benefit. It represents the payment that would be made right now that the hunter views as being equivalent to a payment of $B made 5 years from now. The general formula for an impact of $B that occurs t years from now is

$$PV(B, t) = \frac{B}{(1 + i)^t}$$

This formula works for both benefits and costs. The sum over years and impacts is called the net present value (NPV) of the project and is given by

$$NPV = \sum_{t=0}^{T} \left[\sum_{j=1}^{J} \frac{B_{jt}}{(1 + i)^t} - \frac{C_t}{(1 + i)^t} \right]$$

where T is the time horizon, $j = 1, \ldots, J$ are the different impacted individuals, B_{jt} is the impact (positive or negative) on individual j that accrues t years from now, and C_t is the financial cost to the government in year t of implementing the project. If the project NPV is positive, then it passes the cost–benefit test.

An alternative statistic for judging a project is called the cost–benefit ratio (CBR). This ratio can be calculated as

$$CBR = \frac{\displaystyle\sum_{t=0}^{T} \sum_{j=1}^{J} \frac{B_{jt}}{(1 + i)^t}}{\displaystyle\sum_{t=0}^{T} \frac{C_t}{(1 + i)^t}}$$

If this ratio is >1, then the project passes the cost–benefit test.

Although these two statistics will always agree about whether a particular project generates positive net benefits, the two will give different rankings when there are many potential projects. The NPV statistic is best used in situations when there are many mutually exclusive projects, from which only one (or none) may be chosen for implementation. An example would be a situation where there are several different options for how a natural area should be managed. To maximize the net benefit generated by the area, the option with the highest NPV should be chosen. The CBR is best used in situations where an agency or land conservation organization has several projects available to it, each of which can be independently undertaken, but has a limited budget to spend. An example would be a situation where a land trust has several different wildlife-viewing access sites under consideration and has to choose which to develop.

6.7. REGIONAL ECONOMIC IMPACTS OF WILDLIFE

As discussed in the previous section, a hunter receives some value from being able to go on a hunting trip, measured by his WTP for the trip. But the hunter must pay for expenses related to the trip, such as equipment, gas, food, lodging, and guide fees. The difference between the hunter's WTP for the trip and the amount he must actually pay is the consumer surplus he receives from the trip. This is the correct measure of the benefit to society of the hunting opportunity, but how should we treat the money the hunter spends on expenses?

Wildlife-associated recreation is an important economic driver for many local economies. The money spent by hunters and wildlife watchers visiting a region supports local businesses, incomes, jobs, and tax revenues in that region. Economists have developed methods for conducting a regional economic impact analysis (REIA), which calculates the impact that such expenditures have on a local economy.

To conduct an REIA, the analyst must first define the region to be modeled. Depending on the purpose of the study, the region might be a county, group of counties, or an entire state. The conceptual foundation of an REIA is an accounting table that specifies the linkages that occur among businesses and households within the region. The table, called an input–output table, shows the dollar flows from each industry in the region to every other industry in the region, as well as the dollar flows into the region and out of the region. For example, a dollar spent for breakfast at a hotel by a visiting hunter causes that hotel to spend some money (<$1) on food from restaurants and bakeries. These new expenditures at restaurants and bakeries cause them to spend more money on cleaning services. The cleaning service company must then spend more money on supplies, and so on. The initial dollar spent by the visiting hunter is called the *direct effect* from hunting. The additional expenditures by the hotel, restaurants, bakeries, and cleaning businesses are called *indirect effects* from hunting.

At the same time, the hotel, restaurants, bakeries, and cleaning businesses must hire more workers (or employ their existing workers for more hours). As a result, wages to local workers increase. Elevated sales by hotels, restaurants, bakeries, and cleaning businesses will also lead to increased profits, which go to the owners of those businesses. Local workers and business owners will spend some of their new income within the region, which will generate another round of new sales at local stores and businesses. These increased sales due to higher local incomes are called the *induced effect* from hunting.

This process does not continue forever. At each round of spending, some of the money spent goes outside of the region. Flows of money that leave the region are called leakages. Local economies that are less diverse or smaller will tend to have higher leakage rates than larger regions with more diverse economies.

It is common to calculate a statistic (called an output multiplier) that tells how much total economic activity will occur in a region as the result of $1 in direct effect expenditures. Two types of multipliers are commonly calculated:

$$\text{Type I Multiplier} = \frac{\text{Direct Effect} + \text{Indirect Effect}}{\text{Direct Effect}}$$

$$\text{Type II Multiplier} = \frac{\text{Direct Effect} + \text{Indirect Effect} + \text{Induced Effect}}{\text{Direct Effect}}$$

For a typical small rural area, the Type I output multiplier associated with outdoor recreation expenditures would be in the range of 1.25 to 1.75 (a Type I multiplier near 1.75 is rare). That means that every dollar that hunters or wildlife watchers spend in a region causes indirect expenditures of $0.25 to $0.75. Type II multipliers would be slightly larger.

Multipliers for the number of jobs that would be created, the increase in local incomes, and the increase in tax revenues to local governments can also be calculated. These multipliers are often of great interest to local businesses and local governments. Consider, for example, the purchase of land for hunting access that will allow more hunters to come to a region. If we can project how many hunters will come, and how much each will spend in the region, we can use the REIA multipliers to predict the impact that the new hunting activity will have on local sales, local employment, and local tax revenues. This kind of information influences wildlife management decisions.

6.8. BIOECONOMIC MODELING

A core task of wildlife managers is to integrate biological and social science knowledge in ways that improve decisions about how people affect or are affected by wildlife (Organ et al. 2006). Bioeconomic models offer the wildlife manager a set of quantitative tools to integrate ecological and economic information in ways that promote systems thinking and learning by managers and stakeholders.

Wildlife management policies are typically attempting to achieve multiple objectives, so economic efficiency is rarely the only consideration in decision making; economic information is insufficient as the only basis for decisions. Moreover, models of any real-world system are of necessity a simplification of complex and dynamic interrelationships. When used to reveal the "big picture," however, bioeconomic models are a valuable aid to deliberations. They offer quantitative mechanisms to ask "what if" questions about the economic and ecological consequences of policy choices.

By creating mechanisms to simulate, critique, and evaluate policy options, bioeconomic models can help managers and stakeholders collectively work toward a common goal: management systems that provide economic benefits to society without compromising the sustainability of wildlife populations or natural ecosystems (Woodward and Bishop 1999). Common applications of bioeconomic modeling in a wildlife management context are described in the following sections.

6.8.1. Modeling to Inform Harvest Policies

Most applications of bioeconomic modeling to date have focused on the actual or potential outcomes associated with harvest regulations or other policies to manage wildlife populations and consumptive uses of wildlife. These models are designed to identify the net value of benefits generated by existing or proposed harvest policies, as well as the likely effects of those policies on wildlife populations over time.

A simple bioeconomic optimization model can be constructed for a population of a single wildlife species in a defined area that is subject to harvest. It could be a population that is harvested for personal use (e.g., migratory waterfowl), commercial sale (e.g., furbearers, alligators), or subsistence use (e.g., walrus, caribou). The wildlife population has a natural rate of net productivity, which is often modeled using a logistic growth function. The population will be in equilibrium if the rate of harvest equals the rate of productivity. Absent stochastic variability, that rate of harvest can be sustained each year without affecting the population size. However, the natural rate of productivity of the population will depend on the population size. Figure 6.3 shows an example of how net productivity can vary depending on population size. If the population is at its carrying capacity, *K,* then by definition its natural rate of net productivity is zero. If the population is below its carrying capacity, it will have a positive rate of net productivity. If the population is very small, however, it will have a low rate of net productivity because so few adults are available to reproduce.

We can theoretically evaluate the sustainable harvest (yield) associated with every possible population size (Box 6.1). If the population is at X_H, we can sustainably harvest H animals each year. The harvest will exactly equal the net productivity of the population, and the population size will not change. There is another population size that will support a sustainable harvest of H units each year. If the population size is X_L, the net natural productivity will also equal H, so that H animals can be sustainably harvested each year. The largest possible sustainable harvest is H_{MSY}. This level is called the stock's maximum sustainable yield (MSY), and it is achievable only if the stock size is at X_{MSY}.

It might appear that a harvest of H_{MSY} would be the sustainable harvest that is "best" for society, but this harvest level ignores the cost of harvesting the animals. As the wildlife population decreases, it costs more to find and harvest each animal. To find the sustainable harvest level that provides the largest net benefit to society, we need to consider both the value of the harvest and its cost. Figure 6.4 shows the value of the sustainable harvests associated with different population sizes; it is simply the sustainable harvest levels from Figure 6.3 converted to dollar values. Figure 6.4 also shows a plausible cost function for harvest. At carrying capacity, K, no harvest occurs so there is no cost of harvest. As harvest is introduced and the stock size decreases, it does not initially cost much to harvest

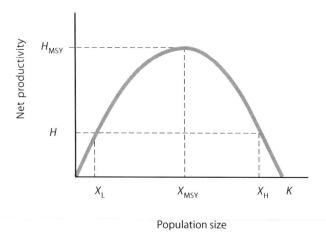

Figure 6.3. Net productivity of a fish stock

Box 6.1 *PRIVATIZATION AS CONSERVATION STRATEGY IN CHOBE NATIONAL PARK, BOTSWANA*

Chobe National Park is a large park in northern Botswana. Hunting is not allowed in the park, but is allowed in areas surrounding the park. Each year, villages near the park are given quotas to harvest different wildlife species. The purpose of these quotas is to allow residents of villages near the park to benefit from the presence of the park, and to compensate those residents for damage they experience from wildlife, including damage to crops, homes, and in some cases deaths of residents from wildlife attacks.

The villages sell the quotas to safari operators. One dif-

Chobe National Park supports a range of charismatic wildlife, including oryx and zebra. Economic analysis can inform decisions about allocation of those resources for the benefit of safari operators and residents of the communities bordering the reserve. (photos courtesy Bret Muter)

Economic Considerations in Wildlife Management   79

the population to a level smaller than K because animals are easy to find. As the population size is hunted to a lower level, however, animals become harder and harder to find, and the cost to achieve the desired sustainable harvest increases. From society's perspective, the "best" level of harvest is where the vertical distance between the value curve and the cost curve is largest. This level of harvest is called the optimum sustainable yield (OSY). It occurs at stock size X_{OSY}. At the OSY, the value of the sustained harvest, V_{OSY}, minus the cost of the sustained harvest, C_{OSY}, is the largest possible. The OSY solution involves a larger stock size than the MSY solution. The impact that stock size has on the cost of harvest (called the stock effect) provides an incentive to society to conserve animals.

Models that seek to maximize or optimize an outcome (e.g., total harvest, profit) help decision makers understand trade-offs associated with policy alternatives (see Chapter 8 for more discussion of trade-offs). Modeling theoretical outcomes such as MSY and OSY can be a valuable learning exercise if the reality of not achieving those concepts is kept in perspective. Their greatest value comes not from direct application, but through indirect use, as a learning tool. Modeling results can

help define useful input for discussions about the net benefit that a given policy choice may provide to society. Importantly, optimization models provide insights that are too costly, time consuming, or risky to obtain through field experiments.

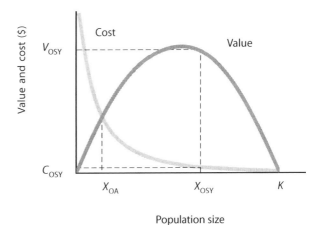

Figure 6.4. Harvest value and cost for a wildlife population

ficulty in setting the quotas is that no one really knows how many animals are in the park. However, the park does have a monitoring program. Each year animal counts are made from airplanes flown along defined transects. If the count is found to be declining, then quotas are adjusted downward. If the count is found to be increasing, then quotas are adjusted upward. As one park ranger explained, "This way, we know we are sustainable."

This approach, however, ignores the fact that any level of sustainable harvest could be supported by two different population sizes. As illustrated in Figure 6.3, when a stable equilibrium is reached, the wildlife managers do not know whether they are at stock level X_H or X_L. Further, there is no particular reason that the initial population size is the best population size. If the initial population size is lower than X_{OSY}, managers should set quotas at a lower level for a few years to allow the population to build up. If the initial population size is greater than X_{OSY}, managers should increase quotas for a few years until the population is reduced. Although it is laudable that managers worry about whether their harvest is sustainable, that is not enough. Managers must also consider which sustainable solution is best for their situation.

6.8.2. Understanding Overexploitation

History is replete with examples of unsustainable wildlife use that led to ecological or economic collapse. Bioeconomic models are used to understand the economic roots of past overexploitation and to analyze the sustainability of proposed policies for wildlife utilization and illegal or unregulated taking of wildlife.

Recognizing and preventing the tragedy of the commons. If a natural resource has open access so that anyone who wants to can join in the harvest, then as long as the resource is generating positive net benefits (value greater than costs) new participants will join in the exploitation of it. This will occur until harvest costs have risen enough, and revenues or values have fallen enough, that no one harvesting the resource is making any money or enjoying any consumer surplus. This occurs at population size X_{OA} and harvest level H_{OA}, where OA designates open access. At X_{OA} the value of the animals harvested is equal to the cost of harvest, so that the wildlife resource is generating no net value to society. This situation, where the potential value to society from the wildlife population is lost due to overharvest associated with open access, is called the "tragedy of the commons," after the seminal article written by Garrett Hardin (Hardin 1968b). The "tragedy" here is that the wildlife population could provide positive benefits to society, in the form of meat or fur or recreation, but is overharvested to the point where its net benefit to society is zero. Exploitation of wild chinchillas is a classic example of the tragedy of the commons (Box 6.2).

The primary approach taken to avoid the tragedy of the commons in wildlife management is to set limits on harvest to conserve the animal population. This can be done by limiting the harvest that each individual can take (season or bag limits) or by limiting the number of harvesters who can participate in the hunt (limited access). Economists have proposed two main economic approaches to wildlife management that try to change the incentives that drive the open access result. The first is to charge harvesters a fee (tax) for each animal harvested. If the fee is set at the correct level, it will push the cost lineup so that the equilibrium population size increases. The animal population would then generate positive net values to society, but these are all captured by the government through the fee. Because it captures all of the net value of the harvest

---

Box 6.2 *EXPLOITATION OF WILD CHINCHILLA AS A TRAGEDY OF THE COMMONS*

Short-tailed and long-tailed chinchilla, species that were both widely distributed along the length of the central Andes and adjacent mountains, were heavily sought after for their high-quality fur. Commercial hunting of both species of chinchilla was widespread in northern Chile beginning in the early 1800s.

The export numbers and prices for chinchilla pelts follow a typical supply–demand relationship. The price per pelt increased as pelt supply decreased. As the resource became scarce and the supply of pelts declined markedly after 1900, pelt prices rose sharply (15-fold during 1902-1909). High

pelt values fueled even greater efforts to harvest chinchilla from a population that had already been decimated. Although killing of chinchilla was banned in 1929, the risk of being apprehended by a warden was low and the financial rewards were high, so harvest continued. The income received by very poor local harvesters, called chinchilleros, for one or two pelts was sufficient to cover their living expenses for at least 2 months (Jiménez 1996).

Analysis of the historical record suggests that the fur trade brought chinchilla to the brink of extinction (Jiménez 1996). Exportation of pelts rose during the nineteenth

Presumed extinct for decades, a small colony of short-tailed chinchilla (pictured) was discovered by Jiménez in 1992. Short-tailed chinchilla inhabit barren tallus slopes at high elevations. (photos courtesy Jaime Jiménez)

through a tax and effectively excludes some people from participating, this approach is politically unpopular and has been rarely applied.

The second economic approach is to convert the unowned wildlife resource into a privately owned resource (Box 6.3). This approach is used in many European countries, where it is most often accomplished through harvest quotas given to landowners. It is also used in some African countries where harvest quotas are given to villages, rather than to individual landowners. The quotas can differ from year to year depending on productivity of the wildlife population. Each quota owner can harvest their quota or sell it to someone else. In Africa, the harvest quotas are typically sold by villages to safari operators, who in turn sell the harvest rights to individual hunters.

6.8.3. Modeling Non-use Values

The basic model described above can incorporate non-use values of the wildlife stock. The top panel of Figure 6.5 shows two curves. The NVH curve represents the net value of harvest and is constructed by subtracting the cost curve in Figure 6.4 from the value curve in Figure 6.4. The OSY that maximizes net value of harvest occurs at population size X_{OSY}. A non-use values curve is superimposed on Figure 6.5. It has been drawn such that the non-use value is negative at higher population sizes. This would occur for species that generate high costs to those who live nearby when the species is abundant. Examples would include game species that cause severe damage to crops or ornamental plants, such as deer in parts of the United States.

The bottom panel of Figure 6.5 shows the net benefit to society generated by the animal population. The net benefit to society is the sum of the net value of harvest and the non-use values. In this case, the socially optimal population size, X^*_{OSY}, is smaller than the optimal size based only on hunting values, X_{OSY}. In such a case, society would benefit if hunters were allowed to harvest more animals, even though total value to hunters would decrease. In an extreme case, there might be a wildlife population where the negative marginal non-use value is so large that the optimal population size is smaller than X_{OA}. An example where this might occur could be a wildlife species that causes economic loss, such as deer in agricultural areas. In such cases, society would benefit if hunters would harvest

century as the demand for chinchilla fur and market prices increased in Europe and the United States. Annual pelt exports peaked at approximately 700,000 in 1900, only to decline by 96% during the next 10 years. This overexploitation was spurred by market prices for the pelts that had increased from 0.088 to 50 Chilean pesos per pelt during an 84-year period. More than 7 million chinchilla pelts were exported between 1840 and 1916, but it is estimated that 21 million chinchillas were actually killed. The resource was depleted so severely that it was effectively economically "extinct" by 1917. The last recorded sighting for a short-tailed chinchilla in the wild was around 1953. Long-tailed chinchilla were once also considered extinct in the wild, but a small population was discovered in 1975. A few scattered colonies of long-tailed chinchilla persist in two separate areas of north-central Chile.

JAIME E. JIMÉNEZ

During the 1980s, a few remaining wild colonies of long-tailed chinchilla were discovered in the desert foothills of the Coquimbo Region of Chile. Although about half of known colonies now live within a protected national reserve, long-tailed chinchilla are critically endangered. (photos courtesy Jaime Jiménez)

Box 6.3 *INFORMING DISCUSSION ABOUT MANAGEMENT OF AN EXOTIC HERBIVORE IN AUSTRALIA*

The utility of bioeconomic modeling to inform harvest policies is well-illustrated by applying it to the harvest policy for an isolated, non-native population of banteng (a wild species of cattle) in northern Australia. Approximately 20 banteng were introduced to Australia in 1849 by European settlers. Banteng have become endangered in their original range (Southeast Asia), but they are thriving in a small peninsular area of Australia that includes a national park and aboriginal land. The area is co-managed by a government–aboriginal council. Aboriginal hunters harvest banteng occasionally and some harvest occurs via permits for trophy hunting (safari hunters pay a royalty to aboriginal owners for hunting permits). Stakeholders in banteng management (aboriginal councils and owners, national park managers, Australian wildlife conservation groups) disagree about policies for banteng harvest. Bradshaw and Brook (2007) developed a bioeconomic optimization model to examine the potential outcomes associated with multiple

harvest scenarios: (1) maximizing sustainable yield (MSY) for meat production, (2) maximizing harvest of trophy males, (3) indigenous sustenance harvest, and (4) scenarios combining two or more options. They also modeled effort needed to achieve partial or complete eradication (a goal for other exotic species in Australian national parks) and minimum viable population size (MVP; necessary to estimate the resilience of harvested population of banteng and susceptibility to unintended eradication). The authors did not offer their results as unequivocal advice regarding policy choices. Rather, they offered the result of model runs to demonstrate that "achieving a balance between population exploitation and control, economic gain and conservation management of the endangered herd of banteng of northern Australia is a realistic proposition" (Bradshaw and Brook 2007:148). That input was offered as a starting point for discussion of banteng harvest policy.

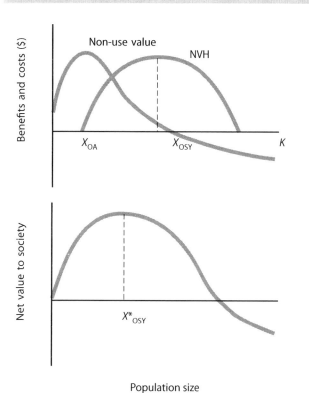

Figure 6.5. Net value to society from a harvested wildlife population with negative marginal non-use value

more of the animals than they otherwise would desire (perhaps especially in urban environments). Hunters would need to be given some incentive to do so because the net benefits they would experience from hunting at that level would be negative.

The opposite situation, where marginal non-use value is positive over the likely range of population sizes, is shown in Figure 6.6. In such a case, the population size that maximizes net benefits to society would be larger than what would maximize net hunting values. Here, society would benefit by restricting harvest below what would be optimal solely from the view of hunting values. An example would be a wildlife species such as bighorn sheep, which is both hunted and enjoyed by wildlife viewers. By curtailing hunting, the population size would be left larger and would generate more non-use value. In an extreme case with large positive marginal non-use values, the socially optimal population size could be K (i.e., it could be socially optimal to restrict harvest completely).

A caveat. Although cost–benefit analysis, cost-effectiveness analysis, regional economic-impact analysis, and bioeconomic modeling can each be used to evaluate the impacts of management decisions on stakeholders and the wildlife resource, they do not provide a complete guide to what decision should be made. A wildlife management agency should use these economic analyses as inputs into its decision-making process but should also consider other factors.

For example, a cost–benefit analysis might conclude that ending hunting for a predator species would generate positive net social benefits, because the non-use values generated by the species outweigh the lost hunting values and the increased damage to livestock. However, we would not want to make this decision based solely on the calculated net benefits. We would also want to consider who is made better off by the change and who is made worse off. Is it fair that hunters and ranchers are made worse off in order to benefit non-users? That is a political decision that cannot be answered using economic analysis. Although economics can calculate who

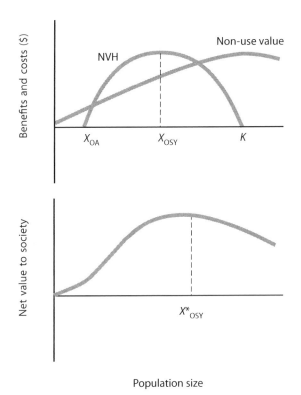

Figure 6.6. Net value to society from a harvested wildlife population with positive marginal non-use value

benefits and who loses, and estimate the size of the gains and losses, it cannot tell us whether the proposed action is fair.

6.8.4. Other Modeling Applications

Bioeconomic and other interdisciplinary modeling aimed at addressing practical problems of wildlife conservation is still in an early stage of development. A few examples illustrate that these approaches can be useful learning tools in deliberations about wildlife policy. Modeling work is used to understand the economic drivers of illegal game-meat harvesting in the Serengeti (Hofer et al. 2000) and to examine the profit motive associated with game ranching in Africa (Hearne and Swart 2000). Modeling is used effectively to evaluate ecological and economic consequences of habitat loss and use of game animals in areas bordering African wildlife reserves (Skonhoft and Armstrong 2005). It also has been used to examine the economic trade-offs in management of bovine tuberculosis in white-tailed deer in Michigan (Fenichel and Horan 2007).

As it is applied to emerging questions about wildlife policy, modeling is not expected to result in detailed recommendations that are followed directly. Rather, modeling efforts—from simple to highly complex—can be expected to clarify issues and deepen understanding of basic processes that should be considered in those policies (Natural Resource Modeling 2000).

SUMMARY

Wildlife agencies are constantly allocating scarce funds across competing programs and projects to achieve multiple purposes: (a) cost-efficiency, (b) cost-effectiveness, (c) benefits for particular stakeholders, and (d) benefits to society as a whole. Economics provides useful, consistent, and objective ways to analyze and evaluate trade-offs that help inform decisions made in wildlife management.

- Economics is the social science that deals with the production, distribution, and consumption of goods and services among competing means to satisfy human wants. Economists study the flow of values, usually measured in monetary terms, through society.
- Economic value, or an individual's willingness to pay, is usually defined as the greatest amount of money an individual is willing to pay for an opportunity, experience, service, or object. Economists, however, estimate many types of values, including consumptive and non-consumptive use, subsistence, non-use, and option values.
- When using traditional economic measures, it is often difficult to calculate a purely market value for wildlife and the activities associated with some human–wildlife interactions. In these cases, economists use non-market valuation methods, such as contingent valuation, to estimate the value of public goods that people value but that are not sold in markets.
- Although wildlife may be a public good that is not traded in markets (at least in the United States), wildlife often provide the foundation on which other markets and economic activity depend; that is, wildlife may have a significant economic impact.
- Economic values and impacts can be evaluated through economic modeling. Although cost–benefit analysis, cost-effectiveness analysis, regional economic impact analysis, and bioeconomic modeling can be used to evaluate the effects of management decisions on stakeholders and wildlife resources, modeling only informs decisions rather than providing a complete guide to what decision should be made.

Suggested Reading

Boardman, A. E., D. H. Greenberg, A. R. Vining, and D. L. Weimer. 2006. Cost–benefit analysis: concepts and practice. Third edition. Pearson-Prentice Hall, Upper Saddle River, New Jersey, USA.

Champ, P. A., K. J. Boyle, and T. C. Brown, editors. 2003. A primer on nonmarket valuation. Kluwer Academic, Dordrecht, Netherlands.

Freeman, A. M. 2003. The measurement of environmental and resource values. Second edition. Resources for the Future, Washington, D.C., USA.

Hartwick, J. M., and N. D. Olewiler. 1998. The economics of natural resource use. Second edition. Addison Wesley, Reading, Massachusetts, USA.

PART III · THE MANAGEMENT PROCESS

In this part, wildlife management is discussed as a cyclical, iterative process that occurs within a management system that consists of ecological and sociocultural components. Comprehensive understanding of the work of wildlife management requires viewing it as a process undertaken to influence a system, rather than as a specific event or a particular technique. A prerequisite for effective management is comprehension of the behavior of the management system and how the system yields negative and positive impacts experienced by stakeholders. A large part of wildlife management work includes understanding the ecological, sociocultural, and institutional components and processes of the system, and learning how to intervene in it to achieve desired outcomes.

Chapter 7 describes wildlife management as an 11-step process, and emphasizes that taking actions (intervention) is only a fraction of the work of management. Although the specifics may differ, the general process we present is similar to others you may have been exposed to in wildlife textbooks. A chief difference in our model is the extent and timing of emphasis on human dimensions. Stakeholders are engaged earlier in the process, rather than later, so that management objectives reflect desired impacts. It is imperative to integrate biological and sociocultural considerations early in the decision-making process.

We also present a recent innovation in Chapter 7, The Manager's Model, developed as a catalyst for situational analyses of current conditions and future scenarios. In anticipating the possible effects of management, developing a Manager's Model encourages identification of intentional and unintentional impacts that potentially could be created as a result of management interventions. Developing a Manager's Model includes the following: envisioning and articulating a preferred future condition with respect to co-existence of humans and wildlife; describing current conditions; analyzing the gap between what is present and what is desired; communicating the new understanding that arises from such gap analyses to partners and stakeholders; and developing ideas about potential management interventions to achieve objectives.

Decision making is an essential task of wildlife management. Decisions create impacts—positive and negative—for stakeholders. Although most wildlife managers probably spend little time thinking about it this way, the bulk of what they do every day is decision making; yet few students take a course in decision making. Chapter 8 aims to build your decision-making skills by describing a structured process for making decisions and identifying psychological traps to avoid when making decisions. Decision making is particularly important in wildlife management because managers are required to address choices involving different dimensions of environmental, sociocultural, and economic impacts, even when they have a great deal of uncertainty about the management system, stakeholders, and the effects of management interventions. Management seldom can wait until the desired level of certainty is attained, so a premium should be placed on learning from management experiences. This chapter introduces active and passive adaptive management as methods of learning while doing management.

WILDLIFE MANAGEMENT AS A PROCESS WITHIN A SYSTEM

SHAWN J. RILEY, DANIEL J. DECKER, AND WILLIAM F. SIEMER

A central theme of contemporary management thought is that change is inevitable and uncertainty and unpredictability are inherent traits of any management environment; this theme applies well to wildlife management. Another aspect of management, regardless of context, is that management is a cyclical, iterative process of analyzing current conditions, describing desired conditions, making decisions to solve problems and create opportunities for achieving desired conditions, taking actions, and evaluating outcomes to learn how to make better choices in the future. Robert Giles (1978:5) aptly described wildlife management as "a decision science"; this is a particular application of the interdisciplinary study of regulatory systems. Systems are defined in various ways, but as used in this book a system is a group of interacting, interrelated, or interdependent components that form a unified whole (Sherwood 2002). To improve effectiveness of wildlife management, it should be viewed as a process within a system rather than as a specific event or particular technique (Box 7.1).

In this chapter, we describe a prototypical management process designed around the concept of wildlife management as a goal-seeking system, and we present a model that emphasizes integration of human dimensions insights into wildlife management decisions and actions. Other concepts and models of wildlife management exist; some are simple, some are complex. The one used here includes key elements of wildlife management that will be helpful as you read subsequent chapters. The process is responsive to considerations of governance and emphasizes management of wildlife as a resource for the benefit of people now and in the future.

7.1. A COMPREHENSIVE PARADIGM

The process of wildlife management has been described by various models, most having key elements in common. Krueger et al. (1986:50) made the point that having a single conceptual model, if it was rational and adaptive, would be preferable:

Resource managers could better guide the management process, especially when the public is directly involved, if they agree to similar concepts about how management should be approached. Such a conceptual model must be rational and logical in sequence and yet be sufficiently robust to accommodate initial errors in decisions. The model must be self-correcting and adaptive to social and biological changes over time, and thus encourage proactive as opposed to reactive management.

In that spirit we present a model of the wildlife management process that has 11 primary elements (Fig. 7.1). As with any model, this one is an oversimplification intended only as a framework to acquaint you with a complex process. Experienced management professionals will recognize that any given wildlife management process may have more or fewer steps than we show. The order in which steps take place also varies, and an agency will almost certainly be involved in more than one step at a time as it juggles its many programs and processes. With those realities in mind, let us consider the overall environment in which the cycle exists.

7.1.1. Management Environment

Wildlife management is a system of governance that functions within an environment having cultural, economic, political, and ecological components. The cultural component includes traditions, religions, values, and philosophies of the general public, stakeholders, and wildlife managers. The cultural component typically contributes most to the establishment of objectives, and the social values inherent in the cultural component provide the principal motivation for wildlife management. Wildlife management exists because its end products are believed to have value to part or all of society and because it results in greater benefits more equitably distributed across society than if wildlife management did not exist as an institution.

The economic component of the management system includes all the assigned and held values discussed in Chapter 6.

Box 7.1 *MANAGEMENT AS A GOAL-SEEKING SYSTEM*

We previously defined a system as a group or community of interacting, interrelated, or interdependent components that form a unified whole. Systems can be as broad as ecosystems, or can be a way to think about organizations, such as wildlife agencies, or even the interactions among and between stakeholders. A common but not exclusive way to think about systems is that they are goal-seeking. For instance, many ecosystems progress toward an equilibrium state (even though most do not achieve equilibrium because of disturbances). Even individual humans are goal-seeking systems—the goal being homeostasis characterized by a body temperature of approximately 98.7° F (37° C). To determine the size of the gap between the desired and current situation requires continual and accurate feedback, which is a critical property of any system. If you venture out into a winter day without a jacket, your body provides fairly rapid feedback indicating the external temperature is lower than your body temperature (you feel chilled). If

you do nothing to insulate your body, then your internal temperature will move toward equilibrium with that outside temperature.

A wildlife management system is, of course, much more complex than deciding to wear a jacket outdoors in winter. Nonetheless, there is a management environment that includes the current conditions existing with wildlife and human–wildlife interactions, and there are many stakeholders who have their own desired condition for wildlife and associated interactions. Later in this chapter we provide a more complex view called a "Manager's Model," which attempts to reduce the gap between the desired condition and the current condition. The array of interventions aimed at understanding and influencing wildlife populations, habitat, and stakeholders, and then gaining reliable feedback on the interventions, constitutes a large part of the day-to-day work of wildlife professionals.

These are the values that influence decisions about wildlife management, including both market and non-market values. The economic impacts created by management in turn contribute to society's perception of the value of wildlife and wildlife management.

The political component of the management environment determines broad policy in wildlife governance. The political component influences management decisions through laws, rules, and personal values of individuals, including those people who enact, enforce, or interpret laws and policies. Legislative statutes and administrative codes define management responsibilities and authority of an agency, but considerable room exists for managers to influence day-to-day management decisions, which is a subject covered in the next chapter.

The ecological component is the ecosystem in which wildlife populations of interest exist. This component defines the physical limits of what management can achieve in terms of production, sustainability, and restoration of wildlife populations as well as the environment in which human–wildlife interactions occur. Because management programs often alter parts of the ecosystem, ecological considerations permeate the decision-making process; they are especially evident in statements of specific objectives and in choice of management alternatives (see Chapter 8).

Simultaneous consideration of these cultural, economic, political, and ecological components of the management environment is important because they influence one another in ways best understood if considered jointly. Although specialists in one aspect or the other of wildlife management (e.g., population ecologists, habitat ecologists, or various kinds of human dimension specialists) tend to perceive wildlife man-

agement through the lens of their specialization, experience indicates that wildlife management is driven mainly by human elements such as culture and norms, institutions, and stakeholder interests. Wildlife managers who consider these elements, as well as the ecology of the system, are better prepared to guide the system toward producing desired impacts. When beginning work on a new management issue or after working on an issue when the complexity is becoming overwhelming, a helpful practice is to conduct a situational analysis. In addition to seeking clearer understanding of the management environment, the types of questions to be asked in this analysis are the following (Riley et al. 2002):

1. What is the range of impacts occurring now and expected in the future?
2. Who are the key stakeholders?
3. What are the operational scales (geographical and temporal extent) of the anticipated impacts?
4. What are the capacity and limits of the resource, stakeholders, and management?
5. What will be the explicit structured-decision processes for making choices along the way?

These five interrelated questions are best addressed simultaneously. Defining impacts to be managed precisely may not be possible if you are at an early point in the process. Nevertheless, it is possible to anticipate the range and relative importance of potential impacts.

A situational analysis uses knowledge about relevant impacts to construct a first-generation "map" of the management system. This step assures managers and stakeholders are in agreement that everyone is working with the same core

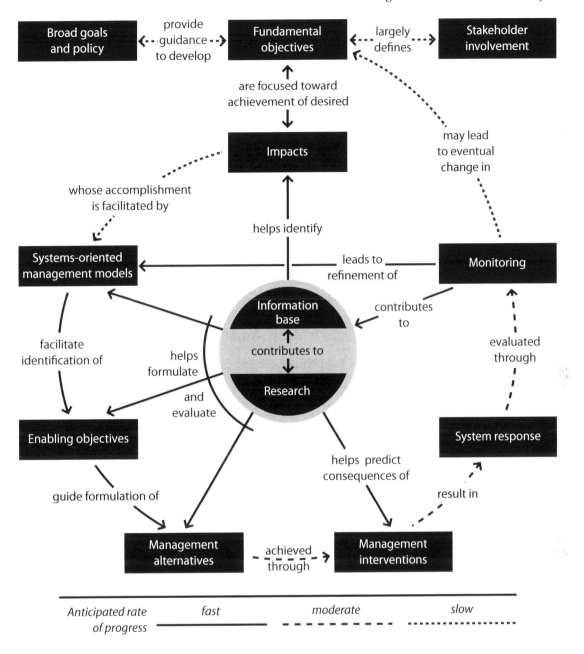

Figure 7.1. A comprehensive model of wildlife management (Riley et al. 2003)

data on the same pertinent issues to be addressed by management. To facilitate thorough situation analyses, we developed a process called a "Manager's Model." The Manager's Model process (Box 7.2) prompts formal responses to the five questions presented above while drawing out underlying assumptions held by managers about the management system. In anticipating the possible effects of management, a Manager's Model also encourages identification of intentional and unintentional impacts created as a result of management interventions.

The 11 primary elements of the wildlife management process are interrelated. Each has an antecedent—a reason or history—and each influences subsequent elements of the process. Each element of the process is discussed in this chapter;

Chapter 8 provides guidance on structured decision-making within the management framework described here.

7.1.2. Broad Goals and Policy

The first elements of the management process—broad goals and policy—emerge from the management environment and reflect the broad values of society. Wildlife policy takes shape through the political process as laws, statutes, and agency directives. Policies are also developed by professional societies and their representatives, which often influence governing institutions. For example, committees of professionals in natural-resource management published broad wildlife policy statements in 1930 and 1973 (WMI 1973) that shaped the practice of wildlife management as we know it today.

Box 7.2 *A MANAGER'S MODEL*

One way wildlife managers can be more effective in their jobs is to improve their understanding of wildlife issues through analysis of wildlife management systems. This includes (1) envisioning and articulating a preferred future condition with respect to co-existence of humans and wildlife; (2) describing current conditions; (3) analyzing the gap between what is present and what is desired; (4) communicating the new understanding that arises from such gap analyses to partners and stakeholders; and (5) developing management interventions to achieve objectives. We devised a "Manager's Model" to help guide wildlife professionals through the process of thinking about and describing attributes of a management system (Decker et al. 2011).

A Manager's Model is a description of the management system from the manager's perspective (i.e., an individual manager or a management team responsible for management of a wildlife resource). The development process of a model includes a rich description of: management purpose, premise, and context; stakeholders and the impacts of management they seek or experience; assumptions; relevant knowledge and knowledge gaps; and management actions and their intended and unintended consequences.

A Manager's Model is one suggested framework for practicing systems thinking, but it is neither a prescription nor a plan, per se. The outcome of developing a Manager's Model is increased understanding of desired conditions relative to existing conditions, factors influencing such conditions and their relationship to one another, and the management interventions needed to reduce the gap between desired and existing conditions.

Building a Manager's Model is most efficient when done in phases:

- *Manager's Model—Phase I:* Describing the context and framing the management issue.
- *Manager's Model—Phase II:* Understanding the management system—questions leading to system definition through identification of constraints, limits–capacity, and opportunities.
- *Manager's Model—Phase III:* Developing the management response—questions to aid development of preliminary impressions about fundamental objectives, enabling objectives, actions, stakeholder reactions, and mitigation.
- *Manager's Model—Phase IV:* Creating, critiquing, and using the preliminary Manager's Model–concept map.

The general template of a Manager's Model is depicted in the concept map. The structured discussion that leads to creating such a concept "map" should yield greater understanding of the management system and awareness of the assumptions and uncertainties that exist about management. Another important but often overlooked aspect of a management system is anticipation of the outcomes from management interventions. Hopefully, a manager's interventions create the intended primary effects, but avoiding unintended collateral effects may be just as important in achieving the overall desired future condition.

Policies are expressed as a purposeful course of government action (Mangun 1992). They are guides to decision making and convey intention about concerns such as enhancing wildlife populations, protecting and restoring wildlife habitats, distributing wildlife-related benefits, limiting human impact on wildlife, and reducing wildlife-related problems. Some examples of legislation that express broad policy positions are summarized in Box 7.3.

7.1.3. Stakeholder Involvement and Fundamental Objectives

Stakeholder involvement is most effective when it matches the geographical, temporal, and social scale of the issue (see Chapter 11). Although, in most cases, stakeholders need to be involved throughout the process, the main purpose of stakeholder involvement at this early point is to gain an understanding of the desired impacts to be achieved. The desired impacts created by management are expressed as fundamental objectives. As you will learn in more depth in the next chapter,

fundamental objectives answer the question, "Why is management necessary?" This question is answered using value-based thinking and precise definition of impacts desired from management based on input from stakeholders.

This way of organizing decisions is a departure from conventional processes. Goals and objectives for wildlife management are most frequently couched in terms of numbers of animals, acres of habitat, and so on, which are referred to herein as targets of enabling objectives (see description below). Because the purpose of wildlife management is to gain benefits and minimize costs in terms of values (impacts) to stakeholders, we find it is more efficient and effective to express fundamental objectives directly in terms of impacts. Engagement with stakeholders for purposes of expressing the fundamental objective of management (i.e., impacts desired from management) is a relatively new way of conducting management for some wildlife agencies and non-governmental organizations. In New York State, for example, stakeholder input groups were convened to help managers reflect on the fundamental

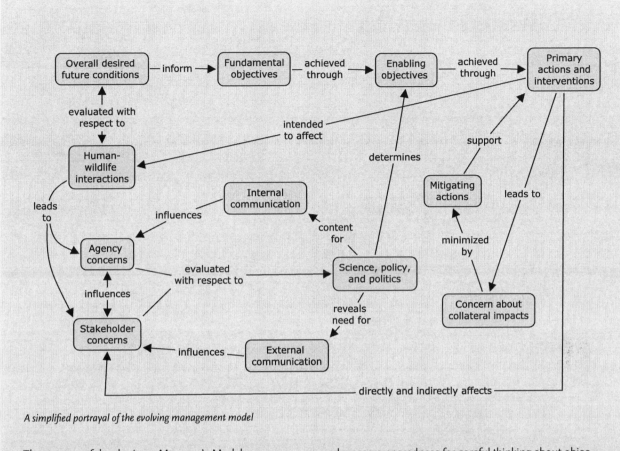

A simplified portrayal of the evolving management model

The process of developing a Manager's Model prepares managers well for the structured decision-making described in Chapter 8. In essence, a Manager's Model process creates a written depiction of the situational analysis and advances preparedness for careful thinking about objectives, management alternatives, consequences (intended and unintended), and trade-offs that occur in choice of alternatives.

objectives for black bear management when human–bear interactions were increasing rapidly (Box 7.4).

Early stakeholder involvement lessens the unproductive situation of having to "sell" agency-defined management objectives to the affected stakeholders. The all-too-often-heard comment, "Now, all we need to do is get 'buy in' on these objectives" usually indicates that stakeholders were left out of the initial stages of the management process. Although at first that may seem like an efficient way to approach management, it usually is not cost-effective in the long run and does not signal good governance practices. Stakeholders neglected early in the process frequently enter later to challenge the implementation of management actions, often in a court of law or in state legislatures. A number of examples exist, however, that exemplify the value of early involvement. The Michigan Department of Natural Resources, for example, recently implemented a new approach for early and ongoing involvement of stakeholders in deer management to define desired impacts from deer management in that state (Box 7.5).

7.1.4. Systems-Oriented Management Models

The purpose of approaching management as a process within a system is to become more effective by (1) adding structure to guide and communicate thinking about complexity of management; (2) building decision-making capacity; and (3) increasing rates of learning (and thus ability to be adaptive). Lee (1993:5) stated, "The essence of managing adaptively is having an explicit vision or model of the ecosystem one is trying to guide." Models of management systems, especially when developed by practitioners (i.e., managers' concept maps of the systems in which they work [e.g., Decker et al. 2006b]), help organize and communicate the complexity of management systems to managers and stakeholders. Model development also exposes important uncertainties and assumptions about the management system. With few exceptions, humans perform poorly at making accurate decisions within a multi-faceted sociocultural–ecological system, such as that in which wildlife management occurs (Kahneman et al. 1982). Models become useful tools for describing and managing a wildlife

---

Box 7.3 *BROAD FEDERAL AND STATE POLICIES*

Some broad federal policies:

- *Endangered Species Act of 1973:* This legislation provides for protection of species officially designated as threatened or endangered. Among other things, it requires that federal and state wildlife-management agencies take steps to maintain or restore viable populations of listed species. It has caused wildlife management agencies to play a greater regulatory role in use of private and public lands.
- *National Environmental Policy Act of 1979:* "This act requires that major federal actions thought to have potentially significant environmental impacts be preceded by an environmental impact statement . . . fully disclosing those effects" (Gray 1993:159). It has caused government agencies and private sector enterprises to give greater consideration to preventing or mitigating negative impacts on wildlife associated with human enterprises.
- *Federal Aid in Wildlife Restoration Act of 1937:* This act (popularly known as the Pittman–Robertson Act) provides funding for the selection, restoration, rehabilitation, and improvement of wildlife habitat; for wildlife management research; and for the distribution of research results. It was amended in 1970 to include funding for hunter-training programs and the development, operation, and maintenance of public shooting ranges. Funds are derived from an 11% federal excise tax on sporting arms, ammunition, and archery equipment and a 10% tax on handguns. Funds for hunter education and shooting ranges are derived from half the tax on handguns and archery equipment. The legislation has had a profound influence on the nature and direction of wildlife research and management for more than 50 years (Kallman 1987).

Some broad state policies:

- *Declaration of Policy* Section 1-0101 of the Environmental Conservation Law of New York declares: "It is the policy of the State of New York to conserve, improve, and protect its natural resources and environment and control water, land, and air pollution, in order to enhance the health, safety, and welfare of the people of the state and their overall economic and social well being."
- *Wildlife Management Authority* Sections 11-0303 and 11-0305 of the Environmental Conservation Law of New York authorize the New York State Department of Environmental Conservation to "enforce all provisions of the laws and regulations relating to fish, wildlife, protected insects, shellfish, crustacea and game." (King and Schrock 1985:184)
- *Conservation Fund Advisory Board* Section 11-0327 of the Environmental Conservation Law of New York authorizes an 11-member board of citizens to make recommendations to state agencies on plans, policies, and programs that affect fish and wildlife. The policy stipulates that all 11 members of the board must demonstrate their interest in fish and wildlife by holding a valid hunting or fishing license while serving on the board.
- *State Environmental Quality Review Act* Section 8-0101 of the Environmental Conservation Law of New York stipulates that protection and enhancement of the environment must receive equal weight with social and economic considerations in public policy. The act requires that New York State agencies prepare (or require the appropriate party to prepare) an environmental impact statement before proceeding with any action they take or approve that may have a negative effect on the environment.

---

management system; they assist the manager in integrating ecological and human dimensions. Models also facilitate identification of proposed management interventions and help define acceptable sets of management options carried forward through the policy process.

7.1.5. Enabling Objectives

Enabling objectives state how fundamental objectives will be achieved. Although fundamental objectives communicate impacts that are the purpose for management, enabling objectives inspire the means to getting things done and establish metrics by which management actions are evaluated. Each enabling objective must link to one or more fundamental objective to create a coherent alignment between desired impacts and the actions needed to attain them. The relationship between fundamental and enabling objectives can be depicted in a linkage diagram, such as the one for black bear manage-

ment in Figure 7.2. Enabling objectives form the basis for selecting management alternatives.

7.1.6. Alternatives

Alternatives are the set of possible management actions available to achieve enabling objectives. The alternatives are the actions to affect

- wildlife populations directly through survival rates (e.g., hunting, ban on taking, reduction of vehicle collisions) or indirectly through manipulation of habitat and of people;
- habitat directly by an agency and indirectly through citizens (e.g., tillage practices, forest management activities); and/or
- humans directly through laws and regulation, or indirectly through communication and education or through economic incentives, recognition, and special opportunities.

Box 7.4 *ENGAGING STAKEHOLDERS TO IDENTIFY FUNDAMENTAL OBJECTIVES FOR BLACK BEAR MANAGEMENT*

An expanding black bear population led to increased human–bear encounters in New York State and prompted wildlife managers to complete a comprehensive, statewide planning process. Black bears have been managed in New York for more than a century, so managers are well aware of how bears can affect humans. Nevertheless, recognizing that their management system was changing, managers undertook a series of stakeholder engagement activities to inform development of fundamental objectives for comprehensive, statewide bear-management program.

Managers began by reviewing input from a series of public meetings on bear management in the mid-1990s. They sponsored a set of three regional scoping sessions to identify the full range of bear-related effects that stakeholders were experiencing across the state. They followed that effort with a statewide mail survey in 2002 that quantified experiences with bears, risk perceptions, and attitudes toward bears and bear-management actions. Input from the 2002 mail survey was supplemented with findings from a stakeholder input group (SIG) process. The SIG process was designed to help New York State Department of Environmental Conservation (NYSDEC) staff identify which effects were considered as impacts that should receive management attention in various regions of the state. University researchers and Cooperative Extension specialists worked with wildlife agency staff to develop and implement the SIG process in three localities in 2003.

NYSDEC created SIGs to be temporary, ad hoc entities. Each SIG had about a dozen members. Process facilitators selected participants to reflect diverse stakes in, and perspectives on, black bear management (i.e., people experiencing different kinds of impacts) and to minimize overrepresentation of any single interest. SIG participants were asked to review two background documents; seek input from others; contribute local experience and knowledge; participate as an individual (not as an official representative of a particular group); and keep an open mind. University and agency staff provided facilitation and subject matter expertise.

Each SIG group met multiple times. In their first meeting, participants were presented with information on bear natural history, were introduced to the concept of impacts, and were instructed to seek input from others in their community or stakeholder group. In the second meeting, facilitators asked participants to review, clarify, and add to the list of bear-related impacts that the Bear Team had

developed from prior stakeholder engagement activities. Participants were then asked to indicate which impacts were most important in their region of the state. Each group was asked to select priority impacts on which to focus further discussion. In the third meeting, facilitators led discussion and ends–means linking exercises that helped participants express their interests and concerns as a set of fundamental and enabling objectives. They also began to describe potential management actions.

Based on a review of all the information obtained from stakeholders, agency staff made a judgment that 12 specific effects were impacts that should be a high priority for management attention (i.e., should serve as the basis for fundamental objectives). Most of the impacts identified were known previously to managers, yet the process created public confidence that the agency was aware of and considering public concerns. The process stimulated articulation of impacts, fundamental objectives, and assumed means–ends connections for black bear management in New York. Having that information in written form made it easier for managers to communicate the essence of their bear-management program to stakeholders and agency administrators. It also created internal and external support for regulation changes proposed by the agency.

A participant in a stakeholder input group process identifies impacts worthy of management attention. Identifying impacts provides input for setting objectives. (courtesy William Siemer)

Humans are affected both positively and negatively by white-tailed deer. Contention about deer management objectives and actions can occur under the best of conditions, and management actions may be halted entirely when stakeholders believe they have not been considered in decision-making processes. Agencies are developing a range of approaches to early and continual stakeholder involvement (see Chapter 11 for more discussion of engagement approaches). In Michigan, for example, managers are implementing a new approach to stakeholder involvement in deer management.

In 2010, the Michigan Department of Natural Resources (MDNR) completed its first strategic Deer Management Plan. The new plan included a commitment to establish Regional Deer Advisory Teams (RDAT) for each of three ecologically based regions of the state. Each RDAT is composed of members with a direct connection to issues in the corresponding region, either as private individuals affected by deer hunting and management or representatives of organizations with members or constituents in the region. Members serve 3-year staggered terms, and participate in biannual meetings co-chaired by a DNR and an RDAT member.

Following a winter RDAT meeting, local open houses hosted by the MDNR Wildlife Division gather additional public input, particularly regarding identified priority topics. At least one RDAT member is asked to volunteer to participate in each open house. Then each RDAT holds a summer (pre–deer season) meeting for finalization of input or recommendations. RDAT members are asked to discuss the nature of this input with the public during the following deer-hunting season in order to aid with communications regarding potential future program directions.

Each RDAT serves multiple functions. In addition to providing recommendations and other input directly to the MDNR from informed positions, each RDAT assists with implementation of the Deer Management Plan, aid in communications with the public and key partners and organizations, and reviews and provides feedback regarding information and education products. Topics for potential recommendations are coordinated among the regions so that deer program areas are comprehensively reviewed. The RDAT approach gives stakeholders opportunities for early input on a variety of topics, such as potential deer hunting regulations; approaches to monitoring deer harvest, populations, and impacts of deer; or concepts and materials for outreach, education, and technical assistance programs to private landowners. The RDAT approach is another example of how wildlife agencies are adopting variations on the types of management processes that include stakeholders early in the process to define impacts.

BRENT RUDOLPH

© *Glenda Powers-Fotolia.com*

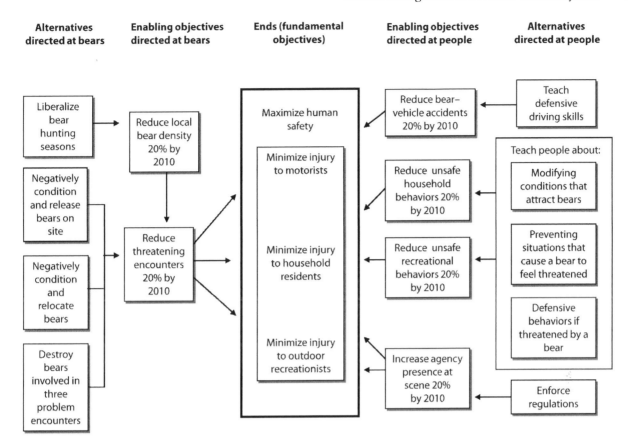

Figure 7.2. A partial means–end linkage diagram for management of safety-related impacts associated with black bears

To be comprehensive, each alternative has a human dimensions component to be considered (Campa et al. 2011). For example, reducing deer populations through hunting requires that people participate, put forth a certain effort, and kill a certain number or type of animal (e.g., adult female deer). This alternative requires human dimensions insights into factors such as hunter motivations and satisfactions, hunters' willingness to use a particular management area, and preferences of hunters for group permits versus individual permits. Those are just a few of the considerations when hunting is one of the actions considered. What about when hunting is not an option? In that case, knowledge is needed about acceptability of alternatives such as euthanasia or sterilization and whether the situation calls for an expert approach, a transactional approach, or perhaps even co-management.

7.1.7. Management Interventions

Management interventions are the actions described in the alternatives that are selected to most likely achieve objectives. As described above, effective interventions involve consideration of all three management dimensions: habitat, populations, and human dimensions. However, interventions ultimately depend on compliance or cooperation of other people to achieve the objective. Furthermore, people determine acceptability of interventions and identify potential unintended collateral impacts.

7.1.8. System Response

Each intervention, if it has an effect, results in a response—the outcome of the intervention—by the system. Those responses occur typically in two or in all three management dimensions. Human dimensions considerations pervade the response stage because, even if it is a change in habitat or population that is desired, stakeholders' behaviors usually are at least the short-term mechanism for achieving objectives. For example, an educational intervention to promote restoration of prairie habitats may focus on landowners holding certain acreage; yet, if effective, the intervention eventually alters habitat and, subsequently, species composition and populations of wildlife.

7.1.9. Monitoring

Monitoring is the effort to measure and evaluate the response(s) of the management environment to the interventions taken. Objectives provide criteria for measuring effectiveness. Whereas most monitoring is focused on the shorter term parameters set out in the enabling objectives, evaluation of the progress made toward achievement of fundamental objectives is more summative and occurs over a longer term. Evaluation provides information for refining system-oriented models or redirecting the management process based on effects of interventions. The model is refined and if changes are needed the enabling objectives or alternatives are reformed to reflect the new knowledge about the effects of management inter-

A process facilitator begins synthesizing stakeholder input as an ends–means diagram. Exercises such as this help clarify the effects that stakeholders find most important and force critical thinking about why they believe managers should take particular actions. (courtesy William Siemer)

ventions. Monitoring (taking into account new developments in the knowledge base) is integral to the feedback that allows a management system to be an adaptive, responsive process. This system of wildlife management recognizes that management interventions change the context for management and may necessitate further adjustment.

7.1.10. Information Base and Research

Information of at least three types is desirable before actions are taken: (1) information about the level of impacts currently existing (prior to intervention); (2) information about the relative effectiveness and costs of interventions under consideration; and (3) information about the acceptability of interventions among primary and secondary stakeholders (i.e., those desiring intended outcomes of management and those affected by collateral effects of management, respectively). Information exists in both published and unpublished data. Sources for the information base include experience and intuition, research and evaluation, inference (through theory and modeling) or experimentation, and culture. The information base encompasses biological, ecological, economic, and social science data and theory; common knowledge; and prevailing professional assumptions and philosophies. This collection of information is the repository of insight from which management processes draw for decision making.

In any type of management, evaluation of actions taken is one of the best sources of information on which to base future management efforts. Adaptability is encouraged by monitoring if each management program or experience is viewed as an experiment. In the next chapter, you will learn about a decision-making process called "adaptive management," where learning about the system is a key objective of management. Although management is not typically a true experiment with controls and randomly applied treatments,

each management intervention is an opportunity to test assumptions, apply theory, refine methods, and contribute to the information base on which future decisions are based. Wildlife professionals benefit from the insights arising from other areas of science, and everyone benefits if wildlife professionals remain inquisitive about their own work with an eye to making improvements. Considerations in planning an inquiry and the methods commonly used by human dimensions researchers are discussed in Chapters 9 and 10, respectively. In practice, management decisions seldom can wait until enough research can reduce all the uncertainties to a comfortable level in the system. That reality accentuates the need for adaptive management and learning by doing.

7.2. DO NOT GO IT ALONE

The complexity of contemporary wildlife management is apparent from the process model described above, which, even though it looks complex, is still a simplification of wildlife management. Wildlife management—whether it is deciding on a management intervention needed immediately at a particular site or planning for the future statewide management of one or more species—must draw on multiple disciplines when addressing wildlife problems or exploring opportunities. The depth of knowledge needed in multiple disciplines often causes wildlife managers to reach out to specialists in those disciplines. Krueger et al. (1986:51) pointed out a reality of wildlife management: "Operating within the management environment described is the resource manager, who is traditionally trained in applied ecology. These individuals often find themselves forced to function within cultural, economic, and political arenas as opposed to the biological foci of their education."

Increasingly, wildlife management is being approached using multi-disciplinary teams whose members bring essential disciplinary strength to the management situation. Depending on the situation, team members may bring expertise from disciplines such as ecology, sociology, economics, administration, political science, educational communication, law enforcement, and management science to conduct comprehensive wildlife management.

7.3. AN EVOLVING MODEL

The case of black bear management between 1970 and 2010 in the Catskill region of New York illustrates well the evolving nature of the wildlife management model. Initially, management of black bears by the New York State Department of Environmental Conservation (NYSDEC) reflected knowledge and concern about a decline in the size of the Catskill bear population.

The broad policy for all wildlife in New York, specified in that state's environmental conservation law, compels managers to maintain wildlife populations consistent with range-carrying capacity and human land use. From that broad policy, the fundamental objective for black bear management in the

Catskill region was developed: to optimize sustained recreational use of bears from populations that are compatible with humans and the habitat. That fundamental objective has been a continuous part of contemporary black bear management in New York, yet the enabling objectives to achieve that end have been modified with each new cycle of management. Decker and associates (Decker and O'Pezio 1989, Decker et al. 2001) described three black bear–management cycles in New York during the period 1970 to 1988. Siemer and associates (Siemer and Decker 2006, Siemer 2009, Decker et al. 2011) described a full management cycle that occurred between 2001 and 2008. In Figure 7.3, we synthesize findings from those studies to illustrate how enabling objectives and human dimensions considerations changed in cycles of bear management between 1970 and 2008. The information base considered in bear-management cycles became more comprehensive and integrative over time, reflecting consideration of all dimensions of wildlife management.

Based on the NYSDEC's evaluation of the situation in 1976, it was determined the Catskill region bear population could be increased. This opportunity became the foundation for an enabling objective to increase the black bear population 60% to 80% during the next 2 years. Previous research identified a significant impediment to meeting the objective; that is, hunting caused most bear mortality in the Catskills. That knowledge led NYSDEC to put in place a 2-year moratorium on bear hunting in the Catskills (this was determined to be the best and perhaps only feasible alternative available). During the moratorium, additional sociological and ecological research was conducted for evaluation purposes, with the intention of adapting the approach using information about the system response (biological, ecological, and human dimensions). Human dimensions research indicated that the bear population before the moratorium was well within the wildlife acceptance capacity of area residents (Decker et al. 1981, Decker and Purdy 1988). The ecological study identified areas where bears potentially could expand into new range.

The system's response to the hunting moratorium was an estimated increase of nearly 80% in the size of the bear population, thus achieving the enabling objective. Wildlife biologists' evaluation of the bears' physical condition suggested the population increase had not exceeded biological carrying capacity. After the population had stabilized at higher levels, further evaluation was planned to determine whether the population remained within the wildlife acceptance capacity of area residents. Additions and modifications to the information base were made, such as the results of the second human dimensions study (Decker et al. 1985b).

In the first three cycles of black bear management shown in Figure 7.3, stakeholders had minimal input to setting objectives or determining the acceptability of management interventions (limited public involvement was typical of bear management across the United States at the time). That changed as a result of pressure by external stakeholders and internal agency initiatives to improve stakeholder engagement and management processes.

Between 1988 and 2001, the cultural, political, and ecological environment in which bear management occurs continued to change in New York and across the country. In New York, stakeholder groups opposed to bear hunting expressed their concerns directly to the governor, who passed an executive order prohibiting bear-hunting with dogs in New York State. The bear population subsequently experienced growth and the geographical distribution of bears expanded across the state into areas closed to all bear hunting. Complaints about problem interactions with bears rose steadily throughout the 1990s, and most of the increase in problem interactions occurred in the Catskill region.

Wildlife managers in New York initiated a new management cycle (from situation analysis through monitoring action outcomes) as a series of linked activities between 2001 and 2008. Five innovative aspects distinguished this process from previous bear-management cycles: extensive situation analysis, an explicit focus on stakeholder-defined impacts, informed transactional stakeholder engagement, use of quantitative systems-thinking techniques, and a conscious decision by the management team to consider the entire experience as a learning opportunity.

Charged to develop a statewide comprehensive bear-management plan, the agency's Bear Management Team worked collaboratively with human dimensions specialists to develop a new framework for black bear–management planning in New York State in 2002. The agency's new bear-management framework was finalized, approved, and released to the public in 2003 (NYSDEC 2003). Based on the precepts of impact management (Riley et al. 2003), the framework established a cyclical process for adapting New York's management program to changing social and environmental conditions. Stakeholder engagement, a focus on impacts, manager–stakeholder deliberation, and adaptive management were incorporated as featured elements of this planning cycle.

Activities conducted in this case (i.e., synthesis of public meeting data from 1992 to 1994, scoping meetings held with small groups across the state, a statewide mail survey) represent consultative or inquisitive forms of public engagement (Rowe and Frewer 2005; see Chapter 11), where stakeholders convey information to policy makers through processes initiated by the policy-making body (NYSDEC, in this case). The final engagement of stakeholders (i.e., the stakeholder input group [SIG] process described in Box 7.4) was a two-way information exchange between stakeholders and NYSDEC. Because the SIG process was designed to encourage deliberation and focus on impacts, it provided opportunities to question both stakeholders' and managers' assumptions and mental models. The SIG process was employed as part of a comprehensive engagement approach, not as a replacement for involvement mechanisms such as public information campaigns, stakeholder surveys, or established regulatory review processes.

Extensive public involvement created public support necessary to take several otherwise controversial management interventions, including the following: a series of hunting

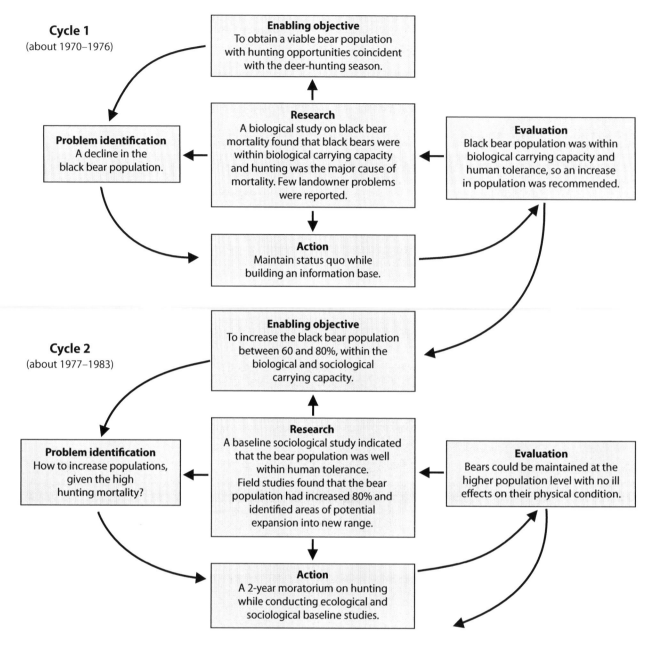

Cycle 1
(about 1970–1976)

Enabling objective
To obtain a viable bear population with hunting opportunities coincident with the deer-hunting season.

Problem identification
A decline in the black bear population.

Research
A biological study on black bear mortality found that black bears were within biological carrying capacity and hunting was the major cause of mortality. Few landowner problems were reported.

Evaluation
Black bear population was within biological carrying capacity and human tolerance, so an increase in population was recommended.

Action
Maintain status quo while building an information base.

Cycle 2
(about 1977–1983)

Enabling objective
To increase the black bear population between 60 and 80%, within the biological and sociological carrying capacity.

Problem identification
How to increase populations, given the high hunting mortality?

Research
A baseline sociological study indicated that the bear population was well within human tolerance.
Field studies found that the bear population had increased 80% and identified areas of potential expansion into new range.

Evaluation
Bears could be maintained at the higher population level with no ill effects on their physical condition.

Action
A 2-year moratorium on hunting while conducting ecological and sociological baseline studies.

Figure 7.3. Cycles of black bear management in New York State, 1970–2008

regulation changes that liberalized bear-hunting opportunity by opening new management zones to hunting and opening hunting seasons earlier; pilot-testing a prevention education program in one Catskill community; and implementing additional education activities intended for a statewide audience. The NYSDEC improved the system they use to monitor citizen reports of interactions with bears. They also began paying even closer attention to bear harvest, especially in areas with new seasons.

Several outcomes associated with cycle 6 of the bear-management system in the Catskills are apparent. Fueled by an injection of financial and personnel resources, energy, and

an innovative approach, a team of managers and human dimensions experts created many bear-management products between 2000 and 2008, including the following: a standard operating procedure manual (Henry et al. 2001); a published operating manual for making bear-management decisions and action recommendations; a publication on black-bear natural history and management; identification of effects that stakeholders in New York regard as impacts (Siemer and Decker 2006); a stakeholder education video ("Living with New York Black Bears: Secrets to sharing the landscape with bears"); a bear-management Web page; bear-management education brochures and billboards; a compact disk of bear harvest rec-

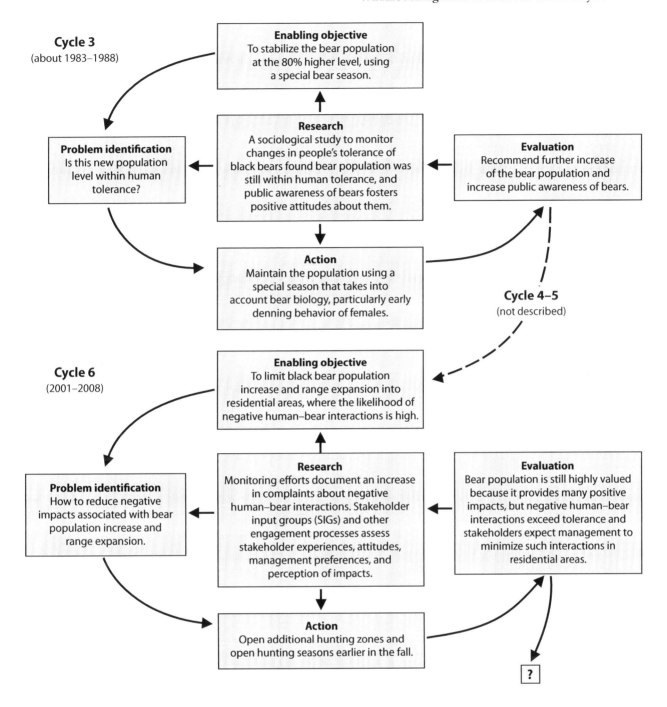

Cycle 3
(about 1983–1988)

Enabling objective
To stabilize the bear population at the 80% higher level, using a special bear season.

Problem identification
Is this new population level within human tolerance?

Research
A sociological study to monitor changes in people's tolerance of black bears found bear population was still within human tolerance, and public awareness of bears fosters positive attitudes about them.

Evaluation
Recommend further increase of the bear population and increase public awareness of bears.

Action
Maintain the population using a special season that takes into account bear biology, particularly early denning behavior of females.

Cycle 4–5
(not described)

Cycle 6
(2001–2008)

Enabling objective
To limit black bear population increase and range expansion into residential areas, where the likelihood of negative human–bear interactions is high.

Problem identification
How to reduce negative impacts associated with bear population increase and range expansion.

Research
Monitoring efforts document an increase in complaints about negative human–bear interactions. Stakeholder input groups (SIGs) and other engagement processes assess stakeholder experiences, attitudes, management preferences, and perception of impacts.

Evaluation
Bear population is still highly valued because it provides many positive impacts, but negative human–bear interactions exceed tolerance and stakeholders expect management to minimize such interactions in residential areas.

Action
Open additional hunting zones and open hunting seasons earlier in the fall.

?

ords, bear-related complaint reports, and other data resources on bear management in New York; and a practitioners' guide to working through bear-management issues (Siemer et al. 2007a).

The level of investment and growth in the NYSDEC bear-management program was unprecedented in the state. Before this period, black bear management was a program in the background of New York wildlife management that received relatively little public attention. With the creation of a permanent staff team to guide the program and implementation of an impact management (IM) approach to bear management, the status and public visibility of the program was elevated

markedly. Evaluation of the project supported assertions by Riley et al. (2003) that a well-implemented IM approach creates a range of outcomes that enhance wildlife agency performance. Future cycles of bear management can build upon this success.

SUMMARY

This chapter introduced a prototypical management process and a model that emphasizes integration of human dimensions in wildlife management. From a starting perspective that wildlife should be managed as a resource for the benefit of people now and in the future, wildlife management is described as a goal-driven, cyclical, and iterative process that functions within *governance* structures and management environments having cultural, economic, political, and ecological components.

- Broad policies reflect broad values of society in the form of laws, statutes, and agency directives. They guide present and future decisions by wildlife management agencies when integrated with stakeholder involvement to define fundamental objectives in terms of impacts.
- Established within the bounds set by policy and stakeholder input in early stages of management, objectives provide a measurable definition of what is expected to be achieved within a particular time frame. Fundamental objectives define the reason for management in terms of impacts, whereas enabling objectives motivate generation of alternatives to achieve fundamental objectives. Systems-oriented management models, such as a Manager's Model, help build a common understanding of the system being managed as well as reveal assumptions and uncertainties that underlie management.
- Alternatives describe the choices of management interventions designed to achieve enabling objectives. The effectiveness of these management interventions is evaluated by monitoring the system's response in comparison to the stated objectives. Knowledge gained through monitoring and evaluation is applied to the information base to inform future decisions and also used to modify system models and objectives.
- Wildlife management systems and processes have been defined, discussed, and refined since the beginning of the wildlife profession. The process described in this chapter is a next step because it advocates early stakeholder involvement and defining the fundamental objectives (purposes of management) in terms of impacts.

Suggested Readings

Checkland, P., and J. Scholes. 1990. Soft systems methodology in action. Wiley, New York, New York, USA.

Decker, D. J., S. J. Riley, J. F. Organ, W. F. Siemer, and L. H. Carpenter. 2012. Applying impact management: a practitioner's guide. Cornell Cooperative Extension, Ithaca, New York, USA.

Riley, S. J., D. J. Decker, L. H. Carpenter, J. F. Organ, W. F. Siemer, G. F. Mattfeld, and G. Parsons. 2002. The essence of wildlife management. Wildlife Society Bulletin 30:585–593.

Riley, S. J., W. F. Siemer, D. J. Decker, L. H. Carpenter, J. F. Organ, and L. T. Berchielli. 2003. Adaptive impact management: an integrative approach to wildlife management. Human Dimensions of Wildlife 8:81–95.

Sherwood, D. 2002. Seeing the forest for the trees: a manager's guide to applying systems thinking. Nicholas Brealey, London, England, U.K.

DECISION MAKING IN WILDLIFE MANAGEMENT

SHAWN J. RILEY AND ROBIN S. GREGORY

Stop for a moment and think about any issue in wildlife management—about any other issue in your life for that matter. The issue might be as simple as selecting the type of boat you will need on a waterfowl refuge in the Midwest, or it may be as complex as multi-stakeholder deliberations about human–wild boar conflicts in suburban Berlin, Germany. What determines whether these situations are problems or opportunities? What can a wildlife manager do about them? If anything is going to happen, someone has to make decisions. Even choosing to do nothing—to make no management intervention—is a decision. Decision making is an essential task of management, including wildlife management; yet few students of wildlife management take a course in individual or group decision making.

This chapter aims to build your decision-making skills. It provides an overview of systematic methods for making decisions through the application of specific decision-aiding techniques. It also identifies common psychological traps that many people fall into when making decisions. Decision making in wildlife management ranges from small, easily described, and uncontroversial choices to those involving more consequential, complex situations about which reasonable people are likely to disagree, no matter which alternative is selected.

An understanding of decision making is particularly important in wildlife management because managers typically are required to address choices involving different dimensions of environmental, sociocultural, and economic impacts, and they are expected to make choices when uncertainty underlies almost every dimension. These choices are often controversial and affect issues and activities about which people are passionate. People involved in decision making frequently voice the opinion that science "makes" good decisions. Science is certainly a cornerstone of good decision making in wildlife management, but science can only inform decisions. Science cannot *make* decisions. People make decisions based on their values and their interpretations of available information.

Although most wildlife managers probably are not aware of this, the bulk of what they do every day involves making decisions. People are prone to do what worked well last time, last week, or last year. They simply react to the crisis of the moment. A lack of structure or rigor in decision making normally means choices are ill-defined in terms of how they are likely to affect the problem (the situation or context) or what the outcomes of various choices might be. Also left out too often are key concerns of stakeholders whose values might be affected by the decision outcomes. Researchers in psychology and judgment suggest there are only two ways to achieve a desirable outcome from a decision: follow a rigorous process or depend on being lucky.

8.1. FOUNDATIONS OF DECISION MAKING

Although it is not always obvious, a method is being applied whenever decisions are being made. In the last chapter, you were introduced to a way to map the management system that included several steps of decision making viewed within the larger context of management. Here, we examine the topic of decisions in more detail, review some common models, and suggest a simple yet effective approach for making better decisions.

Wildlife managers make, or are involved with others in making, more than one type of decision. They face a variety of different decision-making contexts, each of which carries its own challenges, demands, and opportunities. Sometimes managers are asked to recommend the best alternative from among a set of options. At other times, managers need to decide on a preferred sequence of actions, separating decisions into higher versus lower priority choices or decisions made sooner as opposed to later. Sometimes managers are asked to rank different alternatives or to decide whether an action is even within the category of things that they should be concerned about (as opposed to routing the decision to someone or somewhere else). A prevalent question is whether a choice is even theirs alone to make, or whether they need to confer with colleagues, partners, or stakeholders and resource users.

Another common judgment to be made is about information quality and whether there is enough information in hand to make a rational choice. Additional information may need to be acquired through consultation with other professionals or engagement with stakeholders. As pointed out in Chapter 2, to achieve good governance, managers increasingly are sharing decision making with stakeholders rather than only with wildlife experts.

Models of decision making vary with the context and the quality of information that is available on which to base decisions. A purely rational model assumes that managers and stakeholders know precisely what they want, that complete and accurate information about the consequences of management actions is known, and that all relevant management options have been identified. Yet this rational model neglects uncertainties associated with decisions as well as the limits of people's cognitive capabilities. Bounded rationality is the term proposed by Nobel Prize winner Herbert Simon to describe the limited applicability of a purely rational model of decision making. It assumes that not all information needed for a decision is known or can be known; therefore, rationality in a strict sense is bounded by the limits of what is knowable about a system, the cognitive capabilities of humans involved in the decision, and the time frame in which decisions must be made. As Simon (1990:7) wrote, "Human rational behavior is shaped by scissors whose two blades are the structure of task environments and the computational capabilities of the actor (decision maker)."

More recent research on how people make decisions emphasizes the role of heuristics, or mental shortcuts, as part of judgmental processes (Kahneman and Tversky 2000). Rather than being purely rational individuals, as traditionally portrayed in management and economics texts, humans exhibit systematic patterns in their thinking and judgment that work well in some cases, yet lead to misunderstanding or bias in other circumstances. One example is known as the availability bias. People think about an event most often in terms of the more readily available and salient information. This, in turn, misleads people to overemphasize recent or sensational information that they take in and to largely ignore normal conditions. As a result, the number of moose–vehicle accidents is overestimated after reading an account of a fatal crash, or the number of people hurt by accidents with moose is overestimated after seeing pictures of a collision on the news. Another example of bias is overconfidence. Humans tend to overestimate their confidence in our predictions about future events; yet most of us largely are unaware of the extent of this bias (Burgman 2005). The study of heuristics and biases in decision making is the subject of several best-selling books (e.g., *Predictably Irrational* [Ariely 2008]).

A practical model of decision making was proposed long ago by Lindblom (1959) in his paper titled "The science of muddling through." Lindblom proposed that although the literature on decision making promoted and formalized a purely rational approach, in practice (for many of the reasons already identified) it is virtually impossible for managers to achieve a rational approach in complex situations. Instead, Lindblom proposed a description of the process known as "incrementalism" for making decisions and formulating policy. Incrementalism assumes that the hallmark of rational decision making—seeking the single best or optimal solution for a decision problem—is not knowable, let alone achievable. Rather, incrementalism builds on clearly defined objectives to select and to implement alternatives such that changes in a system are achieved in small, incremental steps, thus leaving room for learning and revisions to management actions over time. Although incrementalism as a mode of decision making seldom keeps pace with the rates of change in a management system, it reflects the way bureaucracies function. Later in this chapter you will read about new models of decision making, such as adaptive management, that are evolving to stay abreast with changing management systems.

From a human dimensions perspective, a purely rational approach should be viewed with caution because the approach ignores management's inability to control the behavior of some people or entities whose actions will affect the outcome of decisions, regardless of what management or other stakeholders choose to do. In addition, the search for the perfect or optimal objective and management alternative before making a decision frequently is a factor in delaying action, and in some cases (particularly when wildlife populations are low or threatened, or stakeholders are incurring negative impacts from wildlife) a delay itself carries a cost. Scientists may not want to proceed before additional information is collected, but (as discussed later) managers may prefer to proceed on the basis of what already is known and to carefully incorporate new learning into ongoing actions.

A purely rational decision-making model may ignore how emotions affect decision making. Chapter 4 reviewed the role of emotions in social psychology, especially as an explanatory variable in the behaviors of stakeholders. Research in the decision sciences over the past 20 years reveals a powerful role for emotions in influencing decisions of all types made by all kinds of decision makers (Slovic et al. 2005). This research found that decisions generally are made using parallel but distinct ways of thinking. One is experiential and is based on intuition and beliefs; the experiential mode usually operates quickly and automatically. The other mode involves reasoning, operates more slowly, and is based on rational ways of making sense of information. A distinction between the two systems of thinking also has been explained in a number of best-selling trade books (e.g., Gladwell 2006). Reliance on dual modes of processing information is used by technically trained experts as well as members of the lay public. Although the distinction does not minimize the importance of careful analysis, the distinction does imply that models that do not consider the role of emotions are likely to provide an inaccurate view of decision making. The result can be a decreased ability of wildlife managers to anticipate how emotions affect stakeholder acceptance of decisions (Peters 2006).

8.2. STRUCTURED DECISION MODELS

A useful response to critiques of the rational decision-making model is to introduce a standard for good decision making: a set of steps that, if followed, provide a defensible basis for making good decisions. Effective decision-making processes for wildlife managers should account for two key aspects of choices in wildlife management: (1) values associated with wildlife and wildlife management are multi-dimensional and arise from different interests and concerns (e.g., values exist about environmental, economic, and social interests and concerns; see discussion of values and impacts in Chapters 1 and 4); and (2) usually a diverse set of participants or stakeholders may be affected by a wildlife management decision (Gregory et al. 2002).

One useful way to engage in value-based decisions is through structured decision-making (or SDM), a systematic way of decision making based on theoretical work in multi-attribute utility theory (Keeney and Raiffa 1993) and behavioral psychology (Hastie and Dawes 2001; Box 8.1). The highly readable book *Smart Choices* (Hammond et al. 1999) introduces a decision analytic approach as a sequence of five steps that form the acronym "PrOACT": define the Problem context, clarify Objectives or concerns, identify treatment Alternatives, distinguish Consequences in light of uncertainties, and evaluate key Tradeoffs. Attending to these elements can guide you through a logical sequence of steps for an informed and defensible wildlife management decision-making process, serving as a useful aid to managers and policy makers engaged in selecting and communicating choices that lead to wildlife management actions.

8.2.1. Problem Definition

Wildlife managers try to ensure that actions they take achieve a desired, or at least a more desirable, end state in the management system—the condition of wildlife, habitat, or humans after a management intervention occurs. To determine what decision may lead to a better condition, it is imperative to know the current conditions, to have a desired condition identified, and to understand how and why any particular change might be helpful. This first step in decision making is often described as problem definition, problem bounding, context analysis, or what we called situational analysis in the last chapter (see description of the Manager's Model in Chapter 7). Prior to taking action, managers should have a clear idea of which components of the ecological and human systems might be altered, who might be affected by the decision, and to what extent these changes are likely to be perceived as beneficial or harmful. What you call it—situational analysis, problem definition, context analysis—is less important than making sure you do not follow the human tendency to skip this important first step of effective decision making.

The desired purpose of a situational analysis is to frame the decision-making situation by identifying potentially relevant impacts and describing the management environment in which these impacts occur. You read in Chapter 3 how dif-

ferent stakeholders are likely to frame a situation differently. A government wildlife agency might emphasize needs of achieving regulatory standards, such as those accompanying the Endangered Species Act. A local community might focus on jobs or restrictions in recreational opportunities. An industry might emphasize changes in profits or shifts in the production of certain goods. Conflicts in resource management frequently originate because of a lack of careful thought and explication at this initial problem definition and bounding stage.

A chief goal of most decision makers is to attain sustainable decisions. Sustainable decisions are ones that hold up through time and gain support of stakeholders. Sustainable decision making depends on abilities to summarize the most important elements of a problem, to assemble relevant facts, and to set priorities among the possible outcomes of their choices. The Manager's Model introduced in Chapter 7 is a process for generating a concept map of the management system. Development of a Manager's Model (Decker et al. 2011) or other device for articulating assumptions about the current situation and stating a desired future condition seeks to insure that managers are in agreement about what the problem or opportunity is and that everyone is working on the same core issues. If management processes (such as those described in Chapter 7) are followed, then stakeholders will play important roles at this early stage in identifying and obtaining data in a situational analysis as well as in defining objectives.

8.2.2. Objective Setting

Goals are general statements of intent about the purpose of management. In Chapter 2, we discussed governance as a framework within which wildlife management occurs as the act of orienting, steering, and adjusting organizations. Just as steering a ship in the absence of a chosen destination results in aimless wandering, governance and management cannot be effective (except by chance) without clear goals and objectives (the destination!). An example is the specific goal of bear management in New York, which is grounded in the five major goals of the overall state wildlife program (e.g., "assure that people are not caused to suffer from wildlife or users of wildlife"). Goals often are established through legislation and tend to be vague or abstract. That is, goals provide direction but are generally not quantifiable, nor is there always a realistic expectation that goals will be achieved. They provide guidance for defining a desired future condition, but leave wildlife managers the task of developing specific management objectives that are defined in sufficiently precise terms to direct and evaluate alternative actions in terms of achieving desired outcomes.

Objectives therefore describe what matters in a given situation, in terms of what might be affected with a realistic set of actions over a reasonable time frame. Decision-relevant objectives for any given decision situation are characterized by describing an object and a direction of preference (Hammond et al. 1999). Examples include "increasing the abundance of moose for viewing in a given area" or "increasing the amount of employment from wildlife tourism" or "minimizing the in-

Box 8.1 *STRUCTURED DECISION-MAKING*

Structured decision-making (SDM) is an approach to identifying and evaluating objectives and a set of alternatives to achieve those objectives based on trade-offs, assuming predicted consequences of alternatives. In environmental decision making, SDM combines the analytical methods of decision analysis (Keeney 1992) and applied ecology (Gunderson et al. 1995) with behavioral insights from psychology and the decision sciences (Kahneman and Tversky 2000). SDM is most commonly used to address environmental planning and resource allocation decisions characterized by multiple dimensions (that is, impacts that could occur over a variety of economic, environmental,

social, health and safety, and cultural effects) and multiple participants (in many cases, involving different levels of government, industry, resource users, local citizens, and a variety of interest groups) in cases where there is uncertainty about the likely impacts of management actions. SDM approaches have been used to address a variety of wildlife and fisheries issues (Conroy and Carroll 2009, Gregory and Long 2009), implementation of adaptive management (Failing et al. 2004, Gregory et al. 2006), planning for climate change (Nichols et al. 2011), and consultation strategies involving diverse stakeholders as part of environmental planning (Gregory et al. 2012).

Structured decision-making has been used extensively in adaptive harvest management for mallards and other North American waterfowl (courtesy USFWS)

convenience of obtaining antlerless moose hunting permits." Usefulness of objectives is improved when they define (1) the direction (e.g., increase, decrease, stabilize), (2) an object (e.g., moose viewing opportunity), (3) a specific location (e.g., "Upper Peninsula of Michigan"), (4) the extent (e.g., 10%), and (5) a timeframe for achieving the objective (e.g., within 5 years). Thus, an objective stated as, "Increase moose-viewing recreation days in the Upper Peninsula of Michigan by ≥10%

within five years" is more informative and measurable than a simple objective such as, "Increase the moose population." Developing a set of informative and measurable objectives is essential because they form the basis for evaluating possible management interventions. Objectives formulated with the participation of diverse parties are more likely to be broadly supported and easier to implement, which results in sustainable decisions (Gregory 2000). Our experience suggests the process of formulating clear, acceptable objectives normally receives inadequate attention compared to its importance, although numerous techniques exist for determining objectives (Keeney 1992).

Two types of objectives support making smart choices. Fundamental objectives characterize the reason for management in terms of desired impacts. A fundamental objective answers the question, "Why is management necessary?" and is most effective when it reflects value-based thinking through defining impacts of concern. For example, a fundamental objective of black bear management could be to improve the psychological well-being of a community in which negative human–black bear interactions are frequent events. The second type of objective, an enabling objective, states how fundamental objectives will be achieved. A set of enabling objectives guides development and evaluation of management alternatives and interventions. An enabling objective in the black bear example could be to raise the level of awareness about ways to avoid conflicts related to black bears being attracted to food waste in garbage in that community.

Keeney (1992) provides a method for linking fundamental and enabling objectives through a listing of means–ends relationships. For each objective, participants in the decision-making process should ask, "Why is this important in the specific situation?" The answer either will be that the objective is an essential reason for management (fundamental objective), or the objective is important because it helps attain another objective (means). Each fundamental objective typically will have several means or enabling objectives linked to it. Enabling objectives similarly should support one or more fundamental objectives.

Consider the context of multiple objectives for a regional land-use planning process, involving several state and federal management agencies as well as local residents and resource users. The planning process seeks to improve economic and ecological objectives associated with local timber harvests as well as the quality of hunting experiences in local watersheds. One of the participants, who is an avid recreationist and operates a local lodge, believes it is important to fertilize meadows to encourage productivity (abundance) of moose. Clarifying means and ends is achieved by asking two deceptively simple questions: Why is productivity important? How could we achieve the objective of productivity? The "why question" motivates thinking about impacts to be achieved, while the "how question" motivates thinking about how to achieve those impacts. By asking these questions in tandem, the dialogue proceeds from the expression of means to the articulation of fundamental objectives:

• You want to fertilize meadows. Why?
• So we can increase nutrients. Why?
• So that primary productivity is increased. Why?
• So that moose population growth is increased. Why?
• So that hunting experiences can be improved. Why?
• So that greater economic return will be achieved for the local tourism industry.

At this point, the conversation stops. Economic return for the local tourism industry is identified as an important management objective. When you have arrived at an objective that appears to be important in its own right, regardless of how it is achieved, then you have created a fundamental objective. After a fundamental objective is defined there may be other ways to achieve it than the very first idea suggested. In the case above, fertilizing meadows was suggested first; a risk at that point is that participants in the decision "anchor" on that suggestion (that is, assign it more importance than is warranted). There may be other ideas (such as limiting human access) that might have the same desired outcome without as many side-effects. When setting objectives, remember to ask, "Why is that important?" in order to move from means to ends. Ask, "To achieve the fundamental objective what conditions or processes have to be in place?" in order to move from fundamental to enabling objectives.

8.2.3. Performance Measures

To make objectives clearly understood and operational, it is helpful to develop performance measures or attributes that clearly express the most important and desired characteristics of the outcomes from decisions. For example, stating that higher employment is a benefit of a proposed wildlife tourism initiative may be insufficient for selecting among alternatives because it fails to adequately describe the types of employment that will be provided. Will the jobs be full or part time? Will they be well-paid or minimum wage? Will they be temporary or permanent? Will they be open to local residents or likely to be filled by people from outside the region? Similarly, to make decisions accountable within a program designed to increase abundance of a sparse moose population, the manager needs to distinguish among improvements in adult survival, hunting opportunities for local versus non-residents, or income available to local hunting guides. Any set of performance measures needs to be understandable and easily communicated to stakeholders, who should be informed about the specific outputs and outcomes expected to occur (Keeney and Gregory 2005).

Performance measures used in SDM typically incorporate input from one of three types of measures: natural measures, proxy measures, and constructed measures. Natural measures are in general use and have a common interpretation by everyone. For example, a natural measure of an objective to "increase profits from wildlife tourism" could be in dollars gained per year. Similarly, the number of moose counted over a specified time period within a designated area might be a natural measure for the objective to "increase abundance of moose." Natural measures are best used whenever possible because they are easily understood and thus serve well to communicate a decision.

There are times when it is difficult, if not impossible, to either find or assess a natural measure. In these cases, proxy measures may be helpful. A common example is using the number of pellet groups per hectare as a proxy for abundance of a moose population. A proxy for the objective of increasing profits from wildlife tourism might be the number of vehicles traveling from 4:00 p.m. to 10:00 p.m. on Friday on a road used by tourists to reach a wildlife viewing area. Proxy measures, however, generally are less accurate and less informative than natural measures because proxies indirectly measure progress toward an objective.

A third type of performance measure, constructed metrics, is used when no suitable natural measures exist, or when relevance or accuracy of a proxy measure is tenuous. One example is development of a scale to measure trust stakeholders have in a wildlife management agency to achieve objectives. In this case, a questionnaire could be developed that asks stakeholders to indicate on a scale of 0 to 10 how much trust they have in an agency, where 0 is no trust and 10 is 100% trust in the agency. Another example is a scale to measure community support for a proposed management practice. Because no natural scale exists to measure support, an index can be created (e.g., 1–5 or 1–10) with each rating representing a different level of support. These sorts of human dimensions inquiries are common in wildlife management (see Chapters 9 and 10 for how to design and implement such an inquiry). Many such constructed scales are in widespread use in other aspects of society. For instance, the Gross National Product is a constructed economic measure, as is the Dow Jones Industrial stock average in the United States or the Apgar score used to track the health of newborn babies. When thoughtfully designed, constructed indices facilitate choices by precisely defining ways the objective will be evaluated.

8.2.4. Alternatives

Alternatives represent the range of potential management actions to achieve enabling objectives, which, in turn, are the necessary building blocks to achieve fundamental objectives. When faced with a problem (especially a familiar one), it is common for humans to act on the basis of habits and to turn quickly to familiar alternatives rather than make an effort to clarify or reaffirm the alternatives that will best achieve objectives (i.e., values or benefits to be achieved from wildlife management). Keeney (1992) characterizes this decision error in terms of the distinction between value-focused and alternative-focused thinking. Every management decision can be an opportunity to create and to consider new alternatives. Options should not be limited to those believed to be easily available or previously developed. Linking choices of alternatives back to the enabling objectives, and then rigorously questioning constraints concerned with time, money, or personnel help to determine whether they really apply in a given situation, and can improve the chance of achieving objectives.

Stakeholders often have creative ideas for alternatives and can offer perspectives about expected impacts that may help managers to understand the effects of a proposed management action (Gregory et al. 2012). The experiences of other resource professionals also provide valuable information about traits useful in evaluating candidate alternatives. Access to the Internet makes it relatively easy to find out what others have done in similar circumstances or how other professionals might approach a given situation. Putting effort into identifying and evaluating alternatives is worthwhile because all the due diligence and skillful implementation in the world will not produce the best possible solution to a management problem if important alternatives are omitted from consideration. The key is continual analysis of any proposed intervention in terms of how well the actions will achieve the enabling objectives necessary to meet the fundamental objectives (kinds and levels of impacts) expressed by stakeholders. Whether the proposed alternatives will achieve the enabling objectives is evaluated by estimating their consequences.

A state wildlife manager and a human dimensions specialist (standing) answer questions as an environmental council in Woodstock, New York, discusses results from a quantitative simulation of the consequences of alternative actions to manage problematic human–bear interactions. Quantitative simulations are a structured decision-making tool that can help stakeholders and managers understand decision trade-offs. (courtesy Dion Ogust)

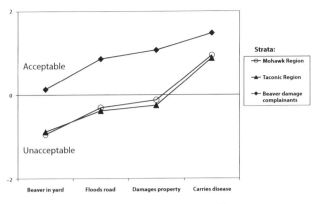

Figure 8.1. Stakeholder acceptance of lethal control of beaver (in 2002) in two regions of New York, and among residents throughout the state who previously had filed a complaint about beaver damage.

Researchers have found that public acceptance of lethal control of species (including beaver, coyote, and mountain lion) depends on how those species affect people (© Tomasz Kubis-Fotolia.com)

8.2.5. Consequences

All actions have consequences, both those intended to achieve enabling objectives and unintended consequences that should be identified and assessed before alternatives are chosen. Estimates of consequences resulting from each alternative—intended, collateral, or subsequent (see Chapter 7)—are informed by available data, new studies, and the judgments of technical specialists or stakeholders. An astute decision maker will remember that information about the consequences of actions may come from stakeholders such as residents of local communities or from knowledge holders within Native populations (indigenous or aboriginal communities). Managers may be particularly interested to learn about the acceptability of management actions from stakeholders; this learning often is achieved through human dimensions inquiry that asks stakeholders about how acceptable various actions are under different scenarios and management responses (Fig. 8.1).

In the case of New York beaver management, stakeholder acceptance of lethal control is greater among stakeholders who experience damage than those who do not (Siemer et al. 2004). These data, collected from a systematic survey in two regions of New York and of stakeholders throughout New York who previously had filed a complaint about beaver damage indicated that, regardless of the situation, a greater proportion of stakeholders who had filed complaints about beaver expressed greater acceptance of lethal control regardless of damage type or place of residence. During the study, 2,400 people in three sub-groups or strata were contacted. Stratum 1 was a random sample of 900 listed households in portions of Rensselaer and Washington counties labeled as the Taconic Region. The Taconic stratum is representative of rural upstate areas with a low beaver density. Stratum 2 was a random sample of 900 listed households in portions of Fulton, Herkimer, Montgomery, Oneida, Saratoga, Schenectady, Schoharie, and Washington counties. This area was defined as the Mohawk Region. Beaver occupancy was estimated to be greater in the Mohawk Region than the Taconic Region during 2002. It is important to note that these study sites were selected to facilitate hypothesis testing, not to provide a representation of the state as a whole.

Stratum 3 was a statewide sample of 600 people who had contacted the New York Department of Environmental Conservation with a beaver damage complaint in 1999 or 2000. The complainant stratum is representative of residential complainants statewide. Members of this sub-group were selected from agency records of complaints filed in 1999 and 2000 (the most recent years for which these data were available from both states). Only private residents were included in the complainant sample (i.e., complaints on behalf of a highway department, municipality, railroad, or place of business were excluded). This is important to note because those sources account for a substantial proportion of the total nuisance complainants in New York. This is just one example of how human dimensions inquiry can help estimate the consequences of various decisions in wildlife management.

Findings about consequences can usefully be summarized in a consequence table, which shows the agreed-upon objectives or performance measures in rows and the agreed-upon alternatives in columns. This provides an easily communicated visual reference that helps to ensure that all participants have a common information base with which to evaluate the alternatives. For example, let us turn from beaver in New York to consider a wind farm proposed for a western state that is on the flyway used by several species of migrating birds. The number of turbines proposed for the site varies from a high of 200 to a low of 50. Impacts on songbirds and, therefore, community opposition are likely to be larger when turbine numbers are higher, even though more electricity can be generated and the per-kilowatt costs of a transmission line connection will be lower. After appropriate studies have been conducted, decision makers might be presented with the following consequence table (Table 8.1), which presents the key trade-offs facing decision makers.

Whenever consequences are uncertain, as shown in this example by the anticipated range of effects for bird fatalities, the quality of decisions can be improved by clearly represent-

Table 8.1. Consequences table displaying trade-offs of three alternatives for wind development

Objectives	Alt A: 200 wind turbines	Alt B: 100 wind turbines	Alt C: 50 wind turbines
Objective 1: minimize bird deaths/yr measured by no. of dead birds found (natural scale)	850 (400–2,000)	300 (150–800)	100 (50–250)
Objective 2: increase local employment measured by the number of different jobs in the community	25	15	10
Objective 3: reduce community opposition to wind development measured by questionnaire (constructed scale: 5 = high conflict to 0 = no conflict)	4	2	1

© Rafa Irusta-Fotolia.com

ing the nature and extent of the uncertainty in a manner that highlights key concerns. This can be done in a variety of ways. A starting point is to compare the consequences of different management alternatives in terms of whether some are more uncertain or less understood than others. In the table, clarification is shown for Objective 1 as a simple range of consequence values along with the best estimates of expected impacts. A more detailed explanation of uncertainty could include a visual presentation of the full probability distribution (showing the endpoints and associated levels of confidence) or, at minimum, additional information such as the mean or median of the distribution. At times, qualitative descriptions are also used to communicate the uncertainty of impact estimates, although this typically is not recommended (at least, not without quantitative uncertainty estimates as well) because the interpretation of terms such as "highly unlikely" or "probable" can vary among individuals, which leads to misunderstandings and confusion.

Different typologies of uncertainty have been developed by statisticians, philosophers, economists, and ecologists. From a practical perspective, two major sources of uncertainty are common to issues in wildlife management (Regan et al. 2002).

One is epistemic uncertainty, arising as a result of a lack of knowledge about facts. Epistemic uncertainty arises from variation in the species, events of interest, environmental variation, measurement errors, imperfect control over human compliance with regulations, or uncertainty in the underlying models (i.e., we never have perfect models of environmental or human systems). The other prominent source is linguistic uncertainty, which arises because people fail to communicate precisely. Linguistic uncertainty emerges from ambiguity in how uncertainty is expressed (e.g., what does it mean that an event is "likely" to happen or that a consequence is "probably" not going to occur?), lack of clarity in terms that are used (e.g., what is "good habitat" or an "unhealthy wildlife population"?), or lack of context (e.g., does risk from wildlife disease characterized as "high" in one location have the same meaning in another?). The influence of uncertainty on the choice of management actions can be profound. To communicate with precision about uncertainty is also difficult, especially when communication involves stakeholders with varied backgrounds and who receive communication through varied channels (see Chapter 12 on communication).

Two other questions related to uncertainty in consequences are important to consider when making decisions in wildlife management. First, ask how differences in the uncertainty of outcomes across alternatives might be addressed. For example, potential ways to gather more information (and thereby reduce uncertainty) or to mitigate the effects of uncertainty (for example, by undertaking incremental and sequenced decisions) should be explored. In some cases, the selection of a preferred management action may be made, at least in part, on the basis of whether it is sufficiently flexible to incorporate new information that is learned over time. Second, ask what the anticipated value of the information expected to be learned is, and whether it might change any aspect of managers' plans. If the answer is no, then the information might be useful from a scientific perspective, yet less relevant from the standpoint of the decision at hand.

8.2.6. Trade-offs

Selection of a preferred alternative is based on achieving a balance among multiple, often competing, values. This usually involves searching for win–win outcomes and facing up to difficult choices when a win–win outcome is not available. Evaluation tools, such as consequence tables (along with other supporting technical information), help to inform decisions with difficult trade-offs, but those devices do not "make" the decisions any more than science can make decisions—people make decisions! Decisions are made with reference to the underlying values and to how impacts are affected by each of the alternatives.

Wildlife managers often need to address trade-offs that require balancing considerations of outcomes and process. Although it is hoped that participants in any management decision agree about the likely consequences of actions (science plays an important role in providing reliable information), there is no reason to anticipate (or to seek to achieve) complete agreement with respect to their values. For example, some hunters might feel that a local forest should be selectively cut because they believe that thinning will improve the habitat for a selected species, and they are anxious for resource managers to get on with the job. Others living in the community may believe that the desires of hunters should be balanced against those of other individuals who collect mushrooms or firewood in the same area, and who might want thinning to be done in a different way or at a different time or not at all. As a result, issues of fairness or trust in the wildlife manager or management decision-making process arise. Other stakeholders may feel that no forest thinning should take place because the dense forest helps provide needed cover for Nordic ski trails in the area. The point is that no decision-making process exists whereby trade-offs either disappear or everyone ends up being happy. Instead, often all that wildlife managers can do is to develop a defensible and systematic decision-making process, follow it transparently, and then provide clear information about what was decided and why. Toward this end, managers may want to engage stakeholders (see Chapter 11 on stakeholder engagement) to develop an advisory committee that helps define and address conflicting preferences.

Table 8.2. Swing-weighting data used in decisions about wind-tower placement

Objective	Best	Worst	Importance ranking	Swing wt (normalized)
Minimize bird deaths	100	850	1	10 (0.55)
Increase local employment	25	10	3	3 (0.17)
Reduce community opposition	1	4	2	5 (0.28)

Several techniques have been developed to help clarify trade-offs, as discussed in Clemen and Reilly (2004) and other texts (von Winterfeldt and Edwards 1986). A common approach, known as swing-weighting, compares objectives in terms of the anticipated range in their values under all alternatives (from worst to best outcomes), proposes a hypothetical alternative with all attributes at their worst value, and then asks participants to identify which attribute they would most like to "swing" from worst to best (Table 8.2). Pertinent information from the earlier wind-farm example is summarized below. First, the different effects are ranked (in this case, from 1 to 3 in terms of importance) and then assigned "value" points, with 10 points awarded to the most highly valued effect and others valued on a ratio basis (i.e., the second most important effect, if valued half as much, would receive 5 points). This same bottom-up process is conducted for each of the attributes with the weights then normalized (so that all weights collectively sum to 1). A top-down approach, in contrast, would ask participants to rank the alternatives in order of preference and then to assign points reflecting how much better one alternative is relative to another. Each approach has advantages and disadvantages. The critical lesson is that methods exist for comparing importance of objectives without obvious commonalities, which can greatly assist in the development of broadly supported management actions.

Whatever weighting process is selected, participants—whether managers within the same agency or a broad-based group of stakeholders—are encouraged to acknowledge and openly discuss trade-offs using explicit criteria for selection among alternatives and review options for achieving an acceptable balance across all objectives. If a decision is significant and multiple stakeholders are affected, managers will need to engage stakeholders (see Chapter 11) to ensure that choices made reflect an understanding of broader community or institutional values. Where consensus on a preferred alternative is not reached, areas of agreement and disagreement (and the associated reasons) should be documented and this information passed on to decision makers. Agreement among participants is therefore not required, but, at minimum, managers should provide decision makers with a clear understanding of the reasons that individuals or groups might support or oppose a particular course of action.

8.2.7. Learning, Iteration, and Monitoring

The final consideration in an SDM process is to recognize that each decision is an opportunity to learn over time and to make better decisions. This learning can be subsequently reflected in many ways: more accurately defining the nature of the problem; more precisely understanding stakeholders' objectives; developing new and more creative management alternatives; more accurate tracking of consequences; greater predictability of stakeholder acceptance of interventions; and evaluating and balancing associated trade-offs. Learning is often enhanced if a clear monitoring program is established to closely follow outcomes of greatest significance and to provide feedback to managers in a form that is readily incorporated into future planning and decision making. This is not easily done; it is often difficult to develop clear measures of outcomes that distinguish the effects of management actions from external developments (e.g., related to land-use changes or climate change), and it is not always easy to maintain flexibility or adaptability within an institutional setting. In the next section we examine some of the successes and failures of one approach to learning in wildlife management, that of adaptive management.

8.3. ADAPTIVE MANAGEMENT

The concept and term "adaptive management" (AM) is firmly fixed in the lexicon of wildlife management and is used with widespread and varied meaning (Organ et al. 2012). The concept, introduced by ecologists C. S. Holling (1978), Carl Walters (1986), and others provides a systematic approach focused on learning. AM seeks to accelerate learning through experimentation and monitoring. Rather than downplaying uncertainty and only using existing knowledge to implement a single "best" plan for ecological management, AM approaches explicitly recognize that uncertainty exists and propose a range of management alternatives to be tested and refined over time. In practice, AM encourages reducing uncertainty, thereby improving long-term performance of wildlife management interventions. AM is linked to effective decision making. The U.S. Department of Interior's *Technical Guide to Adaptive Management*, for example, states that "Adaptive management is framed within the context of structured decision-making, with an emphasis on uncertainty about resource responses to management actions" (Williams et al. 2007:vii).

AM is an attractive concept. Who would not want to learn and be adaptive? Implementation of a scientifically defensible AM approach, however, is intellectually and logistically challenging. Thus, it is not surprising that the case-list of successful applications remains modest. A key benefit of effective AM is a commitment to link learning derived from experimental trials to the wildlife management decisions being made. From a deliberative perspective, however, implementing an AM plan also requires those involved in the overall decision process to hold realistic beliefs about the ability of the proposed AM initiatives to deliver clear results within the desired timeframe (Riley et al. 2003).

Two variations of AM, passive and active, frequently are described in the literature (Walters and Holling 1990). Both passive and active AM recognize uncertainty, and they both design and monitor management interventions to learn while doing management. Choice of approach depends on the expertise and other resources available to the management program. Preferred approach depends on the emphasis placed on learning about the system as opposed to performance or response of the system (how wildlife responds to management interventions). An active approach places full emphasis on learning about the system, whereas a passive approach emphasizes effects of management on resources and stakeholders. With a passive approach, the most widely practiced approach, learning is a useful product of management but not the primary interest.

Passive AM typically involves using historical data and research literature to develop hypotheses and a model of system performance to implement a preferred management action (see Chapter 7 for an example of a conceptual model). An alternative is chosen by comparing potential responses of the system predicted by the model. System response to management is then closely monitored to learn and refine the underlying model through time. Monitoring under this process is similar to normal monitoring of effectiveness. By assessing system changes over time, the intention is to create an improved understanding of responses of wildlife, ecosystems, and stakeholders.

When practicing active AM, managers focus on reduction of uncertainty through the deliberate comparison of different alternatives. For example, consider a situation where there is substantial uncertainty about the relationship between flows within a managed river (e.g., downstream of a hydroelectric dam) and fish populations, which, in turn, affects the health of raptors and bears. A deliberate experimental comparison of the flows in the river might be set up to compare the effects on amphibian populations under flow regimes of 10 cubic meters per second (cms) for 3 years, 20 cms for another 3 years, and then (depending on how these results turn out) 5 cms for another period of time. In this sense, active AM is "management by experiment" as guided by competing models of the environmental system. Planned manipulations of the system through testing a range of alternative management actions or treatments (either simultaneously or sequentially) can achieve substantial learning, but only if accompanied by careful experimental design and monitoring.

It should be noted, however, that the cost of such treatments can be significant and the time required can be long, conceivably measured in decades rather than years (depending on the effect being addressed). As a result, it may be difficult to gain acceptance from participating stakeholders to conduct the experiment (after all, who wants to wait a decade or more to learn about an important issue?) and it also may be difficult to maintain sufficient institutional stability (decision makers have limited life in an agency) to have the experiment run its course and for results to be subject to careful analysis (Enck et al. 2006).

Active AM often is presumed to be more appealing from the standpoint of improving knowledge and contributing to the scientific (and, in turn, management) knowledge base.

This approach, however, imposes higher opportunity costs because other management actions cannot be conducted simultaneously (due to concerns about confounding study results) and typically requires a more sophisticated experimental design; if this is lacking, the experiments might deliver inconclusive information at a relatively high price (Gregory et al. 2006). Active AM also results in a different distribution of risks. Because a broader range of actions is tried, the usual result is that at least some of the manipulations will prove to be unsuccessful and as a result some resource users or other stakeholders may be upset. An active AM approach may be preferred by research-oriented scientists or managers, who tend to place a higher priority on long-term learning. Passive approaches, however, typically are favored by government decision makers and industry, which tend to place a higher priority on short-term results and reduction of risk (Lindblom 1959).

An adaptive approach, which promotes consideration of management actions as experiments, applies to the human dimensions of management (i.e., beliefs, attitudes, and behaviors), as well as to the biological and ecological dimensions. The adaptive part of AM refers not only to learning on the part of scientists and managers, but also to stakeholders learning as they engage in management (Enck et al. 2006). When learning is shared among scientists, managers, and stakeholders, greater support is gained for management actions, which results in more sustainable decisions in wildlife management. A variation on AM that uses stakeholder-defined impacts as the fundamental objective, rather than only the state of a wildlife population or habitat, is termed *adaptive impact management* (Riley et al. 2003). Adaptive impact management proposes that an essential component of any management program—adaptive or not—is determining fundamental objectives, in terms of stakeholder-identified impacts, prior to choosing enabling objectives that are the focus of management actions (see Chapter 7).

SUMMARY

Decision making is a skill critical to attaining success in wildlife management. Decisions range from small, everyday decisions to those that have potential to create enormous impacts (positive and negative) for stakeholders in wildlife management. The quality of these decisions can be improved by following a structured decision-making process.

- Some people argue that more science, logic, and rational thinking result in better choices; yet purely rational models of decisions are usually flawed because they assume managers and stakeholders know precisely what they want, have perfect information about the management system and the consequences of management interventions, and that all relevant management options have been identified.
- A bounded rational approach to decision making is introduced, which can be remembered by the acronym

PrOACT: Problem identification, Objectives, Alternatives, Consequences, and Trade-offs. This structured decision-making (SDM) process encourages rigorous analysis of problems and opportunities; establishment of fundamental objectives in terms of stakeholder-defined impacts; creation of enabling objectives to accomplish fundamental objectives; description of alternatives to achieve enabling objectives; evaluation of consequences of each alternative; and weighing of trade-offs of each alternative relative to the objectives.

- A final component of any well-designed SDM process is learning through monitoring and evaluation. Although often overlooked, this is a vital step in the iterative process of management because evaluation of decisions in one cycle helps inform decisions in the next.
- Whenever uncertainty exists, as it almost always does in wildlife management, there is no guarantee that a decision will lead to desired consequences. Regardless of the process used, reduction of uncertainty associated with decisions remains a key issue in wildlife management.
- Adaptive management, practiced in its many forms, places a premium on reducing uncertainties through learning while doing management. Active adaptive management pursues management as an experiment with the overarching emphasis on learning (that is, reducing uncertainty) about wildlife resources and the effects of management action. Passive adaptive management follows a similar process, yet the chief motivation behind management is to understand the consequences of the decision on the ecosystem and the interests of stakeholders.
- Although decisions are often judged on the basis of outcomes (the quality of the consequences), decisions in wildlife management should be judged by the quality of the processes used to make the decisions. Thorough, inclusive, and transparent public-decision processes are elements of good governance.

Suggested Readings

Clemen, R. T., and T. Reilly. 2004. Making hard decisions with decision tools update. Duxbury, Belmont, California, USA.

Gregory, R., L. Failing, M. Harstone, G. Long, T. McDaniel, and D. Ohlson. 2012. Structured decision making: a practical guide to environmental management choices. Wiley-Blackwell, Chichester, UK.

Hammond, J. S., R. L. Keeney, and H. Raiffa. 1999. Smart choices: a practical guide to making better decisions. Harvard Business School, Boston, Massachusetts, USA.

Lindblom, C. E. 1959. The science of muddling through. Public Administration Review 19:79–88.

Organ, J. F., D. J. Decker, S. J. Riley, J. E. McDonald, Jr., and S. P. Mahoney. 2012. Adaptive management in wildlife conservation. Pages 43–54 *in* N. Silvy, editor. The wildlife techniques manual. Seventh edition. Johns Hopkins University Press, Baltimore, Maryland, USA.

PART IV • HUMAN DIMENSIONS METHODS AND SKILLS

This part contains four chapters that counsel you about working with human dimensions specialists, methods of social science inquiry, stakeholder engagement, and communication planning. They introduce some crucial skills pertinent to the human dimensions aspects of a career in wildlife management.

Chapter 9 reflects decades of experiences and observations in working with wildlife managers to design scores of human dimensions studies. You will find advice about roles for both managers and researchers in planning a study—advice we hope will improve communication between wildlife managers and their collaborators and consultants in human dimensions inquiry. The aim of this chapter is to improve results and application of these inquiries to decisions in wildlife management.

Also contributing toward that end, Chapter 10 is an overview of research methods commonly employed for human dimensions inquiry. Our intent with this chapter is to familiarize you with those methods; we do not expect you to become an expert in them. The chapter outlines the strengths and weaknesses of various methods so you will appreciate the considerations that human dimensions specialists must weigh. We hope this knowledge will facilitate better communication and choices that lead to accurate, reliable results from human dimensions studies.

Gaining stakeholder involvement in decisions is pervasive in wildlife management and fundamental to collaborative governance. Chapter 11 traces the development of professional approaches to gaining stakeholder input and involvement. Stakeholder engagement includes activities ranging from listening to an individual stakeholder to intensive co-managerial arrangements. These types of engagement are being creatively refined through research, experience, and monitoring.

Chapter 12 discusses one of the most crucial of all skills for a wildlife manager—communication. In a world that is becoming globalized at dizzying rates, communication is widely seen as the most important skill sought by employers of wildlife professionals. More diverse stakeholders, more governmental and non-governmental entities involved in governance of wildlife, and the pervasiveness and dramatic speed of electronic media today make well-honed communication skills essential. This chapter describes communication as a social process and explains why skillful communication is so important to being an effective wildlife manager. Communication planning is emphasized, and the use of human dimensions inquiry to inform planning of communication is demonstrated with examples. Important emerging areas of communication, such as risk communication, are emphasized in the context of wildlife management.

9

PLANNING A HUMAN DIMENSIONS INQUIRY

DANIEL J. DECKER

Few wildlife managers will become human dimensions specialists. Mastering human dimensions inquiries requires expertise in advanced social science, with training in both theory and inquiry techniques. It also requires experience in adapting textbook social science methodology to the real-world contexts encountered in wildlife management.

Although wildlife managers do not typically have sophisticated social-science research skills, they do need to have the knowledge and skills to integrate human dimensions into management thinking and action. When wildlife managers are called on to help to develop a human dimensions study, they find themselves on a team of people, often including someone who can use the tools of social science inquiry. Together they must articulate the management problem to be addressed and the decision to be made, and then they must identify what information they will need to have before they can make a good decision.

The manager brings unique knowledge and insight about the context to the study: the history of management actions and public reactions to them, the agency's credibility with various stakeholders, tensions and alliances between stakeholder groups, and information about what the stakeholders need to learn about the issue at hand. Those insights from wildlife managers can greatly enhance the value of the study.

In this chapter, we explore the questions that you (the wildlife manager) should ask yourself before you decide to do a study, the questions you should ask when initially working with the social scientist to frame the study, and the questions the social scientist may ask you. If you have the answers to these questions, you will be able to participate effectively and guide the development of the human dimensions study.

9.1. ATTRIBUTES OF A HUMAN DIMENSIONS STUDY

Human dimensions studies are like biological studies in some respects. For example, the usefulness and quality can be good or poor, depending on the effort that goes into designing and implementing a study.

You need to ask the correct research question to derive results that will yield the insights needed for decision making. That is true not only for field research on habitat utilization of Neotropical migratory birds but also for surveys of birdwatchers to determine their spring habitat visitation patterns and behavior when seeking Neotropical birds.

You need to apply the correct methods of collecting data, and apply them well, if you are to have results you can trust. In a nutshell, you need to apply the familiar standards of scientific rigor to human dimensions studies; these are the same standards you would apply to biological field studies. The specific steps taken to ensure valid and reliable results may be different, but their intent is the same.

Ignoring those steps for the sake of doing a quick-and-dirty human dimensions study would have the same pitfalls as an inadequate biological study: you would waste time and resources and perhaps diminish your credibility among stakeholders. When thinking about quality of data, recall the old phrase used by computer programmers: garbage in, garbage out.

There are distinctions between biological studies and human dimensions studies. Communication with the subject is not a concern in biological studies, whereas human dimensions studies often require direct interaction with stakeholders, which needs to be handled well. How the interaction is handled, both in data collection and in public relations, affects the confidence stakeholders will have in the results and the ultimate decision. A rude and insensitive inquiry can be taken as an indication of the agency's arrogance or confirmation that the study is intended to deflect attention from another issue, support a foregone conclusion, or be a stalling maneuver. Poorly presented studies can cause major public controversies.

Some biological studies certainly evoke public reaction (e.g., when animals have to be killed, when a moratorium on recreational opportunities is required, or when landscapes are altered and aesthetics diminished for some stakeholders). In human dimensions studies, however, there is more potential for controversy; after all, these studies are about people, and the decisions that are based on them may affect people's lives on a daily basis.

Also, people tend to question the methodology of human dimensions studies more frequently than biological studies. That may be because they believe they understand the techniques and primary tool for eliciting responses—written and verbal communication.

Moreover, people can purposefully mislead investigators. Wild animals and plants are better, or at least easier, study subjects. They do not have conflicting political agendas or hidden purposes that can affect their behaviors.

9.2. QUESTIONS TO ASK YOURSELF OR A CONSULTANT

In some situations a human dimensions study is not warranted (the data already exist), is inadvisable (the timing is unreasonable or the political arena dictates the outcome), or cannot be justified (the cost exceeds the likely value to the decision process). We discuss some of those situations below. We also present some questions you will want to ask yourself if you decide a human dimensions study is appropriate.

9.2.1. Do We Really Need a Study?

Is it just a stalling tactic? Is it just an appeasement to a stakeholder group? Would the results of the study even matter?

When an issue is particularly contentious, conducting a study is sometimes a tactic used to create a cooling-off period for the opposing stakeholders or to postpone a decision to a more propitious time. Studies are sometimes used to divert attention from a political reality that will be the only thing that really matters when decision time comes round.

Although those motives are not unique to human dimensions studies—many a biological study has been initiated for the same reasons—they are not good reasons for you to conduct a study. The almost inevitable consequences of such motives are loss of credibility and squandering of resources.

The results of a human dimensions study conducted for the wrong reasons rarely improve the situation, but occasionally the results are so compelling and the public interest so great that the outcome is good. That is a gamble we do not advise you to take.

9.2.2. Does the Information Already Exist?

Is the information already available elsewhere? Could it be adequately approximated?

Perhaps the information you want already exists in a form adequate for decision-making purposes. Perhaps there are secondary data, obtained for another purpose, which could provide the insight you want for the current situation. Your consultant may be familiar with a similar situation in which a study was conducted, and you can discuss whether the results could reasonably be generalized to the current situation. If the information is already available, it does not make sense to incur the cost and delay of a new study.

9.2.3. Is It Worth the Cost?

Is the extra precision that the study results can bring to a decision worth the cost to get them?

Sometimes more or better data would be reassuring, but the extra measure of precision that such data would bring to the decision does not justify the expense of a study. In fact, some decisions you must make are not important enough to warrant a study. In our less-than-ideal world, wildlife managers only have the resources to conduct studies that are essential.

9.2.4. Will It Build Unrealistic Expectations?

Even if the study strongly supported a particular course of action, would we be able to carry out the action? Would conducting the study, in itself, build expectations for an agency action that would be biologically, fiscally, or politically unlikely?

Well-intentioned wildlife managers often find themselves in a situation where they would like to actively identify stakeholder support for an innovative management action. They see a human dimensions study as a way to verify the support or fine-tune an idea. There is nothing wrong with that. However, the decision to undertake such an inquiry needs to be made cautiously; sometimes ideas that make sense to a manager, and may even be acceptable to stakeholders, are too radical to enjoy support within an agency or in the prevailing power structure of state government. Moving ahead with the study might build expectations for follow-through that the agency cannot meet. That can lead to a public backlash.

9.2.5. Is There Enough Time?

Can a study be designed and implemented, and can the results be analyzed, by the time a decision is needed?

Sometimes there is not enough time to conduct an inquiry with proper technique. For example, you want to have the human dimensions data for the decision process, but the human dimensions researcher cannot ethically deliver in the time available. In such a case, do not ask the researcher to do a quick-and-dirty study. It may be possible to launch a limited, but credible, study on short notice with rapid turnaround time, but such studies typically have serious limitations. Do not ask the researcher to compromise on methodology in order to get the job finished on an unreasonable schedule.

In other instances, a good study is designed and implemented, but the results come too late to be used in decision making. At best, the results can be used after the fact to confirm or question the assumptions on which the decision was based.

Sometimes human dimensions researchers are unfairly criticized for taking too long to complete a study. The culprit is often lack of adequate planning; for example, the need for human dimensions information often is not identified soon enough to allow adequate time for a good study.

9.2.6. Who Needs to Be Involved?

Who within your organization will need to support the study? What other organizations need to be involved, perhaps even as co-sponsors?

Chapter 12 emphasizes the importance of internal communication. It is absolutely necessary that agency staff, especially leaders, be supportive of the human dimensions inquiry, for two reasons. First, if the decision is to be made in-house, having agency leaders endorse the study from the outset makes it unlikely that they will ignore its results. Second, engaging agency leaders early in the study design will make it unlikely that they will publicly discredit it.

The decision-making process frequently can be significantly enhanced simply by including other relevant agencies and non-governmental organizations in the design and implementation of a study. They can make valuable intellectual contributions, and through their involvement they become vested in the study and its results and more comfortable with the application of the results in decision making. Moreover, they can sometimes identify ways of modifying the study slightly to obtain data of value to their own organization and will share the cost of the study to gain that benefit. Sharing the cost is not just a monetary arrangement; it is symbolic of close collaboration.

9.2.7. Who Should Do the Study?

Should selection of the study leader be given special consideration? That is, will the stakeholders have more faith in the results if an independent survey firm or a university researcher conducts the study?

One can easily imagine circumstances where the human dimensions study should not be conducted in-house. If an agency has already taken a position on a contentious issue, any study it conducts directly would be seen as an attempt to reinforce its position. A non-governmental organization may simply want to enhance the human dimensions knowledge base on an issue, but because it has a particular position on the issue, the public might not have faith in an in-house study. In such situations you should retain the services of a respected outside specialist, even if your organization has the expertise to conduct the study.

There are private consulting firms that conduct studies for a fee. Universities can also be engaged to conduct studies, especially if it is an opportunity to contribute to theory or a knowledge base for policy. Private consultants tend to have quicker turnaround times than universities, but universities are often seen as less likely to be biased by the funding source. One can, of course, find exceptions. Some university researchers are responsive to time constraints, and some consulting firms have a well-deserved reputation for reporting results without influence of the funder. When deciding where to go for help—a private consultant or a university researcher— think about the likely reactions of those who will be asked to consider the results.

9.3. QUESTIONS A RESEARCHER MIGHT ASK

The human dimensions researcher—whether from within the agency or organization or contracted from the outside— should ask some questions about the context in which the study will be developed and the results used. Think about all these questions, and know what the answers are, before your first meeting with the researcher.

9.3.1. Who Is the Management Decision Maker?

What is the decision that information from a human dimensions study is intended to inform? Who will make that decision? What kinds of data, and in what form, will the decision maker find most helpful?

By focusing on the decision to be made, you make sure that the data you collect will be useful. For example, you may be considering the use of a bait-and-shoot program to manage an urban deer problem; in fact, after working with a citizen task force in a community, you may have decided it is the only feasible way to deal with a situation from a population dynamics viewpoint and in consideration of other variables. What you want to know is whether and why the residents in the trouble area would find that acceptable. Seeking less specific information will not help the decision makers.

On the other hand, perhaps your assessment of the situation does not indicate an obvious preference, and you are considering various management techniques. In that case, you will want to know which of the techniques the residents would find acceptable.

By being mindful of who will make the decision, you also can ensure that you obtain the correct set of information. If you are not the decision maker but have responsibility for collecting and analyzing the information, do not assume that you know what the decision makers want to know. Ask them, and then question them further to be sure. Even if you think bait-and-shoot is the best technique for your situation, the decision maker may need to find out if less controversial approaches would be more acceptable to the stakeholders. In that case, you would have to gather the more general information about acceptability of various techniques, including bait-and-shoot. If a public listening session had been held at which various possibilities were raised, the decision maker might need to have the acceptability of all those options checked.

If you are the decision maker, this process is easier, as long as you have thought carefully about it. Remember that a part of every decision is communicating it to others. Make sure you gather all the information you need to explain your decision, and keep in mind the difference between an *individual's* expression of personal preferences reported in a survey and a *community's* best interests revealed through stakeholder dialogue.

9.3.2. Who Cares?

Who has a stake in the study and its application to decision making?

The human dimensions researcher should ask you to describe the primary and secondary stakeholders in the decision. When planning a study, interest in the stakeholders stems from two concerns: (1) some stakeholder groups may be such an important source of data that they need to be considered a stratum for sampling purposes; and (2) some stakeholders have so much influence on the decision that they need to be represented on an advisory group, or at least involved enough that they will find the inquiry credible.

9.3.3. Will There Be an Advisory Group?

Will a study advisory team be useful? If so, who might be on it? Who will nominate and invite participants? What will their role be?

A human dimensions researcher should be aware of an agency's need to keep internal and external stakeholders informed of the development of a study that will affect the decision they have a stake in. Being informed may be all that is required, but often that is not enough. Interaction and involvement of stakeholders is increasingly typical for human dimensions studies (Mattfeld et al. 1998). Along with the transactional and co-managerial models of public involvement in wildlife-management decision making, we are seeing the desire for more involvement in the study itself.

We have found that study advisory teams are useful

- when multiple entities have jurisdictions relevant to the issue (include people from those entities);
- when the issue is so contentious that organized factions do not trust the wildlife agency, or any other single organization, to design and implement an unbiased study (include people from the various factions);
- when insight from people involved in an issue needs to be on tap from beginning to end and a commitment to an advisory group will help ensure their ongoing input (include local experts—both laypeople and wildlife biologists);
- when others will have key roles in making or communicating a management decision, and when their involvement will build their knowledge (these could be co-workers or outsiders); and
- when others are paying for the study, and their involvement partly or entirely fills accountability requirements (try to identify someone with interest and knowledge in the problem so there will be a real contribution, not just oversight).

In each of those situations the purposes must be clear, and the roles and responsibilities of individuals and the overall group must be spelled out and agreed on at the outset. Through careful planning and clear communication with potential participants as the group is being developed, you will avoid confusion of purpose and mixed expectations.

9.3.4. Who Will Handle the Media?

Are the media likely to be involved? If so, how will media relations be coordinated?

Some studies are low-key undertakings that generate little media interest, and perhaps that is the way the agency or organization wants it. Other studies are focused on such hot topics (e.g., restoring wolves or culling animals in a national park) that the media become involved as soon as they learn about the study.

Media relations can affect a human dimensions study in several ways. First, sensational coverage before and during implementation can affect the results. Second, the researcher can be distracted by having to deal with the media; it takes time to answer reporters' questions, and those of stakeholders that are generated by the media coverage. Third, media treatment of a study can create tensions between the agency and the researcher (e.g., misquotes, or even accurate statements, from the researcher that are not in line with agency policy) and between the agency and the stakeholders.

It is best for the sponsor of the research to be the one that actively deals with media relations and for the researcher to have a role that was agreed on ahead of time.

9.3.5. What Is the Deadline?

You want the study designed and implemented and the report on your desk by when?!

An unreasonable turnaround time for a study is a major challenge that human dimensions researchers frequently face. In the next chapter, on human dimensions research methodology, we outline some of the timing considerations. With the exposure to social science inquiry that we provide in that chapter, you will know when you are asking for the impossible. The human dimensions researcher should raise all kinds of questions about timing and make it clear that compromising study implementation to meet unrealistic time constraints will produce lower quality results. If the prospective researcher does not raise such questions, you might want to look for someone else.

9.3.6. What Information Do You Want?

What information is essential to inform or serve the decision? What information would be interesting for background or perspective, but would not be essential?

Think carefully about the kind of information you need before your first meeting with a human dimensions researcher. It would be an inappropriate delegation of responsibility to leave that task to the researcher, and most researchers would refuse to let you off the hook by doing your work for you.

You should be able to explain to the researcher the kind of information you need and why. Perhaps listing your information needs in three categories would be most useful: (1) essential, serves the decision directly; (2) nice to know, helpful in understanding the stakeholders; and (3) nice to know but has little bearing on the decision. Do not draft a questionnaire, which calls for the technical skills and experience of the researcher. Rather, develop information objectives that the researcher can use to design the questionnaire.

A handy way to organize your information needs is in an information-by-objective matrix. That is simply a table that lists your objectives for the study across the top and the kinds of information needed from the survey down the left margin. You can indicate which information needs relate to which objectives by placing an X in the appropriate cells of the table. Later on, as the instrument for data collection is developed, you can replace the information categories on the table with the number of the question or other measure on the instrument that addresses the objective and is intended to obtain the information desired (Box 9.1).

Box 9.1 *SAMPLE QUESTIONS AND MATRIX*

To show how to set up an information-by-objective matrix, we have excerpted the objectives and some survey questions from a study of public perceptions and preferences about white-tailed deer in Westchester County, New York (Connelly et al. 1987), and we have provided the matrix that indicates which question from this excerpt is intended to fulfill which objective.

OBJECTIVES
A. To determine Westchester County residents' perceptions of the nature and extent of deer impacts on their communities.
B. To determine the support for deer management in Westchester County.

QUESTIONS
1. Overall, how would you describe the amount of deer damage to your property within the past 12 months? (Check one.)
—— none
—— light damage
—— moderate damage
—— substantial damage
—— severe damage
2. How do you feel about the amount of damage your property received from deer in the past 12 months? (Check one.)
—— not aware of any damage
—— the amount of damage was negligible
—— the amount of damage was tolerable in exchange for having deer around
—— the amount of damage was unreasonable

3. Please check any of the items below that represent a concern you or your family have about deer in your area. (Check all that apply.)
—— deer–car collision
—— Lyme disease transmission
—— damage to vegetable garden
—— damage to yard plantings
—— no concern
4. Please circle the one item above that you are most concerned about.
5. Do you feel there is a need for deer management (to increase, decrease, or maintain stable population numbers) in Westchester County?
—— No (Skip to Question 7.)
—— Yes
6. Research and experience have shown that the current hunting of deer by long-bow only in Westchester County is an ineffective method for managing deer herd size. If a situation arises where deer herd size needs to be limited, which of the management techniques used elsewhere in New York State would you support for limiting deer herd size in Westchester County? (Check all that apply.)
—— use of muzzle-loading firearm during a regulated hunting season
—— use of shotgun during a regulated hunting season
—— use of rifle during a regulated hunting season
—— other (please specify: _____)
—— none of these is acceptable

INFORMATION-BY-OBJECTIVE MATRIX

Questions	Objective A	Objective B
1	X	
2	X	
3	X	
4	X	
5		X
6		X
. . .		

9.3.7. Are There Other Relevant Studies?
What studies have been conducted previously, perhaps within your agency or organization, that have a bearing on the information needed?
Sometimes relevant work has been conducted in-house that an outside consultant would not know about. Anticipate a request to share such information, and compile it before meeting with the researcher. It might not be absolutely necessary to have the information available before your first meeting, but you should at least be able to describe the purpose, methodology, and results of the relevant work.

You may be familiar with work conducted by peers in other

states that would be useful. Much relevant human dimensions research that has been done has not appeared in widely circulated journals and is not broadly referenced.

9.3.8. Do People Know Enough?

Is it reasonable to expect stakeholders to be able to provide the information you seek?

A researcher can entice some people to respond to almost anything. The trick is to be sure you are getting truthful responses to reasonable questions. Your expectations of stakeholders' ability to provide information may not be realistic.

Answers to questions about hypothetical scenarios or potential interactions with species that people have not experienced will necessarily be superficial, no matter how sophisticated the inquiry. They may reveal how a stakeholder would react to a proposal, but they would not reveal the stakeholder's likely response to the actual experience.

We have seen studies asking people how they would feel about having a species of wildlife restored to their area when it has been extirpated for the past 100 years, and therefore no one in this generation or the preceding two generations has any experience in co-existing with it. It is similar to asking them if they would like to have mastodons restored to North America along with saber-toothed tigers for mastodon population control. How could the respondents possibly know the consequences on other animals, on plants, and on human land uses? How could they know the personal and community risks and trade-offs? How could they know who would benefit and who would pay? The example may seem absurd, but how different is it from asking people if they would like elk and wolves restored to areas where they have not existed since the 1800s?

If you want information that will be useful in decision making, you need to be reasonable in what you ask people. In most decisions, people are making a choice between two or more courses. Predicting consequences of the action, such as its impacts on stakeholders, is important. Preparing stakeholders to evaluate the likely consequences on themselves and others, by providing them with the information they need, is vital.

9.3.9. How Precise Should It Be?

If sampling is needed, what levels of precision and confidence are you seeking? Will you need to describe responses from sub-groups?

When a sampling strategy is being designed, you should anticipate being asked about how precise the results need to be. Can you live with fairly broad estimates, such as ±7%? Or can you tolerate only a 3% error range? Your answer will make a big difference in sample size, which in turn will influence cost and time.

How confident must you be in the results, statistically speaking? Do you need to have only a 1 in 20 chance ($P = 0.05$) of a measure for the population falling outside the error range? Or, given the level of uncertainty that exists for other parameters in a decision, could you accept a 1 in 5 ($P = 0.20$) chance that the actual value for a parameter might not be within the error range specified and the results obtained from the sample? The more confidence you demand, the greater the sample size

required, and the higher the cost and longer the time required for results.

Sometimes sub-sets of a population of interest need special attention. For example, for a statewide assessment of farmers' acceptance of deer, a sample of 500 might be adequate. However, if you also want data from 10 deer management regions that are specific enough to allow you to be responsive to their needs, the sample obtained in each region from 500 overall might not be large enough. In that case, geographic stratification of sampling would be needed. Each region would be sampled to give region-by-region data, and then they could be combined (with appropriate weighting) to yield a statewide estimate. The total sample size might jump to 5,000 and study costs would rise accordingly.

Or perhaps fruit growers (as a sub-set of farmers) are of special interest because they have the potential of suffering disproportionate levels of crop damage from deer. In that case you might opt for a statewide survey of a sample of all farmers except fruit growers and a separate survey of a sample of fruit growers.

You will need to consider precision, confidence, and stratification carefully to assure that the study yields information of the correct kind and quality for decision making. Typically, you will have to make trade-offs to keep the study within budget. Choosing to have less precision and lower confidence may allow you to add a stratum. Confidence level may be more important than error range in some situations. Every case is unique, and it is the manager, not the researcher, who should make such decisions.

9.3.10. Will There Be Follow-ups?

Is this to be a one-shot study, or are there likely to be follow-ups to monitor change or assess achievement?

This makes a difference. A study designed with the expectation that there will be follow-ups uses a larger sample (to allow for attrition) than does a one-shot study. The content and format of the instrument and the communication associated with the survey are also affected.

Wildlife managers often get wrapped up in the immediate need for information and do not think ahead about measuring objective achievement. The human dimensions researchers should be asking about those possibilities in the initial stages of study design. You should give the question careful thought, because the initial survey of a longitudinal study typically costs more than a one-shot survey.

9.3.11. What about Non-respondents?

How much effort will be put into a follow-up to assess non-response bias?

Forget about the justifications you have read from authors using results from a survey with a response rate less than approximately 70% (e.g., "We experienced a 40% response rate, which is good for single-wave mail surveys."). Unless a response rate is very high, you should be concerned about non-response bias. You will want to know whether the people who did not respond had characteristics markedly different from those who did. If so, knowledge of those characteristics might influence your interpretation of the results and their implications. Previous surveys of the stakeholder groups might give some useful information about non-respondents.

Use proven survey techniques and plan a meaningful non-respondent follow-up. The follow-up is typically a telephone survey of a randomly selected group of non-respondents to a mail survey, and sometimes even of non-respondents to a telephone survey (the procedure is explained in the next chapter). A non-respondent follow-up is necessary in almost every case, and the researcher may well insist on it. If not, you should!

Because a follow-up costs money and takes time, an inexperienced wildlife manager may give it a low priority and opt instead for a larger sample size or an additional stratum in the survey. However, having a non-respondent follow-up that lets you assess the representativeness of your findings is more important than having additional confidence. If the data are not reflective of the population of interest, it is a false confidence.

9.4. SOME FINAL THOUGHTS

Many considerations go into the conceptualization, planning, design, and implementation of a credible human dimensions study. The ability to deal with those considerations makes a human dimensions research specialist a valued member of a wildlife management team; however, you (the wildlife manager) also need to be engaged in the research process from beginning to end. Do not hand off the process to someone else and expect to get useful results in a timely fashion. You can bring insight to the many considerations and decisions that will be made in designing and carrying out a good study.

SUMMARY

When a wildlife manager and a human dimensions specialist collaborate to develop a human dimensions study, they need to ask the correct questions, use appropriate methods, and execute research that meets rigorous scientific standards. Poor-quality inquiry yields poor-quality information and can result in an adverse reaction among the stakeholders, possibly exacerbating a management situation that the research was intended to help resolve.

- Wildlife managers who want to undertake a human dimensions study should ask themselves several questions: (1) Does my agency, or my organization, really need a study? (2) Does the information already exist? (3) Will the answers obtained be worth the cost of data collection? (4) Will a study build unrealistic expectations? (5) Is there enough time to complete a quality study? (6) Who needs to be involved? (7) Who should do the study?
- The human dimensions researcher also should ask questions about the context in which the study will be developed and how the results will be used: (1) Who will decide about what? (2) Who cares? (It will help to know who the primary and secondary stakeholders in the decision are.) (3) Will the study have an advisory group? (4) Who will handle media relations? (5) What is the deadline? (6) Are there other relevant studies? (7) Do people to be contacted know enough to provide informed responses, and if not, is there an information and education plan? (8) What level of statistical precision is needed? (9) Will there be follow-up research, such as a longitudinal study? (10) How will non-respondents to the survey instrument be handled?

Few wildlife managers become human dimensions specialists; mastering human dimensions inquiries requires expertise in advanced social science, with training in both theory and inquiry techniques. Special skills are needed to adapt textbook social-science methodology to the real-world complexities encountered in wildlife management. The ability to deal with these considerations makes a human dimensions research specialist a valued member of any wildlife management team.

Suggested Readings

Creswell, J. W. 2009 Research design: qualitative, quantitative, and mixed methods approaches. Third edition. Part 1: preliminary considerations. Sage, Thousand Oaks, California, USA.
Wholey, J. S., H. P. Hatry, and K. E. Newcomer, editors. 2010. Handbook of practical program evaluation. Third edition. Jossey-Bass, San Francisco, California, USA.

METHODS OF HUMAN DIMENSIONS INQUIRY

NANCY A. CONNELLY, WILLIAM F. SIEMER, DANIEL J. DECKER, AND SHORNA B. ALLRED

In Chapter 9, we discussed considerations for conceptualizing, planning, and implementing a human dimensions study. After you have established objectives for the study, found that you will need to contact stakeholders, and identified stakeholders, then it is time to select a method of inquiry.

We have two objectives in this chapter. We want you to become conversant with the particulars involved in determining the appropriate methodology for a study. We also want you to understand the strengths and weaknesses of various social-science methods commonly employed for human dimensions studies in wildlife management. If those objectives are accomplished, you can contribute to the planning and design of a study. You will be able to discuss methods, ask about the researcher's recommendations, and appreciate how the circumstances within and outside your control constrain the selection of methods.

As a wildlife manager, you will be called on occasionally (and perhaps unexpectedly) to defend studies commissioned to support decision making; you will stand in good stead if you know the reason for selecting one method over another. The information in this chapter should also help you evaluate human dimensions studies conducted by others that may affect a management issue with which you are dealing.

10.1. QUANTITATIVE AND QUALITATIVE METHODS OF INQUIRY

Wildlife managers have always sought to understand what people think and do with regard to wildlife and its management. As the profession has matured, wildlife management has become sophisticated in its approaches to incorporating human dimensions, such as in use of social science inquiry to support wildlife management.

Social scientists usually choose between two overall approaches to collecting information and creating understanding. One emphasizes quantitative techniques where, for example, you might contact a random sample of hunters to determine what proportion would favor a particular manage-

ment action. The other applies qualitative methods to help understand the complexity of human behavior and to capture details and nuances about individuals and groups. The complementary use of qualitative and quantitative methodology can be especially helpful in understanding a situation or phenomenon. Understanding the strengths and weaknesses of both quantitative and qualitative methodologies will help you understand how best to approach the wildlife management problem and context to be investigated.

Applying the standards of scientific rigor to human dimensions research has emphasized the trustworthiness of findings—their reliability, validity, representativeness, and generalizability (Vaske 2008). Although this is not a book on research methods, it is worth visiting those foundational concepts for scientific inquiry before moving to the practical issue of assessing the strengths and weaknesses of specific data-collection methods. We briefly describe the scientific criteria used in quantitative methodology and their parallels in qualitative methodology.

10.1.1. Reliability and Dependability

Reliability in quantitative methodology means consistency or repeatability of findings. It is considered a precondition for validity. Researchers questioning the reliability of a survey of fur trappers, for example, would ask themselves: If we were to conduct our study again in the same way (e.g., a mail survey), with the same sampling scheme (e.g., randomly selecting from a pool of trapping license buyers) and data collection instrument (e.g., a list of questions in a particular order), would we get the same result? How much will situational factors, vagaries in measurement instruments, or other factors affect the repeatability of results?

If a trapper filled out the questionnaire today and again tomorrow, would the answers be the same? If so, the results are reliable; if not, there is a problem. Asking the same question, in the same way, to the same person should yield essentially the same answers.

In qualitative methodology the equivalent notion to reliability is *dependability*. Emphasis is put on improving the research design and data collection method as more is learned during the inquiry. In quantitative methodology, any midstream changes in research design or measurement instrument are viewed as threats to reliability, whereas in qualitative methodology purposeful modifications that improve dependability are encouraged.

10.1.2. Validity and Credibility

Validity refers to the accuracy of quantitative data. A valid instrument is one that actually measures what you intend to measure. A human dimensions researcher, for example, might develop a scale to measure the wildlife stewardship attitudes of private landowners. The researcher would attend to several types of validity concerns:

- *Content validity:* Do the items in the scale cover all aspects (sub-domains) of the general concept of "wildlife stewardship"? Sub-domains in this example might include attitudes about personal responsibility for environmental impact, obligations to future generations, and sustainable use of wild animals.
- *Predictive validity:* Can attitudinal or behavioral differences between landowner sub-groups be predicted based on scores from the wildlife stewardship scale? For instance, will it predict behavioral differences between young and old landowners, peanut and peach farmers, or year-round residents on the land and absentee owners?
- *Construct validity:* Do item groupings identified through content analysis or some other analytic technique actually measure the construct of concern? In our example, the scale would have high construct validity if repeated use of the instrument showed that items in the scale grouped into the expected sub-domains (e.g., personal responsibility, obligations to future generations, and sustainable use of wildlife).

Put simply, validity has to do with measuring what you intended to measure, so that statements of fact based on the data are on target.

Credibility is the qualitative equivalent to validity. Qualitative methods are especially valuable when you are trying to establish what is really going on in the field, and trying to ground the research in the reality of the stakeholders rather than in your hypothesized conceptualization of their situation. The grounding is accomplished in various ways, such as prolonged engagement in the situation, persistent observation of the situation, and (to validate their accuracy) discussion of your interpretations, as they are developed, with members of the stakeholder groups being observed.

10.1.3. Representativeness

Representativeness is an important concern related to sampling in quantitative studies. Quantitative data have little value unless they accurately reflect the population of interest. Take, for example, a human dimensions researcher designing a study to characterize birdwatchers using a particular National Wildlife Refuge. To assess whether the study is likely to produce representative data, the researcher would consider questions such as these: How well will my study results reflect the population of interest (all birdwatchers who visit the National Wildlife Refuge this year)? How representative of the population is the sample of stakeholders queried or observed (does it include different sub-groups of birdwatchers, and does it sample birdwatchers at different times of the day and different seasons of the year)? Probability-sampling designs are one way to ensure representativeness of the sample and inferences about the sampled population within acceptable levels of confidence.

Qualitative methodology does not really have a parallel to representativeness, because the researcher's concern is that all perspectives be discovered and illuminated. *Inclusiveness* is a key element. Rather than making sure that people with certain traits are represented in the sample in the same proportion as they exist in the population from which the sample was drawn, the qualitative researcher tries to be sure that all views that exist in the population are captured in the inquiry. Selecting people to study is not random and probabilistic but purposeful and directed; this has implications for generalizability in the traditional sense, but the qualitative study is concerned with somewhat different attributes of the phenomenon being studied.

10.1.4. Generalizability and Transferability

Generalizability, or external validity, in quantitative methodology addresses the breadth of inferences that can be drawn from the findings of a study. You might look at a study of whale watchers in New England, for example, and ask: Can some traits of these whale watchers be generalized to all wildlife watchers in the region? Or can the characteristics of New England whale watchers be generalized to whale watchers in the Pacific Northwest?

Generalizability is enhanced to the degree it is linked to theory and validity testing (e.g., the predicted relationship or pattern of behavior is actually observed when conditions are right for it). Theory and its refinement are the foundation for development of knowledge in quantitative methodology. A trait of generalizability in quantitative methodology is that the burden of proof for claimed generalizability is on the researcher.

Transferability is a parallel concept in qualitative methodology. It is relative and depends on the degree to which salient conditions found in the situation studied match those in another situation of interest (Guba and Lincoln 1989). The trick to making transferability judgments is to make known all working hypotheses for the study and to present a thorough (extensive and careful) description of the time, the place, the context, and the culture in which those hypotheses were salient and related observations were obtained.

Determining the degree to which findings are transferable is the burden of the person seeking to apply the findings to another situation. The researcher's responsibility is to provide enough contextual detail and depth of description so that others may reasonably make the transferability determination. That often results in lengthy reports!

10.1.5. Objectivity and Confirmability

Objectivity, simply put, is the degree to which the researcher is insulated from affecting the findings of a study. In quantitative methodology, objectivity assurances are "rooted in method—that is, if you follow the process correctly, you will have findings that are divorced from the values, motives, biases, or political persuasions of the inquirer" (Guba and Lincoln 1989:243). Achieving objectivity is the process of eliminating researcher bias. It is an ideal goal that may never be perfectly achieved.

Confirmability, in qualitative methodology, accepts that it is impossible to separate the researcher from the findings. It advocates having data—facts, assertions, constructs—traceable back to their source through clear description and by fully revealing the logic that was used to organize observations into meaningful patterns. The task is to make available both the raw data and the processes used to combine them, so that explanations of the phenomena studied can be developed. The results can also be confirmed by others, perhaps by using triangulated evidence from the same or other studies.

10.2. ASSESSMENT OF QUALITATIVE METHODS

In the following sections we review general strengths and weaknesses of common methods used in studies of the human dimensions of wildlife management. We review qualitative methods first and then quantitative methods and we offer examples of situations that lend themselves to particular methods.

Several texts are available on qualitative methods and data analysis. Among them are Dey (1993), Miles and Huberman (1994), Patton (2002), and Silverman (2004). Each of these books provides in-depth coverage of the methods we briefly outline in this chapter.

Methods for qualitative data collection fall into one of four general categories: (1) in-depth, open-ended interviews, (2) facilitated small-group processes, (3) direct observation, and (4) analysis of written documents. In the material covered below we have selected focus groups as an example of the open-ended interview method. We describe the nominal group as an example of a facilitated small-group process. We describe participant observation as an example of the direct observation method.

We do not discuss qualitative review of written documents because most students are familiar with qualitative literature review (to learn more about conducting an in-depth review of research literature, see Hart [1998]). A quantitative approach to evaluating documents—content analysis—is discussed under quantitative methods.

10.2.1. Focus Groups

The focus group interview is a qualitative research method that dates from the 1920s but is currently popular in market research and elsewhere in the social sciences (Bogardus 1926, Krueger and Casey 2009).

The method has several variations and is more difficult than other social research methods to delimit. Focus groups are used most often (60% of the time in published literature across the social science spectrum) in combination with another data-collection method (Morgan 1996).

Focus groups are considered a qualitative research method because (1) a focus group is limited in size (6–10 people is typical), and (2) individual participants are purposefully selected in a non-random way by the researcher (focus groups are not established through any type of probability sampling and therefore cannot be considered statistically representative of any larger group of society). Focus groups often are used to explore a topic for purposes of gaining initial insights from which a quantitative study (i.e., mail, telephone, Web-based, or face-to-face interview survey) can be designed. Useful perspectives on focus groups can be found in Morgan (1993), Templeton (1994), Greenbaum (1998), and Krueger and Casey (2009).

Although focus groups are conducted in various ways, they typically have these characteristics:

- Each group is fairly homogeneous in interest and stakes in the topic (to encompass other stakeholders, additional focus groups would be conducted).
- A moderator has a prepared agenda of question topics but not the kinds of specific questions that would appear on a questionnaire; the focus group method allows for spontaneity and exploration of a topic or phenomenon.
- The session is typically audio-recorded and transcribed later for analysis and further use in the research process (sometimes focus groups are video-recorded).

Focus groups have been used extensively in wildlife management to reveal how stakeholders may react to specific proposals or management actions—with emphasis on the nature of the reactions and not the preponderance of one reaction versus another. Focus groups have also been used to obtain suggestions for improving services offered by agencies for their "customers" (Box 10.1).

We caution against generalizing from focus groups. They are not designed to be probabilistic or adequate samples of stakeholders. Sometimes wildlife managers find the ideas that surface in focus groups so interesting or so compelling that they ignore the limitations of the method. Avoid that tendency; it is as perilous as it is alluring.

Minnis et al. (1997) reported on the use of focus groups in human dimensions research in Michigan. One application was to gain a better understanding of the reason many deer hunters in certain areas of Michigan had organized to protest the issuance of special deer permits to farmers. From the focus groups, the researchers identified 40 variables related to attitudes and behavior of deer hunters concerning farmers' use of special permits to kill deer and reduce crop damage. The information obtained was used in the subsequent design of a quantitative study. Because of the focus group results, the researchers devoted more attention in the quantitative study to equity and administrative aspects of the special permit system than they otherwise would have done.

10.2.2. Nominal Groups

The nominal group method is a means of generating and prioritizing ideas at a meeting or workshop. It carries that name because the interaction between members of the group is limited; hence, it is a group "in name only"—or a nominal group. In a research setting the method may be used to develop ideas to be pursued subsequently in a quantitative study.

A nominal group commonly uses a facilitator and consists of four steps (Moore 1987):

1. Silent generation of ideas (in response to a question) in writing
2. Public recording and display of the ideas (typically using flip charts)
3. Discussion of the meaning of each idea, taken one by one
4. Voting to prioritize the ideas

Small groups of 5 to 9 are optimal, although up to about 12 can be accommodated. The method can also be used with more people by breaking into smaller working groups (often called breakout groups), each with a facilitator. For example, the Northeast Wildlife Administrator's Association held a workshop in 1998 on cormorant management. Thirty-three participants were divided into three groups to consider biota and natural community impacts, fisheries impacts, and human dimensions issues. The nominal group method was used exclusively in the workshop to elicit participants' concerns and suggestions (Box 10.2).

The nominal group method has been useful in studies where the breadth of stakeholders' concerns was sought. A statewide study of New York deer hunters used the method in its exploratory phase (Enck and Decker 1994). Researchers held the groups in regions of the state that were identified as having different hunter characteristics pertinent to the study. The groups revealed the likelihood that elected representatives of hunters and the hunters themselves had different perceptions of deer hunters' motivations and the satisfactions they sought. Wildlife biologists were also found to have different perspectives from hunters. The subsequent statewide study confirmed the differences that the nominal groups had revealed.

10.2.3. Participant Observation

Participant observation has been used in studies of primitive societies, sub-cultures, organizations, communities, social movements, and informal groups (McCall and Simmons 1969). Participant observation "involves some amount of genuinely social interaction in the field with the subjects of the study, some direct observation of relevant events, some formal and a great deal of informal interviewing, some systematic counting, some collection of documents and artifacts, and open-endedness in the directions the study takes" (McCall and Simmons 1969:1).

The purpose is to develop an analytic description of a social process, group, or organization. It is not usually intended to test theory. It typically seeks to describe organization that is unrecognized (and cannot be directly reported) by group members.

Participant observation is a primary tool in a research tradi-

---

Box 10.1 *A FOCUS GROUP*

STRENGTHS
- Can be assembled quickly
- Is inexpensive
- Entails face-to-face interactions, in which any questions or misunderstandings can be clarified
- Offers opportunities for deliberation of points

WEAKNESSES
- Cannot be used to generalize to a population of stakeholders
- Can be misused to make it appear more complete and conclusive than it was
- Requires a skilled and knowledgeable moderator

Box 10.2 *A NOMINAL GROUP*

STRENGTHS
- Is easy to run
- Can work well with people who do not know each other or have not worked together before
- Gives participants equal status regardless of their position outside the group
- Provides a record of the discussion when results are summarized and distributed back to participants

WEAKNESSES
- May imply a higher level of agreement than was actually reached
- Is properly used only to generate or flesh out ideas

Box 10.3 *PARTICIPANT OBSERVATION*

STRENGTHS

- Is a non-disruptive approach for exploring and describing social processes, group dynamics, and determinants of individual or group behavior
- Allows the researcher to go back and forth between analysis and observation by using new insights and research opportunities to revise analysis
- Is a means of exploring sensitive issues and studying particular social groups
- Provides insights about the context in which social change occurs and the conditions that lead to social change
- Allows the researcher to make use of personal relationships developed with key informants

WEAKNESSES

- Is labor-intensive
- Typically calls for a highly trained and skilled observer
- Can be relatively expensive because of the need to employ a high-level researcher over a long period of time
- Has high potential for observer bias
- May result in observations that are difficult to interpret and synthesize
- Produces results that are not quantifiable and can be difficult to defend to decision makers
- May lead to questions about generalizability of findings because observations focus on a small number of individuals

Source: McCall and Simmons (1969)

tion referred to as ethnography (Patton 2002). Ethnography can be summarized as direct observation of people as they go about their lives, with limited intervention or manipulation by the observer. It involves immersion in the culture or group under investigation and an effort to interpret and apply findings from a cultural perspective (Box 10.3).

Researchers have relied heavily on ethnographic methods to understand Native American communities, including their relationship to the natural environment and interactions with wildlife. For example, anthropologists such as Richard Nelson, Henry Huntington, and James Barker have provided wildlife managers with invaluable insights about some of the indigenous peoples who utilize marine mammals and migratory birds in Alaska and western Canada (Nelson et al. 1982, Nelson 1983, Huntington 1992, Barker 1993).

Although used predominantly to understand indigenous peoples and subsistence issues, the value of participant observation is not limited to those contexts. It has also been employed to learn about the depths of beliefs and cultural significance of hunting within other groups in North America. For example, Enck (1996) used participant observation in developing a theory of identity production in deer hunters, and Marks (1991) used the technique in a 7-year study of hunting culture in Scotland County, North Carolina.

10.3. ASSESSMENT OF QUANTITATIVE METHODS

Quantitative social-science methods are familiar to us all. Most of us have completed questionnaires received via the mail or the Web, and many of us have been interviewed over the telephone or face-to-face. These kinds of inquiries have specific objectives identified ahead of time and perhaps even articulated as hypotheses to be tested. An underlying assumption in this mode of inquiry is that the researcher has enough knowledge of the issue being investigated and the stakeholders from

whom information is being sought to allow the researcher to structure data collection and expect all subjects to be able to respond to a standard set of questions.

10.3.1. Face-to-face Interviews

Face-to-face interviews are data collection events in which a trained interviewer engages a subject in an in-person interview. Interviewees are usually selected to represent the stakeholders who are the focus of the inquiry. The interviewer is trained to be engaging but to avoid interaction with the subject that would introduce bias. Bias can occur inadvertently if the interviewer influences the subject's responses or interprets and records the responses in a way that the subject did not intend.

Face-to-face interviews were employed extensively between 1955 and 1985 for the national survey of hunting, fishing, and wildlife-associated recreation (USFWS 1988). In those important surveys the number and complexity of the questions and the need for visual aids as memory cues precluded other formats of data collection. The sampling strategy placed great importance on verifying that the respondent was the person selected. More recently, however, sponsors of the national survey changed the methodology. More frequent telephone interviews are used to reduce memory-recall bias, a benefit they deemed larger than the benefits of face-to-face interviews (USFWS 1993). Face-to-face interviews are still conducted with segments of the sample that are hard to reach by telephone.

Mail, Web, and telephone accessibility is sometimes a concern in other studies as well, and face-to-face interviews can ensure that certain segments of the population are not being neglected (e.g., inner-city residents, ethnic minorities, and others who have proven more difficult to reach with mail, Web, or telephone surveys).

Face-to-face interviews are well-suited for gaining specific information from users of a particular site. For example, on-

Box 10.4 *A FACE-TO-FACE INTERVIEW*

STRENGTHS

- Allows a lengthy instrument to be administered
- Has a high item-response rate because interviewees usually answer every question
- Can include complex questions
- Can include branching, depending on the answers to screening questions
- Allows the interviewer to clarify questions and probe for a more complete answer
- Allows for field observation of equipment used, game harvested, and others factors of interest
- Can include people who are not likely or able to respond to telephone, Web, or mail

WEAKNESSES

- Is expensive because of staff time and travel costs
- Requires highly trained interviewers
- May require a lot of time to reach potential respondents and complete all interviews
- Has potential for interviewer bias
- Has potential for social desirability bias (when answers are socially acceptable rather than truthful)

As part of a study of human–jaguar conflicts, Dr. Silvio Marchini interviews a landowner (top) and a ranch hand (bottom) on the Amazon deforestation frontier, in southern Amazonia, Brazil. Over 600 adults were interviewed in both rural and urban areas, in order to understand why people kill jaguars. Along with mail survey data, the interview results revealed that fear and personal and social rewards—in addition to the expected retaliation for livestock depredation—are important determinants of the human behavior of killing jaguars (Marchini and Macdonald 2012). (courtesy Silvio Marchini)

site face-to-face interviews have been the method of choice for surveys of motorists at a bighorn sheep observation point in Montana, mountain goat watchers at Mt. Evans, Colorado, bald-eagle watchers at Montezuma National Wildlife Refuge in New York, and backcountry deer hunters leaving a trailhead parking lot in Maine. On-site interviews are convenient for both parties. Memory recall problems are reduced because the activity has just taken place, and the interviewee can supply detailed information. Moreover, the interview can be coupled with other agency functions (e.g., collecting biological data on harvested animals at big-game check stations; Box 10.4).

10.3.2. Telephone Interviews

Telephone interviews share many of the strengths and weaknesses of face-to-face interviews but have additional considerations because of the remote nature of the technique. The interviewer has little idea of, let alone control over, the interview situation. The interviewer cannot be sure that the subject is the person desired or that the opinions of other household members will not be inserted instead of the subject's. Moreover, in these days of increased telemarketing, the subject may assume that the interviewer is actually a salesperson and refuse to be interviewed.

Telephone interviews can be expedient and effective for

situations such as gaining a sense of public support for an upcoming referendum or voter-initiated ballot on a wildlife management issue. Controversies normally have multiple dimensions (see Loker and Decker 1995), and telephone interviews can be used in controversial situations to gain insight into which issues are pertinent to which stakeholder groups, how people are thinking about the issues involved, and what the information needs of the voters are (Box 10.5).

Telephone interviews can reach many people quickly and provide rapid results. The rapid turnaround is desirable for game harvest surveys. They are usually straightforward inquiries; the agency simply wants an estimate of the harvest of particular species. Most state agencies engage in a harvest survey, and because the number of animals taken and the days spent hunting are usually key data, memory recall problems can be minimized by using a telephone survey immediately after the period for which harvest data are desired.

With the increasing use of cell phones and particularly the loss of land lines, surveys of the general public using samples of land-line users are becoming more biased toward older respondents. This bias will increase as more people drop their land lines altogether. Researchers are working on methods to overcome this bias; these methods should be discussed if a sample of the general public is to be surveyed by telephone.

Telephone interviews can augment other primary methods of data collection, as part of an overall research design. They are used for screening to make a preliminary determination of which people to consider in the larger inquiry. For example, telephone interviews were used to identify New Yorkers with an interest in wildlife education programs (Connelly et al. 1995). Those with an interest were sent a more detailed mail questionnaire asking about past program experiences and preferences for specific program elements. Telephone surveys also are used as a follow-up to mail surveys, to reach non-respondents and assess non-response bias.

10.3.3. Mail Surveys

Mail surveys have long been the standard tool of the human dimensions researcher. The wildlife literature contains hundreds of reports based on mail surveys—some good and some poor. Anyone can type out questions, get copies made, insert them in envelopes, and mail them, so many untrained people believe that anyone can do a mail survey. Such surveys seldom produce valid or reliable results.

Even for experts in mail surveys, development of quality instruments and effective implementation of surveys can be difficult. Because of the amount of unsolicited mail people receive, they have become less tolerant of mail inquiries. It takes special attention to all aspects of mail survey design to be successful these days (Box 10.6).

Mail surveys are unsuitable for some applications but are adaptable and effective when the conditions are right for their use. Studies of various wildlife-user populations (hunters, trappers, and wildlife watchers of all sorts) have used mail surveys. Much of what we know about those and other stakeholders (such as farmers, ranchers, homeowners, motorists, and others) has been acquired through mail surveys.

The wildlife management profession is fortunate that stakeholders have been willing and able to provide complex data about important management topics. The data have covered many topics, such as motivations for participating in various wildlife-related activities and satisfactions derived from them, beliefs about wildlife and attitudes about human–wildlife interactions, and preferences for management actions in various situations.

The recipient of a mail questionnaire can take as much time as desired to reflect on responses and to consult records; that leads to confidence in the data collected. The respondent can also be led through a series of questions that encourage consideration of the positive and negative traits of a species (e.g., Canada geese, white-tailed deer, or beaver) in a particular situation (e.g., suburban residential area or public municipal parklands), before reporting a preferred course of action for management.

Although response rates for surveys of the general public tend to be low and require extensive non-respondent follow-ups to assess non-response effects, there are many wildlife management issues in which stakeholder interest is high and response rates are satisfactory. (It is not uncommon to achieve response rates of 60% on well-constructed mail questionnaire surveys on wildlife issues of local importance.)

10.3.4. Web-Based Surveys

Web-based surveys are gaining popularity as a mode of survey data collection. Web surveys can be an ideal method for data collection for populations that have Internet access and technical proficiencies. Advantages of Web surveys are lower costs compared with mail, telephone, and in-person survey methods, as well as the timeliness with which data can be collected. Web surveys also allow branching patterns to be automated, which increases the ease with which surveys can be completed. Web surveys are not without their disadvantages, however. Survey mode experiments consistently show that, when given a choice, individuals prefer to complete surveys by mail (Millar et al. 2009). It also is often difficult to obtain e-mail samples that provide complete coverage of the sample population. It is important to tailor the choice of survey mode to the population being surveyed (Box 10.7).

Design features are critical to Web surveys. First, it is recommended that contacts be personalized because it establishes a connection between the researcher and the respondent that is necessary to invoke social exchange (Dillman et al. 2009). This also reduces the likelihood of survey invitation and reminders being mistaken for spam e-mail. When implementing Web surveys, you should use multiple contacts (3–5 contacts), vary the message across contacts, and keep the e-mails concise (Dillman et al. 2009). The content of the survey invitation should appeal to potential respondents, ensure data confidentiality, explain what is being asked of respondents, clarify the reason that their participation is important, provide clear instructions on how to access the survey, and indicate who they can contact if they have questions (Dillman et al. 2009). When planning a Web survey, be sure to provide adequate time between reminders for respondents to complete the survey

Box 10.5 *A TELEPHONE INTERVIEW*

STRENGTHS
- Can be implemented quickly
- Is highly conducive to branching (with computer-assisted interview instruments)
- Provides more control over who answers questions than a mail survey
- Has a higher cooperation rate than a mail or Web-based questionnaire (but lower than a face-to-face interview)
- Can be implemented with a geographically dispersed group

WEAKNESSES
- Must include questions that are brief and easily understood
- Usually must take only a few minutes to administer
- Has some potential for social desirability bias
- Requires highly trained interviewers willing to work evenings and weekends
- Must consider the potential impact if cell phones are not included in the sample

Staff at the Cornell University Survey Research Institute complete telephone interviews for a study of coyote-related risk perceptions (photos courtesy William Siemer)

(a few days to 1 week apart with surveys scheduled to arrive early in the morning). A non-trivial consideration in Web surveys is what is selected for the "sender name and e-mail" and "subject line" of the e-mail. The sender name should convey to respondents that the message is a formal request from a reputable sender. The subject line of the e-mail should also convey professionalism and key information. Requests for help are often more effective than offers to participate (Trouteaud 2004). For example "Please help X organization with your advice and opinions" is more effective than "Share your advice and opinions now with X organization."

In addition to the design elements discussed above, several techniques can be used in Web surveys to encourage response. The use of incentives in Web surveys has been found in some studies to increase response rate significantly (Tuten et al. 2004, Heerwegh 2006). The incentives that are most effective are drawings for a prize (Bosnjak and Tuten 2003) or cash lotteries. Additionally, informing respondents that they will receive a reminder e-mail if they fail to complete the Web survey increases response rates and has not been found to have a negative effect on data quality (Klofstad et al. 2008).

Respondents are more inclined to complete the Web survey if researchers communicate their desire to send as few contacts as possible. Another method to increase response rates is to send a reminder postcard by postal mail in addition to the e-mail invitation and reminders (Kaplowitz et al. 2004, Broussard Allred and Smallidge 2009). Web-survey response rates rival those of mail surveys when a notice letter is provided by postal mail (Kaplowitz et al. 2004).

10.3.5. Behavioral Observation
Direct observation of animal behavior has long been used as a wildlife research technique, but behavioral observation of people has been employed infrequently to shed light on wildlife management issues. Whether focused on people or wildlife, the technique calls for objective and precise description, recording, and interpretation of observed behavior. Several methods are used. Informal, qualitative methods include narrative description, diary description, and anecdotal records. Each of those methods has its own set of strengths and weaknesses, described elsewhere (Bentzen 1997) (Box 10.8).

Time sampling and event sampling are controlled tech-

Box 10.6 *A MAIL SURVEY*

STRENGTHS

- Can include complex questions
- Can be implemented to a geographically dispersed group
- Allows respondents to reply at their convenience, which results in better memory recall (they can verify the information)
- Has low potential for social desirability bias

WEAKNESSES

- Can include only a limited amount of branching
- Raises problems of non-response bias
- Takes a long time—usually 8 weeks—before all responses are in
- Provides no opportunity to explain questions
- Does not provide certainty about who actually completed the questionnaire
- Does not give the researcher complete control over the order in which the questions are answered

Mail surveys offer a cost-effective technique for conducting quantitative research with large audiences. Declining response rates have prompted researchers to explore alternative data collection methods, such as Web-based surveys. (courtesy William Siemer)

niques used in behavioral observation research. When using these techniques, the observer records only specific behaviors selected for observation and carefully defined ahead of time. The observer typically records characteristics such as the presence or absence, frequency, or duration of specific, predetermined behaviors (Bentzen 1997).

A time-sampling approach would be useful, for example, to quantify the presence or absence, frequency, or duration of visitor behaviors associated with disturbance of shorebird nesting or feeding at a specific site. An event-sampling approach may be more useful to characterize hunter behaviors (e.g., specific shooting behaviors when a waterfowl hunter has an opportunity to shoot at passing birds).

Observing people from afar as they engage in activities,

such as beach use where piping plovers nest or river use where brown bears gather to feed on salmon, can help wildlife managers identify actions that would minimize user impacts on wildlife. Researchers often find that the impacts on wildlife vary by activity, by wildlife species, and even between individual animals within a given species. For example, observation of birdwatchers on wildlife refuges has led to an understanding of their effects on various bird species' use of habitats; some individuals are reclusive and shy, and others are tolerant of human presence and approach (Klein et al. 1997).

For some situations, behavioral observation is the best way to gain the insight needed to understand the human dimensions of a situation. Direct observation has been used in sophisticated and surreptitious ways for understanding poaching and other illegal activity. Though such covert activity should be undertaken only by those trained for that kind of investigation, the expertise of social scientists has been enlisted to shed light on the beliefs of individuals, groups, and communities that lead them to engage in illegal or unethical behavior.

10.3.6. Social Network Analysis

Social network analysis focuses on relationships between individuals, not on individual characteristics (Prell et al. 2009). Characteristics of relationships are analyzed for structural patterns that emerge among individuals, how individuals are positioned within a network, and how relations are structured into overall network patterns. Social network theory and analysis are just emerging as analytic tools for the human dimensions of wildlife management. We envision that it will be very useful in the future in the areas of stakeholder interactions, human–wildlife conflict, and governance.

10.3.7. Content Analysis

Robert Weber, author of many publications on content analysis, provides a succinct definition of this technique: "Content analysis is a research method that uses a set of procedures to make valid inferences from text. These inferences are about the sender(s) of the message, the message itself, or the audience of the message" (1990:9). Human dimensions researchers can use content analysis for many purposes, including coding open-ended responses in questionnaires, synthesizing public input for decisions, describing trends in communication content, or comparing a communication's content with its objectives.

In recent years, multiple software packages have been developed for the specific purpose of quantitative or qualitative content analysis. New software is being developed at a rapid rate, which provides powerful analysis tools to a broader audience of researchers and information managers.

Quantitative analysis software programs (e.g., *General Inquirer, VBPro*) are designed to provide a rapid analysis of large volumes of text (e.g., thousands of public comments on a pro-

---

Box 10.7 WEB-BASED SURVEYS

STRENGTHS
- Cost savings associated with eliminating printing and mailing
- Time saved by eliminating manual data entry
- For populations using the Internet, Web surveys are a useful way of collecting data
- Branching patterns can be automated into the survey rather than relying on respondent to follow instructions
- Responses can be received quickly

WEAKNESSES
- Web surveys of the general public still produce significantly lower response rates compared with mail surveys
- When given a choice, mail surveys are still the preferred survey mode of respondents
- Sometimes difficult to obtain e-mail samples
- Not a good choice for populations with low Web proficiencies or Internet access

---

Box 10.8 BEHAVIORAL OBSERVATION

STRENGTHS
- Provides quantitative data about behavior that could not be collected through surveys or interviews
- Allows the researcher to collect data on a large number of individuals (using sampling techniques)
- Provides direct information about human behaviors of interest
- Provides a means of testing new behavioral theories and better understanding existing ones

WEAKNESSES
- Requires field staff that have extensive training
- Has high potential for observer bias
- Is time-consuming and expensive
- Often results in loss of contextual details because a sampling approach is used

Box 10.9 *CONTENT ANALYSIS*

STRENGTHS
- Provides insights about communication and relationships between groups
- Allows for exploration of social and cultural change over time
- Is unobtrusive and does not interfere with ongoing communication between groups
- Can be used to provide qualitative data, quantitative data, or both

WEAKNESSES
- May not capture the full complexity of rich text passages
- May overlook important information in text passages when techniques such as searches for key words and phrases are used
- May not be able to account for different meanings or emphasis of words at different places in a text passage
- Often results in multiple interpretations of text, which can limit the researcher's ability to use content analysis to test theoretical constructs

Source: Weber (1990)

posed regulation change). These programs are designed for analysis of short units of content (e.g., a word or phrase). They offer an easy mechanism to calculate word frequencies, word clusters, occurrence of word pairs, group or phrase category frequencies, and concordance (an alphabetical list of words used in a document and their context of use). Coding is automatic (researcher judgment is needed to interpret findings, but not to create coding categories).

Qualitative analysis software (e.g., *Atlas-ti, Nvivo*) offers tools to analyze large units of content (e.g., text passages, paragraphs, open-ended survey questions, entire comments or letters). These software packages allow multiple coding of text passages that contain more than one theme of interest to the researcher. The strength of qualitative software is that it allows for subtler and deeper analysis than simple word counts. Such packages are typically costly and complex. They require coder judgment to create units of analysis and coding categories. Purchasing the software and training personnel in its use is generally not cost-effective for small projects or one-time use. Software developers (e.g., *SPSS*) are beginning to offer packages that automate some of the coding process, which is likely to encourage wider use of content analysis to inform natural-resource management decisions (Box 10.9).

Content analysis has had a useful and unique role in providing historical perspective on changing attitudes toward wildlife and wildlife management. Careful document analysis allows the researcher to take a retrospective approach to exploring change in public attitudes over long periods of time. For example, Kellert and Westervelt (1983) used content analysis to trace changes in Americans' attitudes toward wildlife over much of the twentieth century. In a more recent example, Goedeke (2004) analyzed content of scientific and popular literature since 1873 to characterize how public perceptions of Florida manatees have changed over the late nineteenth and twentieth centuries.

Content analysis also can be a valuable tool to characterize public discourse about wildlife management issues. By analyzing content in mass media outlets, researchers can help

wildlife managers understand how those media both reflect public attitudes and influence public actions (e.g., voting behavior). Analysis of media content has been used repeatedly to understand public discourse about carnivore management in the United States and Europe. For example, Wolch et al. (1997:97) analyzed newspaper stories in the Los Angeles Times "to document how coverage of cougar-related issues in southern California's major newspaper shifted between 1985 and 1995." Kaczensky et al. (2001) analyzed newspaper and magazine articles published in Slovenia from 1991 to 1998 to understand public discourse when depredation on domestic sheep was increasing sharply. Zeiler et al. (1999) analyzed content from three prominent Austrian hunting magazines between 1948 and 1996 to assess change in hunters' attitudes toward European brown bears and lynx. Siemer et al. (2007b) analyzed print media content in New York State between 1999 and 2002 to characterize media framing of black bear–management problems, management solutions, and attributions of responsibility for solving bear-related problems.

Content analysis has also been used to help understand how the content of communications influences attitudes toward wildlife management issues. For example, Liu et al. (1997) used content analysis of 124 newspaper articles published in a 5-year period to help understand the "cultural / informational framework" of residents living adjacent to the Key Deer National Wildlife Refuge in Florida. Their analysis helped document that "local media tend to frame environmental issues as conflicts of interest between environmentalists and developers" and that "structure factors affecting both human and Key Deer survival, such as population density, available land, and water pollution and other forms of degradation, are almost never depicted as factors creating common fate for both humans and deer" (1997:113–114).

10.3.8. Secondary Data Analysis

Secondary data are not lower quality or flawed data (unlike "seconds" in merchandise). The term simply refers to preexisting data. Secondary sources of data have been collected by others and archived in some form. These sources include government reports, industry studies, and information in books and journals found in libraries. Secondary information about stakeholders sometimes offers relatively quick and inexpensive answers to questions of interest to managers.

Agencies often find that by mining the wealth of existing quantitative and qualitative data (e.g., data on hunting, fishing, trapping, boat license sales; data on hunter-training courses; internal agency documents and records; Bureau of Census data; market research; the U.S. Fish and Wildlife Service national survey; the U.S. Forest Service national recreation survey), they can avoid the typically greater expense of a primary study. Synthesis and reanalysis of data can often answer new questions with information collected for another purpose.

Secondary research can be used as a substitute for, or a complement to, primary research. At the very least, a thorough review of existing data is the best way for a researcher to identify high-priority information gaps that may be closed through original research. It can provide a starting point by "suggesting problem formulations, research hypotheses, and research method" (Stewart 1984:14; Box 10.10).

Secondary data analysis has been used powerfully in wildlife management decision making. One application is the development of insight about the relationships between socioeconomic and demographic trends and trends in hunting and fishing recreation (e.g., trends in license sales) at the state level. Regression models using various state-level census data and state license sales data over long periods have helped wildlife managers predict program demand and make revenue projections (Brown and Connelly 1994).

10.3.9. Program Evaluation

Program evaluation is not a method per se, but is a useful application of methods described above that can give managers needed information about program impact and can help assess progress toward stated goals. Program evaluation is the activity of systematically collecting, analyzing, and reporting information that can be used to improve the operation of a project or program or the assignment of worth or value according to a set of criteria and standards (Alkin 1990). In a basic sense, program evaluation is comparing what *should be* (criteria, standards, goals, information) to *what is* (evidence, data, information). Program evaluation can be thought of in three distinct steps: focusing the evaluation, collecting and analyzing the data, and using the information (Taylor-Powell et al. 1996). There are two primary types of evaluations. Formative evaluations are conducted to improve programs and summative evaluations are conducted to determine the effectiveness of programs (Patton 2002). It is important to consider the purpose of the evaluation, how you propose to use the information, who will be involved, and whether you will use an internal evaluator or an external evaluator. Next, managers should develop "evaluation questions," which are questions that the evaluation will answer; they are outcomes phrased as a question or set of questions (Posavac and Carey 2006). Examples of evaluation questions include the following:

Does the program or plan match the values of stakeholders?
Does the program or plan match the needs of the people to be served?
Does the program as implemented fulfill the plans?
Do the outcomes achieved match the goals?
Is there support for the program theory?
Is the program accepted?
Are the resources devoted to the program being expended appropriately?

In collecting data for evaluations, measures can be related to performance, process, or both. Types of performance measures are outcomes, efficiency, outputs, and inputs. For those interested in evaluating process, it can be helpful to include measures such as capacity, elements of operations, program participation, and types of service programs. Data can be collected from recipients of programs, program staff or providers of services, observers, and non-users of programs.

Program planning and evaluation models are useful tools that often visually represent the connections between educational and environmental goals. Logic models have been utilized as a way to organize and plan programs and program evaluations. Logic models are a picture of your program or

---

Box 10.10 *SECONDARY DATA ANALYSIS*

STRENGTHS
- Is less expensive than primary data collection
- Takes less time than primary data collection
- Is likely to provide higher quality information than quick-and-dirty primary research
- Increases research efficiency by targeting gaps in knowledge before original research is conducted
- Identifies existing data that can be compared with new data to reveal differences and trends
- Can be used to verify the representativeness of new data

WEAKNESSES
- May be based on a study in which the design or conclusions were flawed
- May make use of a study in which category definitions and means of data aggregation are not well-suited for the new purposes
- May be based on data that are too old to be useful for the new purposes
- May not lead to useful comparisons because of variations in data collection

intervention (a graphic representation of the "theory of action"—what is invested, what is done, and what results; Taylor-Powell et al. 1996). Components of logic models include

Inputs: resources, contributions, investments that go into the program;

Outputs: activities, services, events, and products that reach people who participate or who are targeted;

Outcomes: results or changes for individuals, groups, organizations, communities, or systems;

Assumptions: the beliefs we have about the program, the people involved, and the context and the way we think the program will work; and

External factors: the environment in which the program exists includes a variety of external factors that interact with and influence the program action.

It is also important to consider program ethics in conducting program evaluations. Standards for the practice of evaluation are systematic inquiry, competence, integrity–honesty, respect for people, and responsibilities for general public welfare (American Evaluation Association; www.eval.org).

Managers can utilize evaluation results to improve, better understand, and adapt programs. Examples of potential uses include developing an outreach or communications plan; making changes in and / or improving programs; knowing "what works"; and demonstrating impact of program to legislators, external funders, or internal agency or organization decision makers.

10.4. USING QUALITATIVE AND QUANTITATIVE METHODS IN COMBINATION

Qualitative and quantitative methods can often be matched up in a complementary fashion. Researchers call that a mixed-method approach and they use it to achieve any of several broad purposes (Greene et al. 1989). In the field of human dimensions, mixed methods have probably been used most often for three purposes:

1. *Development:* to inform the development of a future study
2. *Complementarity:* to enhance or clarify the results of another study
3. *Triangulation:* to increase the validity of findings by corroborating the results in different ways

Development. Wildlife managers regularly encounter situations in which they need to understand a new stakeholder group or a new issue for a traditional stakeholder group. When you want to know what proportion of a stakeholder group would support alternative management actions to address an issue, then a quantitative survey is called for; but when the problem is new and previous experience is not relevant to the current situation, then you may not want to make assumptions about what values are operating and how they might manifest themselves. Under those conditions, it is too soon to offer alternative actions in a quantitative survey; you should do some preliminary qualitative work first.

Human dimensions researchers regularly engage in qualitative interviewing with some members of the stakeholder groups that they will eventually be surveying more comprehensively. The interviews are typically open-ended. They may be conducted individually, or they may take the form of a nominal group or focus group (those techniques are described above). Preliminary qualitative interviews are intended to encourage people to reveal—in their own words and way of expression—how they feel about an issue. Some social scientists refer to such preliminary interviews as elicitation studies. Their purpose is to *elicit* views, concerns, interests, or preferences, not to verify preconceived categories of responses. Some general questions are used to guide the interview, but seldom does the researcher assume any response option a priori. Random samples are not used; instead the researcher seeks out people who will give the widest range of responses. Remember, the interviews are conducted to enrich understanding of the scope of an issue (as perceived by the stakeholder groups) and not to yield generalizable statistics.

For development purposes, qualitative and quantitative methods are applied sequentially. First, qualitative methods are used to scope out the issue and learn the vernacular of stakeholders. With that understanding of how stakeholders perceive the content of the issue and communicate about it, the researcher then can develop a questionnaire. A quantitative instrument is likely to be more reliable and valid because of the initial qualitative work done as part of its development.

Complementarity. Sometimes the wildlife manager and human dimensions researcher have good knowledge of a stakeholder group and feel confident about proceeding in the development of a quantitative survey; however, even with carefully prepared hypotheses based on experience and a thorough literature review, unexpected patterns can emerge from a quantitative study.

When that happens, it is logical to question the validity of one's assumptions about the stakeholder group surveyed. Further investigation is then required; but this time the qualitative study follows, rather than precedes, the quantitative study. It may be possible to unravel the mysterious findings in only a dozen open-ended interviews with thoughtfully selected representatives of the stakeholder group that was surveyed (Box 10.11).

Triangulation. Triangulation in the social sciences is achieved by exploring a social phenomenon from more than one methodological perspective (Webb et al. 1966). Using different methods to corroborate the findings of one study with those of another study gives greater confidence in the validity of the findings (see Box 10.11). For example, if a set of focus groups, qualitative interviews, or public meetings suggested that acceptance of a particular deer-management action was associated with gender of the respondent, we would have greater confidence in those findings if a quantitative mail or telephone survey found a similar relationship.

Human dimensions researchers use multiple data sources to strengthen their assertions about the social world, much as

Box 10.11 *USING MIXED METHODS IN YELLOWSTONE NATIONAL PARK*

Researchers at the University of Montana used quantitative and qualitative methods to develop an understanding of the recreational experiences of winter visitors to Yellowstone National Park.

A quantitative study was conducted first to examine visitors' motivations, satisfactions, values, and support for various park management actions. The results revealed some misperceptions about the Yellowstone National Park experience. It described park visitors and illustrated the factors that they considered important to their experience.

The second study, a qualitative exploration of visitor perceptions of the park's winter setting, social conditions, environmental degradation, and management change, provided depth in understanding the meanings and relationships behind those perceptions.

The two studies allowed the researchers to link aspects of the environment to the experience and examine how visitors contemplate their support for management change. Use of quantitative and qualitative approaches in combination allowed the researchers to clarify the reasons that encounters with wildlife are central to the winter recreation experience in Yellowstone National Park.

The qualitative study deepened understanding by providing details and richness that could not have been captured through the quantitative study alone. Corroboration of some findings through studies using different methods also gave researchers greater confidence in their depiction of the winter recreation experience in the park.

Source: Davenport et al. (2002)

Winter recreation in Yellowstone National Park (courtesy USFWS)

trial lawyers use multiple sources of corroborating evidence to support their legal arguments.

10.5. RESEARCH—IT'S NOT JUST FOR PROFESSIONALS ANYMORE

In Chapters 3 and 11, we describe how professional wildlife management has evolved and grown in response to a wider range of stakeholders and pressures for greater stakeholder involvement in every aspect of decision making. One response to those pressures has been an increase in co-management arrangements, wherein wildlife management agencies share responsibilities with stakeholder representatives, local government, and private entrepreneurs.

Research responsibilities are among the duties that stake-

holders are sometimes willing to share. Direct stakeholder involvement in setting the research agenda and executing research is called participatory action research (Greenwood and Levin 1998).

Participatory action research was developed primarily by the international research community working on agricultural, agroforestry, and conservation projects in developing countries. It strives to take advantage of indigenous knowledge, and it promotes adoption of applied research findings by involving key stakeholders in the process of conducting research.

Another method of engaging stakeholders in research is through citizen science projects. These projects, such as the Audubon Society's Christmas Bird Count or Cornell Lab of Ornithology's many citizen-science projects (http://www.birds.cornell.edu/netcommunity/citsci/about), are designed to

Box 10.12 *CITIZEN SCIENCE AT CORNELL—BIRDS, SCIENCE, AND COMMUNITY*

Citizen science at The Cornell Lab of Ornithology has grown dramatically in the past decade; it came into its own as a scientific methodology while striving to integrate science

education with the important goals of getting families outdoors, promoting critical thinking, building conservation communities, and promoting environmental attitudes

Adult and youth participants collect and submit data on their bird observations as part of citizen science inquiries administered by the Cornell University Laboratory of Ornithology (right: courtesy Cornell Laboratory of Ornithology, Diane Tessaglia-Hymes; opposite: courtesy Cornell Laboratory of Ornithology, Susan Steiner Spear)

involve non-scientists in collecting scientific data, which makes large research projects more feasible than would otherwise be possible. The projects promote civic engagement in science and contribute to informal science education (Box 10.12).

Many challenges face stakeholders and wildlife management professionals who choose to use these modes of inquiry, including the need to guard against various sources of bias that can reduce the credibility of research findings. Nevertheless, we foresee exciting roles for wildlife managers and human dimensions researchers who work in collaboration with stakeholders to generate valid and reliable scientific information and create a more informed citizenry.

10.6. WHERE TO FIND MORE INFORMATION

Because social and behavioral science requires specialized training if it is to be done well, and because there are specialists for different kinds of social science inquiry, we do not want to leave you with the impression that taking any single course or reading a book or two would suffice for you to undertake such inquiry on your own. Instead, these modest exposures to social and behavioral science can help you gain an appreciation of the methods wildlife managers are involved with as they work with others to explore the human dimensions of various wildlife issues and seek stakeholder input for decision making.

The following are suggestions about the kinds of courses or references you may find useful.

10.6.1. Courses to Consider
Courses in biologically oriented sampling and statistics cover useful concepts, but many aspects of the design and implementation of human dimensions studies require information best obtained from courses in social science departments. Design and implementation of mail, Web, and telephone surveys (sometimes called survey research methods), public attitude measurement, and qualitative social-science methods are taught in various departments: government, political science, communication, sociology, development sociology, public policy, education (e.g., program evaluation), and business (e.g., market research).

In addition, sampling courses taught in social science departments will emphasize how to draw samples of various human populations and where to obtain samples of various publics. Statistics courses will emphasize univariate and multivariate statistics most often used to analyze data obtained from human populations.

10.6.2. Human Resources
Courses and published references are excellent places to find background information about methods for human dimen-

and stewardship behaviors. It is built around John Falk's concept of free choice learning and engages a self-selected public of all ages to identify and observe birds and their nesting behavior, follow protocols, and enter data online where they are housed in a large, archival, relational database. New Internet tools strive to promote inquiry through dynamic visualizations that literally allow people to "play" with the data.

The proposed impacts of citizen science at Cornell

are broad and evaluation researchers have barely begun to touch upon them. Evaluation of volunteers engaged in projects that involve watching and counting birds at feeders or monitoring nests has demonstrated that citizen science participants gain knowledge of bird biology and engage in scientific habits of mind but do not necessarily develop new attitudes and behaviors, a better understanding of the scientific process, or skills at inquiry. This suggests that attitude change and inquiry are not natural outcomes of nature experiences or data collection even when teachers are involved. Learners may require additional support to develop inquiry skills, perhaps through training of mentors.

JANIS L. DICKINSON
Sources: Brossard et al. (2005); Falk (2005); Trumbull et al. (2000, 2005)

sions research. Nevertheless, there is no substitute for talking with an experienced, skilled, social science practitioner.

Agencies and organizations in some states are fortunate enough to have in-house social science expertise. That is not the norm, however, and even when there is in-house expertise wildlife managers may find that the demands for the expertise are so great that they do not always have access to it on the schedule and to the degree they need. The manager must then turn to external sources.

The primary sources of external expertise are universities and consulting companies. In Chapter 9 we mentioned some considerations associated with each option. A growing number of state universities have faculty members who specialize in human dimensions of wildlife management, and virtually all land-grant universities have applied social science researchers with expertise in survey research methods. At many other colleges and universities one can find faculty members with expertise in social science methods. They may not have any background in wildlife management, but many have the skills to work with wildlife professionals to design a quality survey. The wildlife manager will need to make a commitment to stay engaged throughout the research process, from initial conceptualization to final report.

Consulting companies that do opinion polling seldom publish in peer-reviewed literature, so it is difficult to judge

quality. It is therefore important to check a company's client references. Many companies, especially those that operate nationally, do a good job of carrying out a survey and getting results to the client expediently. Fewer companies can design a survey, and even fewer have any background in natural resources.

If you engage an external consultant, whether private or affiliated with a university, you should shop around. Check references and ask about quality, timeliness, and commitment to wildlife management research. Avoid social scientists that have a history of crossing the line between the responsibilities of being a researcher and being an advocate for a particular policy.

Try to find someone with whom you may cultivate an ongoing professional relationship. You are making an investment to bring the researcher up to speed with regard to agency culture, policy, and processes, and it makes sense to have that investment pay back over more than one study. A researcher who is knowledgeable about the agency's purposes and culture, personnel, stakeholders, and management context will be able to design better studies that yield more valuable results for the expenditures made. Trust can develop and a partnership approach may emerge in such cases, which has proven beneficial (Decker and Mattfeld 1995, Mattfeld et al. 1998).

Develop a relationship with a researcher who is genuinely

concerned with helping you be personally effective and helping your agency or organization to accomplish its mission.

SUMMARY

Human dimensions professionals use many kinds of quantitative and qualitative research methods. Standard social science approaches to inquiries were reviewed in this chapter. Although this chapter was not intended to be a stand-alone treatise on methodology, we hope you gained an understanding of the strengths and weaknesses of social science methods commonly used in studies of the human dimensions of wildlife management.

- Researchers collect qualitative data through in-depth interviews, direct observation, written documents, and organized meetings. Focus groups, participant observation, and the nominal group method are three commonly used qualitative approaches.
- The most commonly used quantitative approaches for wildlife management applications are (1) face-to-face interviews, (2) telephone interviews, (3) mail surveys, (4) Web-based surveys, (5) content analysis, and (6) secondary data analysis.
- Representativeness is the degree to which the data collected represent the population of interest. Qualitative studies do not really have a parallel to representativeness. Qualitative researchers strive for inclusiveness, which refers to the inclusion of all viewpoints that exist in a given population.
- *Validity* refers to the accuracy of quantitative data. *Credibility* is the qualitative equivalent to validity. *Generalizability*, or external validity, refers to the breadth of inferences that can be drawn from the findings of a quantitative study. *Transferability*, a parallel concept applied to qualitative studies, refers to the degree to which salient conditions found in the situation studied match those in another situation of interest.
- *Objectivity*, or *confirmability* in the case of qualitative research, is enhanced when standards of scientific rigor are met. *Reliability* refers to the consistency or repeatability of findings from quantitative studies. *Dependability* refers to the stability of observations from qualitative studies.

- Agencies and organizations in some states are fortunate enough to have in-house social science expertise but that is not the norm. Even when there is in-house expertise, wildlife managers may find that the demands for the expertise are so great that they do not always have access to it on the schedule and to the degree they need. In this case, managers should turn to external sources.
- You may not always feel completely prepared to conduct human dimensions inquiries on your own. A growing number of state universities have faculty members who specialize in human dimensions of wildlife management, and virtually all land-grant universities have applied social science researchers with expertise in survey research methods. Consulting firms also exist with expertise in social sciences, especially in survey design and polling. Avoid social scientists who cross the line between the responsibilities of being a researcher and being an advocate for a particular policy.

Suggested Readings

Dillman, D. A., J. D. Smyth, and L. M. Christian. 2009. Internet, mail, and mixed-mode surveys: the tailored design method. Wiley and Sons, Hoboken, New Jersey, USA.

Fishbein, M., and I. Ajzen. 2009. Predicting and changing behavior: the reasoned action approach. Psychology Press, London, England, U.K.

Kish, L. 1995. Survey sampling. Wiley, New York, New York, USA.

Miles, M. B., and A. M. Huberman. 1994. Qualitative data analysis: an expanded sourcebook. Second edition. Sage, Thousand Oaks, California, USA.

Oskamp, S., and P. W. Schultz. 2005. Attitudes and opinions. Third edition. L. Erlbaum, Mahwah, New Jersey, USA.

Salant, P., and D. A. Dillman. 1994. How to conduct your own survey. Wiley, New York, New York, USA.

Scheaffer, R. L., W. Mendenhall, and L. Ott. 1990. Elementary survey sampling. Fourth edition. PWS-Kent, Boston, Massachusetts, USA.

Silverman, D. 2004. Qualitative research: theory, method and practice. Second edition. Sage, London, England, U.K.

Sirkin, R. M. 1999. Statistics for the social sciences. Second edition. Sage, Thousand Oaks, California, USA.

Vaske, J. J. 2008. Survey research and analysis: applications in parks, recreation, and human dimensions. Venture, State College, Pennsylvania, USA.

Weinberg, S. L., and S. K. Abramowitz. 2002. Data analysis for the behavioral sciences using SPSS. Cambridge University Press, New York, New York, USA.

STAKEHOLDER ENGAGEMENT IN WILDLIFE MANAGEMENT

T. BRUCE LAUBER, DANIEL J. DECKER, KIRSTEN M. LEONG,
LISA C. CHASE, AND TANIA M. SCHUSLER

The state of Nebraska, like all U.S. states during the first few years of the twenty-first century, was faced with the federally mandated task of developing a comprehensive plan for conserving its wildlife resources—a State Wildlife Action Plan. The challenges it faced in trying to do so were considerable. The plan had to address the needs of all wildlife and habitat within the state, so it required significant resources to develop; however, the Nebraska Game and Parks Commission, like many state fish and wildlife agencies, already had its hands full with existing responsibilities. It simply did not have the capacity to complete such an effort on its own. Because Nebraska's landscape is dominated by private lands, the state would need to win the support of landowners to implement its plan successfully. Conservation would have to be voluntary and incentive-based rather than top-down and regulatory.

Stakeholder engagement at multiple scales helped the Nebraska Game and Parks Commission solve both these problems. At the outset of the process, the commission formed a project planning team consisting of federal agencies, other state agencies, and a wide spectrum of private interests (ranging from conservation to agriculture). This team oversaw the development of the plan, provided much of the needed expertise for its development, and helped to build support for it throughout the state. The work of the team was complemented by a series of regional workshops for conservation practitioners to identify priority conservation areas. The scientific expertise and local knowledge of the practitioners contributed to the designation of 40 "biologically unique landscapes" (or BULs) throughout the state.

When the BULs were identified, the Game and Parks Commission had to approach conservation in each in a way that landowners would support. In 12 BULs that were considered particularly high priorities, locally based stakeholder groups were created to develop objectives and strategies that their community would accept. Some of these groups have been remarkably successful and have even established commitments to work toward conservation of species such as beetles (not a typical priority for many private landowners!).

This example shows the power of stakeholder engagement, which often provides the foundation for successful wildlife management. We cover basic information every wildlife manager should know about stakeholder engagement in this chapter. After defining the term, we discuss the reasons for involving stakeholders in management. Next we discuss six primary approaches to stakeholder engagement and describe some specific stakeholder engagement techniques. We then briefly discuss some of the challenges to engaging stakeholders effectively. Finally, we share advice about how you can learn more about stakeholder engagement.

11.1. WHAT IS STAKEHOLDER ENGAGEMENT?

Stakeholder engagement has become a regular feature of wildlife management. People use the term to mean a wide range of activities, and not all people define it the same way. We use the term *stakeholder engagement* to refer to the process of involving and interacting with stakeholders in the making, understanding, implementing, or evaluating of management decisions for improved wildlife management. Recall that a *stakeholder* is anyone who significantly affects or is significantly affected by wildlife or wildlife management.

Stakeholders may become engaged in management in many ways. They can respond to surveys, serve on committees, attend public meetings, or vote on ballot initiatives. Many possibilities exist for involving stakeholders in wildlife management, and new techniques are being developed regularly for specific situations.

11.2. STAKEHOLDER ROLES IN MANAGEMENT DECISIONS

The main purpose of stakeholder engagement is to improve wildlife management. That is achieved by engaging stakeholders in four main aspects of management:

1. *Making decisions.* A common way for agencies to engage stakeholders is to involve them in making decisions, either by providing input or by actually helping to decide on

objectives or a course of action. Task forces, focus groups, public meetings, and surveys are vehicles for gathering information to use in making decisions. Committees are sometimes appointed to address a particular wildlife issue, such as an overabundant resident Canada goose population.

2. *Understanding decisions.* Helping stakeholders understand the factors pertinent to a particular decision and how a decision will be made often requires communication between the wildlife manager and stakeholders and among stakeholders. Presentations at meetings of organizations or public hearings, where managers can respond to questions, are common modes of interaction. Call-in talk shows on television and radio are used to reach large audiences for the same purpose. Interactive Web sites are becoming another medium for interaction and participation to improve stakeholder understanding, but experience has demonstrated that the best understanding comes from actual involvement in the decision-making process. Through such involvement stakeholders learn more than raw facts; they also learn why others hold the views they do.

3. *Implementing decisions.* Stakeholder engagement in implementing wildlife-management decisions is not common, but is increasingly being used in the United States and Canada. For example, the Clifftop Alliance, in southwestern Illinois, is a grassroots group that organized to promote the preservation and restoration of bluff-lands and karst and has implemented management actions on both private and public lands. The Illinois Department of Natural Resources and Illinois Nature Preserves Commission have helped to nurture the Clifftop Alliance by providing it with technical information, equipment, and supplies. As acceptance of such co-management, or shared responsibility for management, increases throughout the United States and Canada, stakeholders are becoming more involved in implementing wildlife management decisions.

4. *Evaluating decisions.* Stakeholder engagement can be helpful for evaluating the management programs and actions resulting from decisions. Wildlife management goals typically set targets for producing benefits for people, so who better to have involved in assessing programs than the people who are supposed to be receiving those benefits? Listening sessions, surveys, and advisory boards representing stakeholders are among the stakeholder engagement techniques used to gain input for program evaluation.

11.3. IMPROVING MANAGEMENT

While the overall purpose of stakeholder engagement is to improve management, wildlife managers often have more specific objectives for it. The three common ways that stakeholder engagement can enhance wildlife management are (1) by improving the information base for wildlife management decisions, (2) by improving the judgment that leads to the decisions, and (3) by improving the sociocultural component of the management environment.

11.3.1. Improving the Information Base
Stakeholder engagement provides information about stakeholders' needs, desires, beliefs, values, and behaviors. Such information can contribute to better management decisions in many ways. For example, Martin (1997) demonstrated how gathering information about the kinds of wildlife-viewing experiences people enjoyed could improve management. He surveyed non-resident visitors to Montana and found widely differing preferences in terms of their interest in using developed facilities and in having contact with other people. He argued that satisfying the wildlife-viewing interests of a broad cross-section of people requires understanding how people's preferences differ and providing a wide spectrum of opportunities to meet those preferences.

11.3.2. Improving Judgment
Obtaining more information about stakeholders' perspectives does not necessarily make decision making easier for wildlife managers, but the added information can reveal how complicated a situation is. Stout et al. (1994) discussed deer management scenarios in which diverse stakeholders hold strong and contrasting viewpoints. The potential for conflict between stakeholders is often high in such situations. Even when a manager is well-informed about the diverse perspectives present, it can be difficult to evaluate the implications of that information in order to reach a decision. Managers face the task of determining the degree to which various stakeholders' needs, wants, and desires will be satisfied and whether some will be addressed at all. It is likely that decisions will be unacceptable to some stakeholder groups under those conditions.

Stout et al. argued that in such situations it could be beneficial to involve stakeholders in the process of recommending a decision that appropriately considers the needs and concerns of all interested citizens. They suggested a model, the citizen task force, in which stakeholders with diverse interests work directly with each other and deliberate to find an acceptable management decision. Such dialogue among stakeholders often leads to alternatives that might not have been identified by managers alone and may satisfy a broader range of interests.

The ability of a task force to produce high-quality, broadly acceptable outcomes depends largely on membership of the task force being representative of the range of stakes at issue. If the individuals serving on a task force do not represent the stakeholder profile of the community, the community will probably be skeptical about the product of the task force.

11.3.3. Improving the Management Environment
Stakeholder engagement can also contribute to wildlife management in less direct ways. Sound management depends on a citizenry that will support, and can contribute to, management. Stakeholder engagement is commonly used to try to improve the social climate in which management occurs by

(1) transforming people's beliefs and attitudes, (2) changing people's behaviors, and (3) increasing management capacity.

Transforming beliefs and attitudes. Stakeholder engagement can have an important influence on public attitudes about management decisions. Mitchell et al. (1997) reported on a community in Indiana that struggled for years with the question of how to manage a locally abundant deer herd. Because it was a controversial issue, agency personnel found it difficult to choose and implement a plan to control the deer herd.

It was not until a study committee (composed of policy makers, ecologists, staff from the parks agency, staff from the wildlife management agency, hunters, and representatives from environmental organizations and an animal welfare organization) was appointed that sufficient public support for management action was generated. The study committee put tremendous energy both into background research and into gathering public input as they developed their recommendations. Their efforts helped not only to produce a sound management plan but to build support for its implementation.

Changing behaviors. Using stakeholder engagement activities to influence behavior can be a critical component of improving the social climate to help meet management objectives. In the early 1980s, biologists became concerned about declining stocks of beluga whales off the coast of northern Quebec, Canada. They recommended that the harvest be reduced and that beluga sanctuaries be established. Hunting beluga whales, however, is an important cultural tradition for the Inuit of northern Quebec, and the harvested whales provide much-needed food for these people. Managers in the Canadian Department of Fisheries and Oceans knew that establishing regulations without the cooperation of Inuit subsistence hunters would be futile. The hunters would follow their traditions rather than externally imposed regulations, and the area was too remote for the department to police effectively (Osherenko 1988).

The department contacted an organization of Inuit hunters, Anguvigaq Wildlife, to discuss concerns over declining beluga stocks. Instead of approaching Anguvigaq with a top-down regimen already established, the department worked with the hunters to devise a plan they could live with. The plan protected beluga females accompanied by calves and called for efficient hunting practices, such as harpooning before shooting to improve retrieval rates. In addition, a small sanctuary was established in an area where the beluga stock was especially depleted (Osherenko 1988). Compliance with the plan was high, and the beluga take was decreased by a third in the first year. The Canadian Department of Fisheries and Oceans was able to reduce the beluga harvest, and the Inuit were able to continue their traditional practice of hunting belugas.

Increasing management capacity. Stakeholder engagement may increase the capacity of citizens, communities, local governments, and agencies to participate in wildlife management. Increasing capacity can mean (1) changing attitudes (e.g., willingness to participate in decision making or receptivity to concerns of others), (2) improving relationships among stakeholders or between stakeholders and agencies, (3) increasing

Participants in the Vermont Habitats and Highways Program discuss wildlife issues during field visits to highway structures that impact wildlife habitat. These experiences give participants opportunities to discuss wildlife needs and highway agency needs and constraints. (courtesy Chris Slesar)

During field visits to highway structures, naturalists from the Vermont Fish and Wildlife Department provide interpretive information that builds appreciation for wildlife and use of highway corridors by wildlife (courtesy Chris Slesar)

knowledge (e.g., understanding the likely consequences of management actions), (4) developing skills (e.g., communication skills), or (5) changing organizational structure (e.g., establishing a new agency office to address particular concerns; Box 11.1).

Building management capacity can lead to a variety of other benefits. As relationships improve, knowledge increases, and attitudes change, stakeholders become more willing to share resources, reach agreement, and support management. For example, as part of an effort to build a more collaborative relationship between agencies, the Vermont Fish and Wildlife Department and the Vermont Agency of Transportation helped to initiate a "Habitats and Highways" training program. This primarily field-based educational program, run by two Vermont naturalists (with Keeping Track and the Vermont Reptile and Amphibian Atlas Project), aimed to build awareness of, and appreciation for, wildlife and habitat and how they relate to transportation. It focused primarily

Box 11.1 *TRANSFORMING CONTROVERSY TO COLLABORATION ON LAKE ONTARIO*

Little Galloo Island in Lake Ontario's eastern basin was donated to the New York State Department of Environmental Conservation (NYSDEC) in 1998. Little Galloo and several nearby islands and adjacent shoals provide habitat for warm-water fishes, colonial waterbirds, waterfowl, and shorebirds, including Caspian terns, black-crowned night herons, several gull species, and a highly controversial population of double-crested cormorants.

The cormorant population on Little Galloo grew from 22 pairs in 1974 to a peak of 8,410 pairs in 1996. Suspicions also grew that cormorants contributed to declining fish populations upon which many residents of local tourism-based communities depended for their livelihoods. Scientific studies implicated cormorants in the decline of smallmouth bass, a popular sport species. Local charter captains, marina owners, and others urged state and federal agencies to aggressively control cormorants' impact on the fishery, while birdwatchers, environmentalists, and animal rights supporters sought the birds' continued protection. The controversy exploded when a group of men from shoreline communities illegally shot nearly 1,000 birds.

In this controversial management environment, NYSDEC engaged stakeholders from local communities in planning for the islands using a process called a "search conference," with organizational assistance from Cornell University (Schusler and Decker 2002). The search method involves collective planning to solve problems and identify opportunities; it typically lasts 2.5 days, includes 25–75

participants, and encourages dialogue through the interplay of plenary and small group sessions (Emery and Purser 1996). The search conference took place in 2000.

Thirty-two stakeholders with diverse interests addressed the question: "What is the ideal future land use and management of NYSDEC-owned islands in the Eastern Lake Ontario Basin, considering the relationship of the islands to coastal communities?" Participants developed action items in five priority areas: community cooperation, ecosystem management, education, recreational resource use, and

Participants in a search conference on cormorant management create a shared history of their community, before and after double-breasted cormorants colonized their region (above: courtesy Tania Schusler; opposite: courtesy Bret Muter)

on Agency of Transportation employees and enrolled nearly 90 participants over 6 years. The most immediate benefit of the training program was improved relationships between individuals of the two agencies and increased appreciation of each other's needs and constraints. These changes in attitudes and improved relationships also resulted in a greater willingness on the part of the Agency of Transportation to consider wildlife and habitat impacts in its decision making and to use its resources to mitigate these impacts.

11.4. HOW TO ENGAGE STAKEHOLDERS

The many benefits of stakeholder engagement are emphasized in these examples, but designing an effective stakeholder engagement strategy is not straightforward. No single strategy works in all situations and no single recipe for success exists. Chase et al. (2002) developed a four-step process that can serve as a framework for helping agencies constructively engage stakeholders.

The first step is to gain an understanding of the situation in question. Local wildlife managers typically know who the

key stakeholders are, the nature of their concerns, and their attitudes. Greater understanding can be gained through interviews of key stakeholders, solicitation of comments, and systematic surveys. The second step entails using this understanding of the situation to define objectives for stakeholder engagement (outlined in the preceding section). The third and fourth steps—selecting a stakeholder engagement approach and designing a context-specific strategy that includes specific engagement techniques—are discussed in more detail below.

11.4.1. Approaches to Stakeholder Engagement
The practice of stakeholder engagement in wildlife management has been pursued by wildlife agencies for many years. The intent and extent of engagement continues to evolve with experience. The trend has been to encourage more stakeholder involvement and engage stakeholders in more aspects of management.

In this section we describe six general approaches that characterize most of the ways wildlife managers engage stakeholders (Leong et al. 2009a): (1) expert authority or authoritative, (2) passive–receptive, (3) inquisitive, (4) intermediary, (5) trans-

sustainable resource-based tourism. NYSDEC incorporated some of these action items in their management plan.

The search process also helped improve the management environment. Stakeholders—some with contentious histories in the cormorant debate—reported learning about the concerns of other stakeholders, identifying common purpose, and developing collaborative relationships. Participants learned through dialogue with others who held different views and from those with expertise on key topics. The search question encouraged stakeholders to focus beyond the cormorant controversy to broader management issues where they found more common ground. The process increased the agency's credibility and provided a foundation for future collaboration between the agency and local communities.

TANIA M. SCHUSLER

actional, and (6) co-managerial (Fig. 11.1). The approach adopted by managers will vary based on the characteristics of the issue they are trying to address, and managers may even switch from one approach to another as an issue develops. We give examples of the implementation of each approach.

Expert authority or authoritative approach. This "top-down" approach, in which wildlife managers make decisions and take actions unilaterally, was the norm when managers served a narrow constituency with whom they identified and shared values. Value differences among stakeholder groups were less of a challenge when fewer and less-diverse stakeholders were considered in wildlife management.

Under the expert approach, game management created notable successes continent-wide. Biologists used their expertise to set and achieve objectives for game populations and hunter use that enhanced hunting opportunities. That goal was embraced by most hunters (the category of stakeholders perceived as the primary clients of wildlife management at the time), so when game populations and hunting opportunities increased, the efforts were viewed as successes.

Even today, an authoritative approach by wildlife managers

(the biological experts) can work when there are few groups of stakeholders and the stakeholders recognize that the experts share their values. When both stakeholders and wildlife managers generally regard an issue as a biological problem whose solution requires biological expertise, the expert model may be the most efficient. For example, under unusual emergency situations, such as when chronic wasting disease (a severe infectious disease of white-tailed deer) is discovered in an area, a rapid response is needed to contain it.

Such situations are rare. Typically each issue has several stakeholder groups that hold diverse values and actively advocate their preferred outcome in a management decision. Managers have learned that taking a paternalistic, "we know what is best for you" approach alienates stakeholders, especially those whose values differ most from the values traditionally reflected by the agency.

Disgruntled stakeholders who feel that they have been dealt with unfairly can delay or derail decision making and wildlife management programs. Ballot initiatives and legal actions are familiar examples of what can happen when stakeholders believe they have not had a voice in decision making.

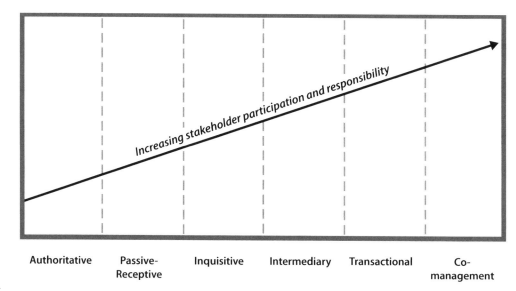

Figure 11.1. Six general approaches that characterize most of the ways in which wildlife managers engage stakeholders (Leong et al. 2009a). Level of stakeholder participation and responsibility increases from left to right.

Passive–receptive approach. In the passive–receptive approach, common in wildlife management, managers simply keep their eyes and ears open and note their observations. They welcome stakeholder input but do not seek it systematically. They listen to the concerns of stakeholders who take the initiative to make their views known, and they informally consider those concerns in their decisions. When making decisions, the wildlife managers (not the stakeholders) determine how much weight to give to concerns voiced by stakeholders.

The passive–receptive approach requires little initiative on the part of the wildlife manager. All public wildlife managers certainly need to be receptive to unsolicited stakeholder input, but managers who do not make an effort to gather opinions are placing little importance on the potential value of such input. An implicit assumption is that if they do not hear from a stakeholder group, then the group must not be interested in the outcome.

When operating in the passive–receptive mode, managers simply listen to whoever gains their attention—whichever stakeholders figure out how to get their message to the manager. Active, organized stakeholders accept this approach because it is to their advantage; they can use it effectively, even to the exclusion of other legitimate stakeholders who have not figured out how to attract managers' attention to their concerns.

Managers who, by their own choice or by circumstances of limited personnel and funds, rely on the passive–receptive approach may experience unanticipated problems when a decision is about to be made (e.g., when citizens speak out against an agency recommendation at a wildlife commission meeting) or after a decision is made (e.g., when citizens initiate legal action to block the decision). Many wildlife management agencies have found it beneficial to use other approaches, described below, together with the passive–receptive approach.

Deer management in the agricultural region of central and western New York operated in the passive–receptive mode until the mid- to late 1970s. Managers gave considerable weight to the complaints farmers made about crop damage and kept the deer population well below range-carrying capacity (Brown et al. 1978). Subsequent inquiry found that managers were missing an opportunity, which we will describe in the next section.

Inquisitive approach. Managers using the inquisitive approach actively seek information about stakeholders to inform an anticipated management decision (i.e., decisions about appropriate goals or methods for achieving them). They also reach out to stakeholders to evaluate programs that are in place (i.e., to refine goals or management policies, regulations, or activities).

The inquisitive approach emerged after wildlife managers gained precision in manipulating wildlife populations through regulated hunting (especially for big-game species, such as deer and elk). As they gained the ability and confidence to fine-tune population objectives, wildlife managers sought more knowledge of stakeholders. They started examining their assumptions about stakeholder values, beliefs, attitudes, behavioral motivations, and preferences for management actions and outcomes (Enck and Decker 1997).

Managers take an inquisitive approach because they recognize that unsolicited input alone can lead to a biased perspective; marginally important stakes can be magnified and some stakes can be missed. Managers using this approach seek input from a broad array of stakeholders and from many members of each stakeholder group. They employ systematic surveys to be more scientific in their efforts to understand stakeholders (Box 11.2).

The inquisitive approach has yielded many benefits. Deer management in central and western New York changed after managers abandoned a passive–receptive approach and adopted an inquisitive approach. They learned from several

Box 11.2 *APPLYING AN INQUISITIVE APPROACH TO MOUNTAIN LION MANAGEMENT*

Mountain lions are among the most resilient of the world's species of large cats. Considered rare 30 years ago, mountain lions have undergone remarkable population increases and have expanded their range into areas where they were historically rare or absent. Expansion of mountain lion populations has occurred at the same time as dramatic increases in the human population of western states.

Growth in the mountain lion population in Montana, especially in areas that coincided with new residential development, contributed to a rapid increase in the number of people–lion encounters where humans were threatened. Dealing with such encounters in residential settings was a new and unexpected challenge for wildlife managers in Montana and other western states.

Montana Fish, Wildlife and Parks had traditionally based lion-management decisions mainly on perceptions of lion population levels and demand for lion hunting. With the emergence of new stakeholders, however, the department began looking for ways to integrate more social information into its decisions.

The department addressed the information need by co-sponsoring research (Riley and Decker 2000) to describe Montana residents as stakeholders in lion management and shed light on the factors that affect residents' attitudes and preferences related to lion population levels. The study included a survey of Montana residents in three geographical areas—western, central, and eastern Montana.

The survey findings gave the department a better understanding of residents' attitudes, perceptions, and management preferences. It revealed that attitudes toward lions were more favorable than had been expected. Several findings suggested the presence of conservation-oriented values that the department could use as a base of support for their efforts to help people share the landscape with lions.

The research provided Montana Fish, Wildlife and Parks with information for developing a statewide education campaign. The ongoing education program seeks to influence stakeholder acceptance capacity for mountain lions through modification of stakeholder knowledge, risk perception, and behavior.

SHAWN J. RILEY

© Megan Lorenz-Fotolia.com

studies that farmers were more tolerant of deer than they had assumed. With that knowledge, they took actions to allow some populations to build, and they systematically monitored farmers' acceptance of deer (Decker et al. 1985a).

After deer populations increased, it was critical that hunters participate in the deer-management permit system. Managers needed to understand hunter motivations in order to ensure that hunters harvested enough deer to control populations. They sought hunter input to determine the best way to encourage hunters to harvest antlerless deer (Decker and Connelly 1989).

Stakeholder inquiry was also helpful in other kinds of people–wildlife interactions. Residents of Denver's western suburbs had voiced concerns to the Colorado Division of Wildlife about the number of elk in the area, and in the late 1990s wildlife managers held public meetings to learn about residents' attitudes toward elk and elk management. The meetings provided information from stakeholders with extreme positions on both sides of the issue, but managers also wanted to hear from the silent majority. They conducted a mail-back survey of a representative sample of area residents in 1998 (Chase and Decker 1998), and they used the results, along with input from the public meetings, to develop elk population objectives.

In South Dakota, the Division of Wildlife frequently conducts surveys of the general public and specific groups of stakeholders, to gather information for decision making. The studies have been on various topics, ranging from recreational interests in wildlife to the agency's image throughout the state (Gigliotti 1996). Missouri has a long tradition of inquisitive activity to help it better serve its citizens. Gaining such systematic input adds to the reservoir of knowledge about state residents obtained through meetings and other public contacts.

Intermediary approach. Inquisitive techniques often rely on one-way communication, where stakeholders respond to surveys or other requests for input. While this provides managers with good information about stakeholder positions, in most situations there are a diverse range of interests behind the same positions (Leong et al. 2009b). The intermediary approach emerged as managers recognized the value of more dialogue with stakeholders.

For example, one common survey question used to gather information on stakeholder opinions about wildlife management asks respondents whether they want a wildlife population to increase, decrease, or stay the same. Although this type of question provides a gauge of stakeholder tolerance for impacts that they are experiencing from wildlife at current population levels, it does not reveal why people hold their opinions, and these opinions could be important in management decision making. For example, people who want deer populations to decrease may be concerned about agricultural damage, car accidents, Lyme disease, or forest regeneration. Those who want populations to stay the same or increase may be concerned about hunting opportunities, viewing opportunities, or the ethics (and related problems) of using lethal techniques

to reduce deer numbers. Management responses may differ depending on key stakeholder concerns.

In addition to conducting surveys that provide a one-way flow of information from samples of stakeholders to agencies, managers also reach out to stakeholders in person and engage them in dialogue about issues. Typically managers regularly attend scheduled meetings of stakeholder groups (e.g., sportsman clubs, homeowner associations, agricultural groups), and take part in one-on-one discussions at open-house-style public meetings.

This approach encourages two-way communication between individual stakeholder groups and the wildlife management agency but does not emphasize dialogue among stakeholder groups with different concerns. Instead, managers act as *intermediaries* in deciphering similarities and differences in stakeholder positions and interests. In the world of international relations, the process of sending an outside party to serve as an intermediary between principal parties is called shuttle diplomacy. Wildlife managers playing the role of an intermediary are in a delicate position. As managers attempt to weigh and balance competing stakeholder concerns, they may find that they are caught in the middle and are being pulled in many directions by various stakeholder groups (Fig. 11.2).

Transactional approach. Since the mid-1980s, wildlife management has expanded to include attention to a diverse array of different types of people–wildlife interactions. Hunters, farmers, and rural landowners have been joined by other vocal groups that expect benefits from wildlife or relief from costs associated with wildlife.

In urban and suburban environments, for example, residents are finding that living with burgeoning numbers of deer, coyotes, and Canada geese, and increasingly frequent encounters with black bears and mountain lions has its challenges. The great increase in people–wildlife interactions in such environments is requiring more attention today than ever. As a result, managers are dealing with populations of stakeholders who have diverse beliefs and attitudes about wildlife management. In many states, wildlife agencies have limited experience working with urban and suburban stakeholders. These "new" stakeholders may not be familiar with or have reason to trust the wildlife agency. In such cases, inquisitive approaches used alone are inadequate because managers need to understand stakeholder beliefs and attitudes and also to consider the likely consequences of management alternatives on the diverse stakes.

Managers take a transactional approach when a choice must be made about how to prioritize different stakes when making management decisions, and they want to engage stakeholders in that prioritization. Managers and stakeholders need to find objectives and actions that are acceptable and this can be a daunting task for wildlife management issues where a wide range of stakeholder views are encountered (Box 11.3).

During the 1990s, wildlife managers (in their attempts to balance competing interests) were being pulled in every direction (see Fig. 11.2). It became an untenable situation. As a result, there was an explosion of experimentation in approaches to solving difficult people–wildlife problems. Managers gravitated toward processes where stakeholders engaged one another directly to articulate their values and stakes (Fig. 11.3).

Managers found they could get out of the middle; they did not need to convey the values of every stakeholder group to every other group. They applied their ingenuity to the need for innovative solutions and in so doing created a new approach to managing people–wildlife problems.

Those developments in the evolution of wildlife managers'

efforts to obtain stakeholder input in decision making resulted in the transactional approach. In this approach, stakeholders describe their stakes to each other rather than having the manager act as intermediary, and they collaborate to prioritize these stakes. By learning about the issue, conducting inquiry, discussing viewpoints, debating the issue, and compromising, stakeholder participants frequently reach consensus about appropriate objectives and courses of action (Nelson 1992). Depending on the nature of the decision, the confidence of the wildlife manager, and the policy of the agency, the

Figure 11.2. Managers being pulled in every direction by stakeholders (illustration by Annie Campbell)

Figure 11.3. Stakeholders interact directly with one another in transactional approaches by managers who also have input but mainly facilitate interactions among stakeholders (illustration by Annie Campbell)

The gopher tortoise is a charismatic, long-lived, and widespread species in the Southeast United States. Gopher tortoises occupy upland habitat throughout Florida, including forests, pastures, and yards. They dig deep burrows for shelter and forage on low-growing plants. Gopher tortoises share these burrows with more than 350 other species and are therefore referred to as a keystone species.

Gopher tortoise management is a longstanding, statewide management issue in Florida. The well-drained upland habitats preferred by gopher tortoises are also in high demand as sites for residential and commercial building, which brings tortoises and humans into frequent contact and conflict.

In 2004, social conflict over tortoise management decisions devolved into a disruptive management issue that received extensive media coverage. Conservation groups voiced deep dissatisfaction with Florida Fish and Wildlife Conservation Commission (FWC) efforts to protect tortoise habitat. Landowners, the regulated community, and an industry of consultants advising them raised protests about undue or even unconstitutional interference with private land use and development rights as well as the costs of complying with FWC regulations. Animal welfare groups and the media expressed outrage at some elements of gopher tortoise mitigation, such as entombment of tortoises by permit.

The need to address this disruptive issue gave FWC the impetus to develop structures for improving communication among FWC staff and various stakeholder groups. Beginning in July 2005, FWC utilized its contracted-facilitation leadership initiative to assist stakeholders in (1) creating their own forum for discussions, (2) developing an effective governance structure to facilitate equity and communication among stakeholders, and (3) transmission of stakeholder views and recommendations to FWC. This stakeholder group now operates effectively to discuss issues, review FWC proposals, and develop alternative or additional possibilities.

This forum evolved over time, beginning as a process for information exchange and then developing into a program of active inquiry into stakeholder views. Mechanisms for transactional forms of engagement were later added, and the process eventually developed into a cautious, co-management approach (i.e., the group serves as a citizen consultation body as FWC and other partners implement management actions associated with a statewide management plan).

The evolution of stakeholder engagement in gopher tortoise management was not a linear process. It was set back several times due to reversions to unproductive behavior by stakeholders and agency staff, shifting expectations that needed to be regularly recalibrated, and unbridgeable divisions of attitude and values among some stakeholders. Yet, with persistence and dedication, managers and stakeholders overcame these difficulties; over time, the process improved trust among stakeholders and between them and the agency and allowed the exploration and development of novel solutions to tortoise management problems. The use of a dedicated facilitator for both stakeholder and agency groups and the clearly defined and transparent governance process were factors in this success.

JAMES PERRAN ROSS

Gopher tortoises inhabit upland habitat throughout Florida (courtesy Cliff Leonard, Florida Fish and Wildlife Conservation Commission)

transactional approach may even permit the stakeholders to make a binding decision within some bounds set by the wildlife management agency.

One important role of wildlife managers in the transactional approach is to facilitate productive interactions between the agency and stakeholders and among the stakeholders. The process should include people who reflect various key stakes in the wildlife management decision, but the participants need not be formal representatives of interest groups organized to represent those stakes in a political sense (Box 11.4).

Another role of wildlife professionals in this approach is to ensure that stakeholders are well-informed about biological and socioeconomic facts and legal considerations pertinent to the issue. This establishes the groundwork for informed discussion and decisions. Implementing this approach often requires a wildlife management team that includes professionals with various skills and areas of expertise.

For example, around 1990, the New York State Department of Environment Conservation instituted citizen task forces as an integral part of deer management in New York. The managers designed an approach that removed them from the "middle" or intermediary role; they are in charge of managing the process and act as deer management specialists. Essentially, they provide a forum and some ground rules that allow task force members with various interests to explain their stakes to each other. The goal is consensus about deer population objectives.

Citizen task forces in New York exemplify the transactional approach. All important stakeholder perspectives ideally are included, an educational element ensures informed discussion, and an external, skilled facilitator handles the meetings. The facilitator encourages discussion and deliberation among stakeholders to achieve an understanding of the stakes involved. In the vast majority of cases, the task forces determine management objectives, stated in terms of desired changes in deer population numbers (Nelson 1992, Stout et al. 1996).

The transactional approach has gained favor for a variety of applications. In the late 1980s, the Cameron County Agricultural Coexistence Committee was organized in South Texas to contribute to the resolution of a dispute centered on the restoration of aplomado falcons to the Laguna Atascosa National Wildlife Refuge. Northern aplomado falcons are listed as federally endangered in the United States. To benefit falcon restoration, the U.S. Environmental Protection Agency had banned the use of many pesticides in the cotton-growing area. The new federal regulations had substantial negative impacts at the local scale, jeopardizing the livelihoods of the cotton farmers who provided the foundation of the local economy. The Coexistence Committee, composed of a range of stakeholders, was organized to try to reach consensus on recommendations that would protect both the falcons and the farmers and satisfy both local and national needs. The group eventually arrived at a solution that involved banning some pesticides but specified different application rates and procedures for others. In this case, rather than serving as the conveners of the commit-

tee, agency staff members worked side by side with farmers, environmentalists, and other stakeholders. This type of novel agency role is becoming much more common, as we next describe under the co-managerial approach.

Co-managerial approach. The challenges of increasingly complex wildlife issues have stretched agency approaches to decision making further. Several trends have hastened this evolution: (1) growth of management demands, including conservation of rare and declining species at one end of the spectrum and managing conflicts with abundant wildlife at the other, (2) the authorities and resources necessary for management being split among a variety of government agencies and stakeholders operating at local, state, national, and international scales, (3) broadening of agency responsibilities, and (4) limitations on agency funds and personnel. These pressures have led agencies to share responsibility and authority for management with other government agencies that have mandates unrelated to wildlife as well as with non-governmental organizations and local communities.

A fundamental distinction exists between the co-managerial approach and the approaches discussed previously. The expert authority, passive–receptive, inquisitive, intermediary, and transactional approaches to stakeholder involvement are approaches in which wildlife management agencies grant a role to stakeholders in their decision-making processes. Stakeholders may have greater or lesser degrees of control over the decision, but it is the purview of the wildlife agency to decide how much that will be. In co-management scenarios, however, the authority and resources necessary for effective management are so fragmented that wildlife management agencies must work in partnership with others. Specifics of these partnerships are negotiated on a case-by-case basis and the responsibility, as well as the authority, for wildlife management is shared. These partnerships vary in form because they are tailored to individual circumstances.

One increasingly common form of co-management into which state wildlife-management agencies enter is partnerships with their sister-state transportation agencies. The negative impacts of roads on wildlife and habitat range from habitat fragmentation to vehicle strikes. Wildlife–transportation agency partnerships help to mitigate these impacts. The Vermont Fish and Wildlife Department (VFWD) and the Vermont Agency of Transportation (VTrans) work collaboratively to improve the connectivity of fish and wildlife habitat, increase roadway safety by reducing wildlife collisions, and avoid unnecessary delays and unanticipated costs in transportation projects. This collaboration began in earnest in the late 1990s; in an effort to improve their relationship, the two agencies created a state-level interagency steering committee, which meets quarterly and serves as a forum for discussion on a wide variety of topics relevant to transportation and wildlife. The role of the committee has been formalized through a Memorandum of Understanding.

One of the primary values of the committee has been the cultivation of personal relationships between employees of the two agencies, which then influence activities at the local scale.

Box 11.4 *MANATEE FORUMS*
A Transactional Approach to Management Planning

The Florida manatee is a sub-species of the West Indian manatee native to Florida. Humans radically altered the manatee's habitat (coastal areas, rivers, and estuaries) throughout its historical range, which contributed to population decline during the twentieth century. Manatees received federal protection through the Marine Mammal Protection Act (1972) and the Endangered Species Act (1973). The Florida Fish and Wildlife Conservation Commission (FWC) classified manatees as a threatened species in 1974 and reclassified them as endangered in Florida in 1979. Manatees were afforded additional protection in 1978 under the Florida Manatee Sanctuary Act (Fla. Stat. § 379.2431(2).

In response to endangered species designation, tremendous resources from local, state, and federal government agencies and the private sector have been directed toward research, protection, and conservation of manatees. Little was known about manatees when they were first listed, but they are now one of the best-studied marine mammals. The manatee conservation effort has become a case study in endangered species public policy and is chronicled in books, periodicals, and virtually all types of media. Manatees have become a Florida icon and engender international public support.

Florida manatee numbers have increased in recent decades, which suggests that conservation actions have produced the ecological effects that managers intended; however, many of these actions have produced negative economic and social impacts (i.e., collateral and subse-

quent effects) as well. Public debate on how to balance manatee conservation with other social values has resulted in one of the nation's most contentious environmental issues. As a response to this contention, the FWC and the U.S. Fish and Wildlife Service (USFWS) began working collaboratively to improve stakeholder engagement in decisions about manatee management and conservation.

One improvement in stakeholder engagement was the development of a transactional approach called Manatee Forums. The goal of the Forums is to improve communication and understanding among key stakeholder groups and participating agencies. The Forums are designed to help stakeholders identify areas of common ground, identify problems or conflicts, and develop ideas about how to resolve those problems and conflicts. The Executive Director of the FWC and the Director of the Southeast Region of the USFWS were instrumental in developing and implementing the Forums.

The first Manatee Forum was convened in the summer of 2004. Representatives of 22 selected stakeholder organizations participated, and the Forum met 10 times from July 2004 through June 2007. Meetings lasted from 1 day to 3 days. Initial meetings served to define the group's mission and operating guidelines. The next several meetings focused on presentations of the latest available information regarding manatees and their habitat. The next two meetings concentrated more directly on conflict resolution and the goal of finding common ground. The last three meetings centered on explaining and discussing FWC's draft

These relationships have had tangible impacts on a variety of transportation projects, including modification of plans to upgrade an important state route in northern Vermont by elevating a portion of the roadway through a critical habitat area, a switch from plastic matting to natural fiber matting for erosion control to protect snakes, and modification of a project on an interstate highway to protect the state's only known population of black racers. VTrans' staff, funding, and support of these projects were absolutely necessary to their implementation. VFWD could not have done this work on its own.

Partnerships between wildlife and transportation agencies have just two primary organizations involved. Other partnerships are more complex and involve multiple government agencies, non-governmental organizations, and even private landowners.

Another common co-management scenario involves community-based wildlife management to address local people–wildlife conflicts. One well-documented case of the

co-managerial approach focused on deer management in Irondequoit, New York. Beginning in the 1980s, citizens of the Rochester suburb of Irondequoit were divided over management of the deer population. The deer population had been growing unchecked for decades due to archery and firearm restrictions. Local citizen groups with different viewpoints on deer management had organized and were demanding attention.

In autumn of 1991, the state wildlife agency decided to apply a modified version of the citizen task forces it had been using successfully in rural areas. The charge to the task force was to set a deer population objective and recommend methods for achieving that objective. The success of the modified task force has been debated (Baker and Fritsch 1997, Curtis and Hauber 1997), but one outcome is clear: the transactional approach of the citizen task force led to increased responsibility on the part of the community and a shift toward the co-managerial approach.

The task force recommended culling deer through a

management plan (drafts of the plan were revised based on extensive public comment received during two open public-comment periods; the Commission approved the plan in Dec 2007). This structure allowed Manatee Forum members to provide input on the content of a management plan early in the process of its development and as it was revised and finalized.

Although the effort required a significant time commitment from both agency staff and stakeholders, both FWC and USFWS believed the work was instrumental in moving the manatee discourse forward as well as reducing the intensity of the conflict among the stakeholders. The agencies plan to sponsor the Forum meetings into the future as long as they continue to be productive and are valued by the stakeholder participants.

R. KIPP FROHLICH
Source: Florida Fish and Wildlife
Conservation Commission (2007)

Courtesy USFWS

bait-and-shoot program and bow hunting in designated areas, coupled with research on contraception as a long-term method of population control. An interagency task force composed of a representative from the state wildlife agency, the Irondequoit town supervisor, a town board member, a couple of county legislators, the head of the county transportation office, and a representative from the county parks department took the lead for implementation. In addition to sharing decision making and implementation with the state wildlife agency, the community funded the bait-and-shoot program. The New York State legislature paid for the contraception research.

Development and widespread adoption of a co-managerial approach requires rethinking the role of state wildlife agencies and their professional staffs. It also necessitates empowerment of other agencies, organizations, and local communities for greater responsibility in solving wildlife problems. Operational guidelines for partners (standards, criteria, and requirements for specific management efforts), as well as oversight, accountability, and evaluation processes and assignment of responsibility for them, need to be negotiated and often spelled out in a written agreement (Box 11.5).

The role of the wildlife management agency varies widely with the context and might include providing biological and human dimensions expertise, managing processes, training representatives of communities or other organizations, approving management plans, certifying private consultants and community wildlife managers, and monitoring management activities. The approach also calls for educational communication programs on a level seldom seen in wildlife management.

To operationalize the co-managerial or transactional approach on a broad scale requires a professional management team of specialists in several areas such as wildlife biology, human dimensions, citizen participation, educational communication, and community leadership development and capacity building. One model is a wildlife management team that is made up of professionals with appropriate expertise, which operates on a regional basis within a state.

Box 11.5 *PRAIRIE RESTORATION IN THE GRAND RIVER GRASSLANDS*

The Grand River Grasslands is a 70,000-acre (28,328-ha) area straddling the Missouri–Iowa border. It is an important site for native prairie restoration, which is a key regional habitat priority. Local landowners had long used the area for cattle grazing but it retained a core of native grasslands. Conservation work in the Grand River Grasslands began in earnest in the late 1990s, when The Nature Conservancy (TNC), the Missouri Department of Conservation (MDC), and the Iowa Department of Natural Resources (Iowa DNR) each acquired parcels of land with significant potential for native prairie restoration. TNC and MDC now own approximately 4,500 acres (1,821 ha) of land on the Missouri side of the border, and Iowa DNR owns nearly 1,400 acres (567 ha) of land on the Iowa side of the border. Each has been engaged in habitat restoration on their properties.

Despite their significant landholdings in the area, these organizations recognize that returning the full complement of grassland wildlife species will require restoration of native vegetation on private agricultural lands. Some prairie restoration practices are beneficial to cattle grazing, but these practices require funding. Additional govern-

ment agencies, including U.S. Department of Agriculture's Natural Resources Conservation Service and the U.S. Fish and Wildlife Service, have entered into contracts in which the agencies share in the costs of implementing these new practices with many private landowners. Neither Missouri's nor Iowa's state wildlife-management agencies would be able to achieve the conservation gains that have been seen in this region without active engagement with the full range of partners operating at multiple levels ranging from local landowners to federal agencies and national non-governmental organizations (such as TNC).

Prairie restoration in the Grand River Grasslands is a true co-management effort. Communication and coordination among the partners have been facilitated by regular meetings. These meetings have helped the partners recognize common needs and interrelated activities, which has contributed to refining priorities and coordinating activities. Partners have developed innovative ways to share funding and equipment, which has benefited their common conservation agenda.

Pawnee Prairie Natural Area is one of three core areas designated for prairie restoration within the Grand River Grasslands area of Missouri and Iowa (courtesy Michael Lykins, Missouri Department of Conservation)

11.4.2. Grassroots Participation

All of the approaches to stakeholder engagement that we have discussed involve wildlife management agencies either as the leader or as an active participant. In reality, it is often the stakeholders themselves who initiate much of their participation. Grassroots organizations and individuals advocate

for and against wildlife activities as diverse as reintroduction of wolves, rehabilitation of small mammals, and control of white-tailed deer in suburban areas. Ballot initiatives, litigation, and legislation to influence the authority of wildlife management agencies also have become increasingly common throughout the United States.

We call those efforts the grassroots, or bottom-up, approach. The citizens may initiate their own participation for any number of reasons, and the agency may or may not embrace their involvement. When wildlife agencies discourage citizen involvement by appearing unreceptive and uncooperative, stakeholders may feel the only way to have an effect is to circumvent established channels, which may leave the wildlife manager entirely out of the decision-making process.

For example, the Colorado Division of Wildlife announced a change in the elk-hunting license policy in certain parts of western Colorado in 1998: unlimited either-sex elk licenses were to be sold over the counter. In Gunnison, a coalition of hunters, environmentalists, and other stakeholders concerned about overhunting formed to oppose the agency's action and sue the Division; they claimed that there had not been enough opportunity for public input. That situation illustrates that stakeholders expect the wildlife agency to include citizens in its decision-making process; they are willing to go to court to ensure that they have input in agency decisions.

11.4.3. Stakeholder Engagement Techniques

Imagine that you have settled on taking a particular approach to stakeholder engagement. The next step is to flesh out your strategy for stakeholder engagement by choosing specific techniques. You have many to choose from. Here are brief descriptions of some of the major forms.

- *Information dissemination* refers to techniques that distribute information intended to reach out to large audiences. One avenue is to use mass media outlets, such as press releases, newspaper inserts, and press conferences or interviews. Other print media, such as fact sheets, newsletters, brochures, issue papers, or progress reports, may be used to disseminate information through mailing lists or at agency offices and visitor centers. These materials also are readily disseminated electronically by using agency Web sites, listservs, electronic mailing lists, and even social networking sites, such as Twitter and Facebook.
- *Open houses* are interactive public outreach and education efforts where stakeholders gather information, ask questions, and offer feedback in informal settings. They typically last a few hours and feature a number of stations with multi-media presentations, exhibits, or posters addressing various aspects of wildlife management. Managers staff each station so that participants can ask questions or communicate one-on-one with agency personnel. This format provides an atmosphere that is more comfortable for individuals who dislike speaking in front of a crowd or desire the opportunity to engage in dialogue with the agency without interruptions and with the expectation of getting answers immediately.
- *Public meetings,* like open houses, are short one-time events that afford stakeholders an opportunity to learn about and provide input into management. Participants hear relevant information and then have an opportunity to ask questions and comment in a public forum. In the past, many

public meetings followed strict "public hearing" formats, where participants were allowed a limited time to state their position before agency representatives and commissioners. Although this approach fulfills legal requirements for public input, it is now largely recognized as fueling controversy. As a result, agencies are experimenting more with dialogue-based approaches to public meetings.
- *Solicitation of comments* from stakeholders is common. A formal solicitation of public comment may be issued, which requests that written comments on a specific topic be sent by a specified date (e.g., the comment period for new regulations). Comments are typically sent by letter, e-mail, or the Internet. Agencies may also let stakeholders know they are open to feedback through less formal means; for this, they might use technological innovations such as blogs and interactive Web sites. When agencies solicit comments, they typically hear from the most passionate stakeholders or those with time to respond to the solicitation. This must be considered when managers decide how to incorporate comments into decisions.
- *Surveys* are typically employed when wildlife managers want to hear from all perspectives regarding an issue, including the "silent majority." A systematic survey to collect data can be conducted by telephone, mail, e-mail, Internet, or in-person. The sampling frame (target audience of the survey) may be the general public, or the survey may focus on specific stakeholder groups. Survey responses are usually kept confidential, which may encourage input from stakeholders who do not typically communicate with agencies. Surveys can be an important source of information for decision making; they provide data on beliefs and attitudes that may be overlooked when only the most vocal stakeholders are heard.
- *Focus groups* bring together a small group of stakeholders to discuss issues of concern. They are frequently used in market research and political analysis to provide information on stakeholder opinions. Focus groups typically meet only once and stakeholders share information from their own perspectives without expecting to educate themselves about a topic or fulfill a particular task (such as making a management decision). The discussion of a focus group can provide in-depth information beyond that attainable through a survey, but focus groups are limited in terms of the number of people who can participate and share their views. They are not generalizable from a scientific perspective.
- *Workshops* are typically one-time events that last from a couple of hours up to a full day. Agencies usually ask workshop participants to complete some type of task to further wildlife management. The particular task may be developing a list of information needs, generating a set of management alternatives, identifying possible concerns about a particular alternative, or any one of a variety of others. By engaging stakeholders in these efforts, agencies broaden the thinking that informs management and ensure that management fairly considers the needs and concerns of

different stakeholders. Participants also improve their understanding of management issues and other stakeholders' concerns.

- *Task forces* are stakeholder committees that typically meet multiple times over weeks or months to accomplish a task. Usually the task is larger and more complex than the tasks addressed in workshops. For example, many agencies have asked task forces to develop management proposals that take into consideration multiple stakeholder interests. Completing this task may require task-force members to gather information about wildlife and management impacts, reach consensus on goals and objectives, study management actions that could help to achieve these goals and objectives, and recommend a particular set of actions.

- *Large-group planning and decision-making processes* encourage direct dialogue and interaction among many different stakeholders. One example is the search conference (see Box 11.1), which is typically a multi-day planning event involving 25–75 stakeholders, in which participants collectively envision and plan for a desirable future. This method considers wildlife within the broader human–environment system and promotes the shared construction of knowledge, open dialogue, and democratic decision making. Planning a search conference and following up afterward require substantial time. Large-group processes can be effective in transactional and co-managerial approaches to stakeholder engagement. They are best implemented with the assistance of professional facilitators.

- *Advisory committees* do not restrict their activities to one particular task but rather advise a management agency on an ongoing basis. Advisory committees may focus on a particular management program, such as deer management or wildlife diversity. They are often required by law, with rules for selecting committee members and holding meetings spelled out in the legislation. Advisory committees provide stakeholder feedback on regulations, policies, and other topics. They serve as a check on agency judgment about how to balance management concerns, although how well they perform this function is influenced by the interests represented on the committee.

- *Negotiated agreements* often accompany a co-managerial approach. Written agreements specify stakeholder roles, responsibilities (including cost-sharing), and mechanisms to ensure accountability when wildlife agencies share responsibility for the implementation of management actions with other agencies, local communities, resource user groups, or other stakeholders. Examples include memorandums of understanding and project-based agreements. Each agreement must be tailored to its specific subject, context, and scale; negotiated with and deemed legitimate by the relevant actors; and comply with existing law. Agreements also include mechanisms for resolving disputes, procedures for monitoring and evaluation, and provisions for periodic revision as change occurs in the management environment.

11.5. CHALLENGES OF STAKEHOLDER ENGAGEMENT

Stakeholder engagement is a mixed bag. It has the potential to be a powerful aid to management, but it also presents many challenges.

11.5.1. Internal Challenges

Some of the challenges are internal; they are related to the management agency and how it operates.

Resistance. Some management agencies resist stakeholder engagement. They may view it as a threat because it means they must relinquish some control over decision making and management. They may view it with suspicion because it seems to be a catalyst for change. Manring (1998) reported that the U.S. Forest Service resisted some forms of citizen participation for many years out of a concern that it might interfere with timber production, which the agency viewed as its central mission.

Agency structure. The structure of the agency may not be conducive to stakeholder engagement even when wildlife managers support it. In an example about bear hunting discussed in Chapter 3, Colorado Division of Wildlife staff had conducted a study of public attitudes and, on the basis of the study, had recommended to the Colorado Wildlife Commission (a politically appointed group of laypeople) that spring bear hunting be eliminated. The commission did not adopt the recommendation. Loker et al. (1994) concluded that the composition of the commission, which reflected predominantly consumptive interests in wildlife, discouraged input that was more representative of the voting public.

Lack of time and money. Stakeholder engagement (public meetings, citizen task forces, mail surveys, educational campaigns, and other techniques) can be time-consuming and expensive and sometimes diverts staff from other important work, so managers may prefer not to involve stakeholders extensively in management processes.

Nevertheless, *not* involving stakeholders can be costly, too. Stakeholders who feel their needs were not seriously considered in decision making may delay or block the implementation of management plans through litigation and ballot initiatives. The short-run costs of incorporating citizen participation into decision making may be less than the long-run costs of litigation and unfavorable legislation.

11.5.2. External Constraints

Some of the challenges to stakeholder engagement are external; they are related to characteristics of stakeholders and how the agency interacts with them.

Difficulty getting the word out. Getting stakeholders involved is not always easy. Many wildlife management issues today involve a wide variety of interested and affected people. It can be difficult to figure out who they are, and it can be even more difficult to find stakeholder engagement techniques that are accessible to all. Stakeholders may be dispersed over broad geographic areas, which makes it difficult for them to attend particular events, such as public meetings. Managers may need to devote particular attention to encouraging involvement by non-traditional audiences, such as minorities, people with disabilities, or non-English speakers.

Complexity of weighing input. Managers have to weigh different, and often conflicting, input from stakeholders before reaching a decision. When considering whether to accelerate the restoration of moose to northern New York, for example, the New York State Department of Environmental Conservation had to (1) balance the wishes of local residents living in the restoration area with those of other citizens who had a general interest in state wildlife resources, (2) compare the value some people placed on restoring a native species with the concern others had about moose–vehicle collisions, and (3) weigh the wishes of a small segment of the public who knew much about the issue against those of the vast majority who did not.

Weighing different types of input is not easy. Even when managers use a transactional approach and share the locus of control, they choose which stakeholders to involve, and those stakeholders' perspectives are emphasized by virtue of their participation. Weighing stakeholder input is a problem in every approach and the responsibility of doing so differs depending on the approach used.

Poor relationships. Some agencies are plagued by poor relationships with certain stakeholder groups. They may range from hunters' organizations that want more voice in decision making, to animal welfare organizations that are philosophically opposed to wildlife management, to advocates of private property rights who resent the intrusion of government in their lives. When poor relationships and a lack of trust thwart attempts to involve citizens in management, agencies may have to take a long-term view; they might first create or restore relationships with certain stakeholder groups and only then try to secure the contribution of those groups to management decisions and actions.

11.6. SUCCESS IN THE PROCESS

No sure-fire secret to success exists in the business of managing people and wildlife for peaceful co-existence across the broad and varied landscape of human–wildlife interactions. There is no pat solution and no single human dimensions technique that can solve the many human–wildlife issues that wildlife managers face. Rather the solution is in creatively adopting an approach to stakeholder engagement that is sufficiently robust to encompass the breadth and ever-changing nature of stakeholder interests. A key ingredient for success is a wildlife manager's informed decision about whether and how to initiate and manage stakeholder engagement in wildlife management. Setting clear goals, establishing transparent expectations about roles and responsibilities, and maintaining good communications all strengthen the endeavor.

Wildlife management is successful when the process that yields decisions about management objectives and actions ac-

commodates the diversity of stakeholder interests. Stakeholders' overall satisfaction with those objectives and actions and the impacts they achieve largely depends on their satisfaction with the process used to incorporate their concerns into decision making and the degree to which their changing beliefs and attitudes are reassessed and integrated into an adaptive management strategy. Wildlife managers who recognize these facts will be better equipped to face the human dimensions challenges of wildlife management in the future.

SUMMARY

Stakeholder engagement at multiple geographical and temporal scales is becoming a regular practice in wildlife management. This intervention represents expanded attempts to increase transparency of decision-making processes and improve accountability (both important components of good governance). Stakeholder engagement has come to include a wide range of activities ranging from listening to individual stakeholders to intensive co-managerial arrangements. These types of engagement are being creatively refined through research and experience.

- Stakeholder engagement is the process of involving and interacting with stakeholders in making, understanding, implementing, or evaluating management decisions for improved wildlife management.
- The main purpose of stakeholder engagement is to improve wildlife management. That is achieved by engaging stakeholders in four aspects of management: (1) making decisions, (2) understanding decisions, (3) implementing decisions, and (4) evaluating decisions.
- The three common ways that stakeholder engagement can enhance wildlife management are (1) by improving the information base for wildlife management decisions, (2) by improving the judgment that leads to the decisions, and (3) by improving the sociocultural component of the management environment.
- Six general approaches to stakeholder engagement are characterized by increasing levels of stakeholder involvement: (1) expert authority or authoritative, (2) passive–receptive, (3) inquisitive, (4) intermediary, (5) transactional, and (6) co-managerial. The approach adopted by managers will vary based on the characteristics of the issue they are trying to address.
- Stakeholder engagement techniques include (1) information dissemination, (2) open houses, (3) public meetings, (4) solicitation of comments, (5) surveys, (6) focus groups, (7) workshops, (8) task forces, (9) large-group planning and decision-making processes, (10) advisory committees, and (11) negotiated agreements.
- Challenges to stakeholder engagement include both those that are internal to agencies and the way they operate and external—related to characteristics of stakeholders and how the agency interacts with them.

Suggested Readings

Arnstein, S. A. 1969. A ladder of citizen participation. Journal of the American Institute of Planners 35:216–224.

Decker, D. J., C. C. Krueger, R. A. Baer, Jr., B. A. Knuth, and M. E. Richmond. 1996. From clients to stakeholders: a philosophical shift for fish and wildlife management. Human Dimensions of Wildlife 1:70–82.

Fisher, R., W. Ury, and B. Patton. 1991. Getting to yes: negotiating agreement without giving in. Second edition. Penguin, New York, New York, USA.

Susskind, L., and P. Field. 1996. Dealing with an angry public: the mutual gains approach to resolving disputes. Free Press, New York, New York, USA.

Thomas, J. C. 1995. Public participation in public decisions: new skills and strategies for public managers. Jossey-Bass, San Francisco, California, USA.

Wondolleck, J. M., and S. L. Yaffee. 2000. Making collaboration work: lessons in innovation from natural resource management. Island Press, Washington, D.C., USA.

12
COMMUNICATION FOR EFFECTIVE WILDLIFE MANAGEMENT

JAMES E. SHANAHAN, MEREDITH L. GORE, AND DANIEL J. DECKER

Communication is a process we engage in daily—at school, on the job, and at home. We cannot avoid it—one cannot *not* communicate. We are all experienced with communication as both the originators and the recipients of messages. Professionals engaged in various roles in wildlife management spend much of their time communicating with various colleagues, staff, supervisors, partners, and stakeholders. It is an essential part of their jobs.

Individuals are involved in hundreds to thousands of communication events daily. Our routine exposure to "purposeful" communication includes Internet Web sites and "social media," e-mail listservs, television commercials and newscasts, radio programs, magazines, newspapers and on-line news, billboards, junk mail, graffiti, and even course lectures and textbooks such as this one. All those communication vehicles are trying to deliver messages to us, trying to get us to buy, believe, or act in a particular way. Some messages are effective; some are not. Do you ever wonder why?

Apart from messages with specific persuasive intent, we are constantly engaged in sending, receiving, and storing messages relating to all aspects of personal, family, community, and professional life. Disentangling specific messages of interest from all the background noise can be challenging. That is as true in wildlife management as in any other endeavor.

12.1. COMMUNICATION AS A MANAGEMENT IMPERATIVE

A wildlife manager's job is not simply to persuade people to adopt someone's notion of preferred behaviors. Previous chapters described how stakeholder involvement in wildlife management has evolved since the 1970s. The proliferation and diversity of stakeholders expands the number of audiences for communication. Two major associated changes accentuate the importance of communication in wildlife management: (1) the complexity of goals sought in wildlife management, and (2) extensive public participation in management. Basically, stakeholders often expect clear and complete agency com-

munication (transparency) about management decisions, can articulate their views rapidly and with great effect, and may anticipate certain agency responses.

No matter how difficult, how challenging, and even how frustrating communication can be, nothing happens without communication; it is a fundamental social process. That is why effective communication among staff in a wildlife management agency or conservation organization and between them and their various partners and stakeholders is an essential process in wildlife management.

This chapter discusses why successful wildlife management is accompanied by good communication and why ineffective management is often the result of poor communication. The chapter provides an introduction to communication models and theories, communication planning, and effective communication with stakeholders.

Effective communication has the potential to be a powerful tool of wildlife management, but how can you, as a wildlife professional, realize some of that vast potential? How can you recognize and practice good communication? Let us begin with a practical example.

12.2. COMMUNICATION PRACTICE—HUMAN DIMENSIONS OF URBAN DEER MANAGEMENT

The imperative to communicate effectively creates both need for human dimensions inquiry and demand for application of research findings. Students often are exposed to communication primarily from the standpoint of skills development—improvement of their writing, speaking, and facility with electronic communication technologies. Such skills are necessary, but to communicate well also takes strategic thinking about the "who, what, when, and how" of human interactions and about ways to integrate informative and educational communication into comprehensive wildlife management.

The importance of communication often is acknowledged in wildlife-management case descriptions, but key points of communication success or failure are seldom analyzed systematically. The vital role of communication typically plays out in a variety of ways from start to finish of a wildlife management issue. This is demonstrated in a case of urban deer management in Cayuga Heights, New York.

Cayuga Heights is similar in several ways to other middle-class suburban communities with wildlife management issues. As in many suburban locations in the eastern and north-central United States, a locally overabundant white-tailed deer population (approx. 200 animals in <2 square miles [<518 ha]) is firmly established among the human residents' manicured gardens, lawns, parks, cemeteries, and greenways. Negative interactions between the humans and deer residing in Cayuga Heights, including property and plant damage, deer–vehicle collisions, and concerns about disease, prompted calls for management actions. A grass-roots citizen "deer committee" was organized and eventually was legitimized by the village trustees. The committee sought human dimensions research assistance to better understand community sentiment about deer, knowledge of deer management, and preferences for providing citizen input into management decision making.

An early community survey showed that most residents of Cayuga Heights were negatively affected by the presence of deer and wanted a reduction in the deer population. Although surveys indicated that a majority of residents agreed that something should be done to reduce negative health, safety, economic, and aesthetic impacts of deer in their community (an agreement over *ends*), they also affirmed that residents disagreed about the *means* that might be used to achieve desired outcomes. Some residents supported lethal control of deer, but a sub-set favored accomplishing this only by sharpshooting, whereas others opposed sharpshooting but supported use of regulated hunts. Still others opposed lethal means of control and supported non-lethal methods (such as immuno-contraception or capture followed by surgical sterilization). The wide range of costs of these different solutions also was an issue among taxpayers in the village. Impassioned debate among residents and local officials about deer management in Cayuga Heights captured local media attention.

Research also found that residents thought they should have the main "say" in deer management decisions in preference to other special interest stakeholders (such as deer hunters), or other less affected persons (such as village visitors and people living outside the village; Chase et al. 1999*a*). Village residents wanted communication channels and opportunities created that would allow them to learn about, hear about, and *be heard* about deer management.

The study informed the village leaders' approach to communicating about the deer situation and the process for addressing it. Educational meetings and a Web site were established to improve residents' access to information. Public hearings, expansion of membership on the village-appointed "deer committee," resident surveys, and other vehicles were used to facilitate one-way and two-way communication that would reach various segments of the village population and allow them to express their opinions.

Overall, a variety of research (human dimensions and bio-logical–ecological studies of deer) has directly and indirectly informed deer management in Cayuga Heights. Human dimensions research on human–wildlife conflicts points to the fact that stakeholder opinions change over time and lean toward less tolerance of negative impacts. For example, after examining more than a dozen survey studies conducted over 15 years, Butler et al. (2003) concluded that (contrary to the assumption that suburban residents were increasingly resistant to lethal forms of local wildlife population management) residents were sensitive to wildlife impacts around their homes and many were willing to support lethal control of wildlife to achieve tolerable levels of impacts and perceived risks of impacts. Surveys in Cayuga Heights showed this as well; many village residents were intolerant of garden damage and driving hazards caused by deer. Over time, surveys of the village residents have documented that most have become more accepting of the need for deer population management, but communication about management options and about the management decision-making process that led to a bait-and-shoot program has not satisfied all stakeholders that their voices have been heard. Why? In large part because some people who are dissatisfied with the management decision and wish to present dissenting opinions have used electronic media options to magnify public perception of their view about lack of fairness in the management decision-making process.

Communication and education specialists advising the community of Cayuga Heights were not surprised by this development. Research on citizen task forces, commonly used to inform deer management in New York State (Pelstring et al. 1997), showed that controversy and conflict would be inevitable in the new communication environment for managing wildlife. Although creating communication forums that are oriented toward expanding how voices are heard in the management process may increase the likelihood of consensus, it also presents the possibility that disagreement might be amplified.

Cayuga Heights, where a dozen years of management have been informed by human dimensions research, highlights the ever-present issues of communication. In particular, it highlights the question, "How can a small community make an important decision that is informed by biological, ecological, and social science while also allowing for appropriate levels of citizen input and public expression of potentially conflicting views?" Although innovative science and wildlife management advice are being used in this community, controversy about what to do with the deer persists. Some portion of the residual disagreement cannot be resolved by either state-of-art management technique or excellent communication because basic value differences (and, therefore, differences in perceptions of impacts) endure among affected stakeholders. In this case, much research on attitudes toward wildlife in the suburban environment guided the committee that formed recommendations on how to deal with the deer problem. Wildlife managers need to understand that being well-informed may be necessary but insufficient for community-based decision making in wildlife management.

Currently, a plan to manage the deer herd that includes lethal options has been adopted, and organized opposition to the plan is already formed. Based on problems that occurred in other communities where lethal methods of deer control were implemented, local police are undergoing training to deal with public responses that could include civil disobedience and organized protest (Gashler 2010). With the potential for confrontation of some type, managers' attention to communication about deer management has never been keener in Cayuga Heights, and will continue to be an essential aspect of deer management in the village.

Despite tenacious controversy in this community, there has been an attempt to balance the positive and negative impacts of deer and deer management via broader understanding of the human dimensions of suburban deer management in general, and of the Cayuga Heights residents' beliefs, attitudes, and preferences in particular. Continuous communication to inform and educate stakeholders has been a key part of the management strategy. Cayuga Heights presents a case study of how human dimensions inquiry can be used to inform communication about a wildlife issue in a community.

Cayuga Heights also offers a practical lesson for wildlife managers about the multi-edged nature of communication in any context. Communication is a very powerful tool for wildlife management, but one must remember that it can be wielded by people who oppose, as well as those who support, a particular course of action (which can range from the general governance approach taken to specific actions chosen to achieve management objectives). This is not a new phenomenon, but it has become a greater concern because of the availability of electronic communication to anyone who chooses to use it. This reality accentuates the everyday necessity for effective communication between wildlife managers, their partners, specific stakeholders, and the public.

12.3. THE CENTRAL ROLE OF COMMUNICATION

The importance of communication in wildlife management is difficult to overstate. Governance and management of wildlife resources do not occur in the absence of extensive and periodically intensive communication effort. The opportunities for communication successes and failures (and, therefore, management successes and failures) have never been greater, largely because of the convergence of three factors: (1) more diverse stakeholders are interested in wildlife and its management; (2) more governmental and non-governmental entities are involved in governance of wildlife; and (3) the media available for communication are changing dramatically.

The variety of communication channels available to managers (and others) is impressive. The manager can pick and choose among them for the most effective way to reach particular audiences. The variety of stakeholder involvement tools used in wildlife management that has developed during the past two decades also has expanded options for two-way, face-to-face engagement with diverse stakeholders. These tools are not always intuitive in their application, so managers need to understand which tools to use, and when and with whom to use them. They also need to consider carefully why they will engage in communication; that is, what purpose is communication to play? (We later address purposes when we describe informative, educational, persuasive, and performative communication.)

For a long time, the common attitude among wildlife managers was that more "information" or "education" ("I and E") was needed to gain consensus among stakeholders about contentious issues. Without discounting the importance of "I and E," we strongly encourage you to adopt a quite different mindset about communication, which extends beyond just I and E. First, "information and education" assumes a predominantly one-way, hierarchical, linear flow of information from manager to "client," which does not reflect contemporary approaches to wildlife resource governance in most cases. As we will see, human dimensions research in general, and the communication research that has been done within the wildlife management field in particular, both point to more complex relationships and effects (Box 12.1). Communication occurs within a complex and multi-layered array of intersecting publics who have different interests in and concerns about wildlife management. Different publics often require different kinds of information and different educational approaches to create the knowledge, belief, and attitudinal and behavioral effects desired. Different theories and models of communication may be required for different situations. In the next section we review some of the more useful theories and models for wildlife managers.

12.4. COMMUNICATION MODELS AND THEORIES

Theories of communication help us understand how communication works. Theory does not always predict communication outcomes precisely, but it sensitizes us to the factors one should consider when thinking about communication strategically. Wildlife management problems, such as those in Cayuga Heights, can become provocative and contested. Any tool to improve understanding of how communication can be used to improve wildlife governance—that is, to increase both the fairness of the management process and quality of decisions that come out of it—is of potential value to wildlife managers. In this section, we review some theories in communication that have proved valuable for wildlife managers.

12.4.1. A General Linear Model of Communication

Ideally, wildlife managers do not communicate haphazardly; they organize their thinking around communication models that help them achieve intended outcomes. Lasswell's (1948) communication model can be very useful; it asks five questions: (1) "Who? (2) Says what? (3) To whom? (4) In what channel? (5) With what effect?" This "linear" model guides managers to consider parts of the communication process that need attention (Table 12.1).

Insofar as the model focuses on a communication "outcome," it helps managers interested in communicating to

Box 12.1 *MASS MEDIA EFFECTS AND PUBLIC PERCEPTIONS ABOUT BLACK BEARS IN NEW YORK*

The degree to which management strategies designed to reduce risks associated with human–bear interactions are effective depends on myriad factors, including mass media coverage. Multiple studies highlight the importance of considering the mass media's role in public perceptions of wildlife and wildlife management and the utility of human dimensions inquiry to increase understanding about media effects.

One study examined effects on risk perception of media coverage of New York's first documented black bear–related human fatality. Based on the social amplification of risk framework, researchers anticipated that media coverage of the incident would affect perceived bear-related risk among residents living within New York's black bear range. They compared results from a preincident mail survey and a postincident telephone survey to determine whether perception of personal risk had changed as a result of the death. Additionally, they performed content analysis of news stories referencing the incident. They found that most respondents had heard about the incident on the news but the fatality did not serve to influence risk perception enough that stakeholder groups were motivated to promote change in wildlife management policy (Gore et al. 2005).

Siemer et al. (2009) explored the extent to which print media use and television viewing could influence people's concern about risks posed by black bears. Using a mail survey of New York residents to assess public perceptions of risk and media exposure, the researchers found no evidence that print media coverage was amplifying risks associated with black bears, but they did find that television viewing was related to elevated concern about the risks from black bears.

Another study used content analysis to review television, newspaper, and radio reports about black bears to better understand how stories differed in relation to identification, attributions of responsibility, and proposed solutions to black bear–management problems. Researchers found that mass media identified few bear-related problems, suggested few solutions to problems, and tended to attribute responsibility for solving problems to individuals rather than government agencies. In framing interactions with black bears as personal problems and not public issues, the media left a void in public communication that was of concern to wildlife management agencies because the coverage

oversimplified the complexity of black bear management (Siemer et al. 2007*b*).

Gore and Knuth (2009) explored mass media effects on the operating environment (i.e., sphere of activity within which a program operates) of a black bear–related outreach program, the "New York NeighBEARhood Watch Program" (NYNWP). By using pretest and posttest mail surveys and a content analysis of media coverage on the NYNWP, they found that the media did affect the operating environment of the program by expanding the intended geographic range of the program, by reinforcing messages from program materials (e.g., both a refrigerator magnet and newspapers discussed the need to remove food to remove bear problems), and by slightly influencing risk perception.

A biologist with the New York State Department of Environmental Conservation handles a bear trapped and immobilized in a suburban backyard. Human–bear interactions in New York increased markedly during the 1990s, generating public concern and media attention. (courtesy NYSDEC)

influence knowledge, attitudes, and behavior. It is reasonable, for example, that a manager talking with members of the public who are concerned about human–bear interactions will try to communicate agency "advisory notices" or regulations effectively, hoping to get people to behave in a certain way (for example, not to feed the bears). When thinking in that way,

the manager is concerned with persuasive communication—influencing what people think and do. In other instances, the manager may simply want to distribute information with intent to have people become knowledgeable about a topic (e.g., a new invasive species threat) but with little interest in how people act. The manager is then engaged in informational or

Table 12.1. A basic communication model

Question	Research focus	Examples
Who?	Characteristics of the sender	Credibility, trust
Says what?	Qualities of the message	Urgent, fear appeal, vivid
To whom?	Nature of the audience	Active, latent, inactive
In what channel?	Effect of the medium	Audiovisual, print, multi-media
With what effect?	Evaluation	Informed, participatory

educational communication. Both of these aspects of communication can be understood with Lasswell's five questions.

Along with the questions, Lasswell identified five basic "components" to the communication process: (1) communicator, (2) message, (3) audience, (4) channel, and (5) feedback. It is reasonable to think of wildlife managers as the "communicators" and publics or stakeholders as the "audience"; however, in two-way communication those roles often switch. In fact, during the past two decades, the trend toward greater citizen participation in wildlife management has had the effect of putting the manager in the audience box with increasing frequency (see Chapter 11).

At one end of this linear model is the communicator—typically the wildlife manager in communication about wildlife management. If the audience is to accept the message and act on it, the communicator generally needs to be perceived as accurate and credible. Perceptions of trustworthiness and honesty are important components of credibility (Cutlip et al. 1985). At the other end of the model is the receiver of the message—the audience. Knowledge about various socioeconomic, demographic, attitudinal, and knowledge traits, as well as communication habits, of the audience is considered key to communicating effectively about any topic, including wildlife. Audiences for wildlife management consist of an incredible array of groups and individuals, which makes reliance on untested assumptions about them precarious and the need for audience analysis imperative. Additionally, managers must intentionally plan most communication activities to be successful with such audiences.

Lasswell's communication model points us to questions that need to be addressed, preferably with answers supported by data. Planning questions include: What kind of message should I use? What channel(s) is best? What effect am I hoping for? What approach will give the communication credibility with the various audiences? There are many more possible questions to think about. The importance of giving such questions careful consideration can be illustrated by thinking about any single communication characteristic. Credibility, for instance, requires considering more than technical competence. Think of the possible consequences if, during an informative public meeting, an audience is put off by an expert whose speech is too technical or who is unnecessarily condescending. Witness the effects of such treatment of an audience just once and one quickly learns that technical competence does not equate to communication effectiveness.

12.4.2. Other Models of Communication
Lasswell's model is a useful starting place both for informative and persuasive communication in wildlife management (also see sections 12.7.2 and 12.7.3), but it is linear and focused on simple "transmission" of messages. More sophisticated theories are available to guide more sophisticated planning for strategic communication. These are especially valuable for communication associated with stakeholder engagement, where two-way (at minimum) and multi-way exchanges of information are the norm. For example, in co-management of wildlife, where communication is multi-way (with many partners and players in a give and take of ideas, information, and preference negotiation), then the wildlife manager would need to go beyond Lasswell's model. Fortunately, other conceptual models exist to help the manager understand and guide communication in such cases. These are not explored in detail in this chapter, but we recommend that prospective wildlife managers become familiar with the more practical of these models, perhaps through coursework in communication theory and application.

12.5. COMMUNICATION AND COMPLEXITY OF MANAGEMENT GOALS

The goals of wildlife management are complex for many unavoidable reasons discussed elsewhere in this book. For example, managers often attempt to optimize benefits—defined in terms of values and satisfactions—among various stakeholders. This means that a simple measure, such as the number of animals harvested by hunters, cannot be a primary indicator of accomplishment. Stakeholder perception of, or satisfaction with, the economic and sociocultural effects that they perceive will be achieved by actions such as harvest is what counts, and these impacts will vary across types of stakeholders. That is the root of the complexity of wildlife management.

Wildlife managers can engage in inquisitive communication to set goals for economic and sociocultural benefits (i.e., positive impacts from wildlife). Inquisitive communication in this application seeks a clear understanding of the benefits desired by the various stakeholders. Some policy manuals would also refer to this as "formative research." Inquisitive communication takes many forms, such as when managers create the forums in which stakeholder views can be expressed openly and become more widely known (Box 12.2).

12.6. PUBLIC INVOLVEMENT AND COMMUNICATION

As noted in Chapter 11, a defining characteristic for contemporary wildlife management is greater public involvement in management decision making. Many wildlife management programs include public participation throughout the planning and decision-making process. Participation requires effective multi-way communication with stakeholders (between wildlife managers and stakeholders and among various stakeholders).

Box 12.2 *USING THE INTERNET TO PUBLISH PUBLIC INFORMATION ABOUT CHRONIC WASTING DISEASE*

The Internet is a vital source of information for many wildlife stakeholders and a useful tool for state wildlife agencies (SWAs) that seek to communicate timely information to the public about serious issues such as Chronic Wasting Disease (CWD). A project at the University of Wisconsin examined SWA Web sites in Colorado, Utah, Wisconsin, and Wyoming to determine which of four approaches to public information publication (based on theories from risk communication and stakeholder participation) was used for CWD. The "private citizen" approach assumes agencies make decisions with minimal feedback from citizens; SWA Web sites on CWD using this approach could include details on human health risks and technical descriptions of the disease. The "attentive citizen" approach suggests that decisions are made by an agency with some feedback from stakeholders; SWA Web sites on CWD using this method could include data about management plans, current and proposed regulations, and agency research projects. The "deliberative citizen" approach involves broad stakeholder input in agency decisions; Web sites using this tactic could highlight arguments from all sides of the debate, links to informative Web sites, and outcomes of research. Finally, the "citizen publisher" approach assumes that stakeholders actively participate in decision making; SWA Web sites on CWD using this strategy could serve as a conduit through which stakeholders share research findings or promote

remote deliberations about management alternatives. The researchers concluded that the private citizen approach dominated Web sites; attentive citizen approaches were meager and deliberative and citizen publisher approaches were absent. Within regulatory constraints, SWAs may have more flexibility in regard to publishing public information on the Internet (Eschenfelder 2006).

Wildlife pathologists process deer heads as part of a chronic wasting disease surveillance project in New York State. Surveillance programs that require emergency response and lethal removal of deer are controversial; communication is critical to successful implementation. (courtesy NYSDEC/Jim Clayton)

The move from one-way to two-way communication in wildlife management was a novelty only a couple of decades ago, but it was only the beginning of a major shift in how wildlife managers approach communication commensurate with the shift toward participatory governance of wildlife that has unfolded. We quickly moved into "multi-way" communication, where multiple channels and networks of communication are readily available and utilized. Expanding social networks, together with increased reliance on "social media" for wildlife management, is the next frontier of research into communicative possibilities for managers. Regardless of what type of communication wildlife managers are engaged in, experience amply demonstrates that communication opportunities and challenges are always changing.

12.7. COMMUNICATION PLANNING

The secret to consistently effective communication is good planning, guided by a clear idea of how various kinds of communication fit in a comprehensive approach to the wildlife management issue at hand. Ideally, communication planning is informed by good data about the audiences with whom managers want to communicate. Communication planning normally begins with an evaluation of how communication

will contribute to overall program objectives in a particular wildlife situation; linking communication objectives (a type of enabling objective) and actions to other program objectives and actions needed for a wildlife management issue is a critical early step, which (perhaps surprisingly) is often neglected. Communication goals can include information sharing, education, persuasion, dialogue, and behavioral change, among others. An inquisitive communication goal can be for the manager to learn how stakeholders want the agency to proceed on a certain issue.

Planning for communication is, therefore, part of planning for wildlife management overall—it is an essential part of management that, at minimum, seeks to (1) gain intelligence about the management environment through multiway communication with stakeholders and others (e.g., part of situation analysis per Manager's Model), (2) keep internal and external stakeholders informed about the management process and activities, and (3) influence knowledge, belief, attitudes, and behaviors as integral management actions (along with wildlife habitat and population alteration objectives). Effective communication—meaning effective message development, treatment, and delivery through the correct channels—often is supported by specific human dimensions inquiry that characterizes the intended audiences and their communication habits or preferences. Understanding the fundamentals of communication planning is helpful, and the manager can easily gain command of the basics.

Wildlife management organizations engage in performative, informative (or educational), and persuasive communication activities on a regular and recurring basis. The following sections provide an introduction to those types of communication and their application to wildlife management.

12.7.1. Performative Communication

One practical kind of communication essential for wildlife managers to consider is "performative" communication (Austin 1962), or the "symbolic" use of communication to demonstrate a particular trait or maintain a relationship. It is vital for maintaining agency image. For instance, wildlife managers cultivate professional relationships with the public through simple communication that shows that the managers respect members of the community who have legitimate interests in management. Managers attempt to uphold the relationship with performative communication even when they disagree with some stakeholders. The sum of past interactions between an agency and its publics result in an overall "image" that is a background for all future events and interactions that occur in any given situation (Box 12.3).

12.7.2. Informative Communication

Most wildlife managers are well-versed in informative or educational communication. Facts and ideas need to be transmit-

---

Box 12.3 COMMUNICATION AND AGENCY IMAGE

A positive agency image enhances open communication and fosters public support for citizen involvement processes that the agency initiates.

Studies conducted for the New York State Department of Environmental Conservation (NYSDEC) illustrate the importance of agency image. Public perception of a wildlife agency—the agency's public image—has three elements: (1) perception of agency personnel, (2) perception of agency function, and (3) perception of communication behavior of the agency (Decker 1985). Several studies have determined that perceptions of agency personnel are frequently the strongest element of an agency's image. Studies have shown that the more contact stakeholders have with agency personnel, the more positive their regard for these professionals.

For example, a study of northern New York deer hunters and their beliefs and attitudes toward NYSDEC reported that the agency was not effectively communicating with them. Nearly half of those questioned said that the department did not listen to their views. In two other studies, leaders of organizations with interest in deer in northern New York reported that although they had a positive image of the agency's wildlife personnel and management programs they had poor regard for its communication behavior. A common theme for all three studies was the need for improvement in communicating with the public.

The public image of agencies and conservation organizations is conveyed through interpersonal and mass communication. In this photo, a state agency biologist gives a television interview during a visit to a black bear den in New York State. (courtesy William Siemer)

ted to stakeholders regularly. Common means for this type of communication include informational brochures, bumper stickers and other notices (Fig. 12.1), presentations at public meetings, workshops, print mass media, electronic media, and the Internet.

Informative communication becomes more difficult as the complexity of issues and the diversity of audiences increases. Informative communication can be limited in its effectiveness when management objectives (1) are controversial (e.g., lethal control of an overabundant species, such as double-crested cormorants), (2) involve uncertainty (e.g., how many cougars live close to town), or (3) require more than simply providing the best available information to publics.

The informational objective often is obvious and straight-forward. For instance, if a wildlife agency or wildlife conserva-tion non-governmental organization (NGO) acquires a wet-land and restores habitat for birdwatching, then the associated communication goal may be as simple as informing birdwatch-ers of the availability of the site for their use and enjoyment. When the goal is simple and non-controversial, as in this case, the manager can move on to other issues. That is, with the message defined, the manager then considers the question of how to position the message for maximum informative im-pact within the resources (e.g., budget, expertise) available for the purpose. That brings us to media planning and message design.

12.7.2.1 Media Planning

Selecting appropriate channels for messages is called media planning. In mass communication, that means using market-ing and demographic data to reach targeted stakeholders. In an agency, however, media planning refers to the informal process of determining the best way to reach an audience. For example, the manager may know that birdwatchers typically read the state birdwatching organization's newsletter and one or two commercial wildlife publications. That may be enough information about media use by birdwatchers to develop a media-use plan. Media planning for simple informational messages is often relatively automatic because the manager typically develops a list of regular publicity contacts: outdoor writers, sportsmen's clubs, bird conservation organizations, public officials, and so on.

Nevertheless, the manager should also consider whether audience and media research would help in making media choices aimed at reaching desired audience segments. Using surveys, focus groups, and interviews, for example, managers can identify attitudes of audience segments and the media they commonly use (Box 12.4). Research on stakeholder ex-periences with wildlife and their knowledge, beliefs, and at-titudes about wildlife can yield insight vital for effective media planning. With such information in hand, plus research-based knowledge about the communication channels used by dif-ferent groups of stakeholders, you can develop a functional media plan.

Ideally, managers should use multiple channels (e.g., news-papers, Internet, TV) to reach an audience. It is advisable to use channels "in parallel" (several channels that reach the audience

Figure 12.1. The logo for New York State's "NeighBEARhood Watch" program was the touchstone for a research-based media plan to edu-cate stakeholders about preventing negative human–bear interactions (courtesy NYSDEC)

directly) rather than in series (one channel that reaches another channel that then reaches the audience). "Gatekeepers" of in-formation in the channel series approach can pose challenges to managers by restricting information flows, whereas using multiple channels reinforces ideas and reaches more people.

For example, if a wildlife manager wants to inform guides who take people on birdwatching tours about the results of an agency-funded study of birdwatchers' motivations, then pro-ducing a 100-page report and posting it on the agency Web site may not be sufficient to reach the entire target audience. For that to occur, all guides would have to be aware that the report was in the agency's site, and they would need to read the entire report to receive the intended message. The agency would be-come a gatekeeper of information. A more effective approach would include multiple actions such as the following: writ-ing a media release with pertinent points for newspaper, TV, podcast and radio use; preparing an article for the tour guide organization's newsletter; e-mailing a notice or blog about the report's availability to a list of tour guides; or having the study leader make a presentation at the tour guides' annual meet-ing. The former approach relies on communication working in series; the latter approach has many elements working in parallel.

When selecting and using channels of communication, a manager should pay attention to

- the purpose of the message,
- the content of the message,
- the characteristics of the audience,
- the available channels that will reach the audience,
- the personnel available with the skills needed to use the channel effectively,
- how to combine channels and use them simultaneously,
- how to minimize the use of channels in series,
- the cost of a channel in relation to its effectiveness, and
- the time constraints.

Note that, in some situations, communication cuts across different "scales" that affect the wildlife management process.

Box 12.4 *STAKEHOLDER COMMUNICATION NETWORKS*
Double-crested Cormorant Management in Michigan and Ontario

An example of determining channels for communication about wildlife can be seen in an effort by the Great Lakes Fishery Commission to explore a social network of stakeholders involved in double-crested cormorant management. Social networks can serve as important channels through which people share, receive, and exchange information about wildlife and wildlife management issues, especially when those issues are complex or contentious. Researchers found that the more frequently two individuals communicated about cormorants in northern Lake Huron the more likely they were to share attitudes about the birds and their management. Beyond providing insight into network effects on stakeholder attitudes, social network analysis produced sociograms (i.e., maps of the social network), which illustrated the structure of the network. The sociogram clarified which stakeholders are central in the

network, which are disconnected, and who may serve as a bridge between groups (Muter 2009).

Double-crested cormorants on Lake Huron, Michigan (photos courtesy Bret Muter)

In the example mentioned in Box 12.4, social network analysis revealed a complex array of social connections that bisect different levels of government and jurisdictions ranging from local to state, interstate, and even international. These types of social networks are common and require wildlife managers to navigate the complexity of intra- and interorganizational communication as well as to reach people experiencing negative impacts from wildlife.

12.7.2.2. Message Design
Even with a simple informational message, managers should pay careful attention to how the message is phrased, the de-

Box 12.5 *DESIGNING A COMMUNICATION CAMPAIGN USING AUDIENCE SEGMENTATION*

Hanoi, Vietnam is a center for wild animal consumption in the form of both health and beauty products and bush meat. Many species of wildlife consumed in Hanoi are critically endangered (e.g., the Saola, a rare forest-dwelling bovine). The World Wildlife Fund's TRAFFIC program un-

dertook a study to gather information on people's attitudes, motivations, awareness of legislation, and receptiveness to different types of media for the purposes of designing a communication campaign to ensure sustainable levels of wildlife consumption. After surveying Hanoi residents of

Forest rangers in Vietnam and Laos receive classroom training, during which instructors from the Saola Working Group explain the concept of social marketing. In one exercise rangers create educational posters, like the one displayed in this photo, to internalize social marketing concepts. (courtesy Serda Ozbenian, Saola Working Group)

sign elements surrounding it, and other attributes of the communication package. In most cases, the manager should assess how the intended audiences are likely to interpret the message. Doing so can be as simple as trying it out on a few people (e.g., "pilot testing").

The message contains the information that the manager wants the audience to understand, (usually) to accept, and often to act upon. Messages should be easily understood by the recipients and perceived as being useful to them. Two traits of messages are important (Cutlip et al. 1985):

- *Content:* The message must have meaning for the recipients and must be relevant; people listen to messages that offer them something concrete and useful.
- *Clarity:* The message should be couched in simple terms that mean the same thing to the audience as they do to the communicator. Complex issues must be simplified and clarified.

An effective message is (1) *matched* to the intellectual, social, economic, and physical capabilities of the audience; (2) economically, socially, or culturally *significant* to the needs, interests, and values of the audience; (3) *concise;* (4) *timely,* especially when there are seasonal factors and issues are current;

(5) *balanced* by factual material covering both sides of an issue; (6) *applicable,* presenting information the audience can use; and (7) *manageable* by the communicator within his or her time constraints and resources (Box 12.5).

12.7.3. Persuasive Communication

Informative communication is relatively simple in comparison to persuasive communication. It is more difficult to get someone to change either an attitude or a behavior than it is to raise knowledge and awareness. Persuasive communication campaigns can be more elaborate to plan than informational communication campaigns, but wildlife managers can be well-prepared to engage in persuasive communication by following some fundamental tenets of persuasion to guide planning.

Research on persuasion is one of the oldest and most active fields of investigation within communication research. Due to the organizational imperative to communicate "good" information to the public *and to get the public to act on such information,* the wildlife manager has long been oriented toward achieving attitude or behavior change through communication; however, real persuasion—measured by change in attitude or behavior toward a particular end—is difficult to

varying professions, age groups, and income levels, researchers learned that demand for wild animal products is prevalent and widely accepted; many respondents identified themselves as future users. The most common sources of information about animal products were television, the Internet, and newspapers. Few respondents were aware of legislation protecting endangered animals. This inquiry

helped wildlife professionals learn which specific messages could be delivered to the public (e.g., illegality of wild animal trade), what media could be used to communicate messages, and the types of messages that should be delivered to the most frequent consumers of wild animals (e.g., government and business employees; TRAFFIC Southeast Asia 2007).

A forest ranger conducts community outreach to raise awareness about Saola conservation in the Quang Nam area of Vietnam. Saola are threatened by poachers' snares and local hunting for bush meat. (courtesy Serda Ozbenian, Saola Working Group)

A street sign created by member organizations in the Saola Working Group raises awareness about the vulnerable status of Saola (courtesy Serda Ozbenian, Saola Working Group)

achieve. Short-term persuasion—or a rapid change in attitude or behavior—is what wildlife managers often want, but this goal is fraught with difficulty, seldom assured, and may not be long lasting.

One-shot, matter-of-fact, informational messages alone rarely result in attitude or behavior change, so more needs to be considered when such change is desired. Change typically requires long-term, intensive and extensive efforts to influence attitudes and behaviors. Change is not based only on polished communication from wildlife managers to stakeholders; it requires many opportunities for feedback from stakeholders and a social context that rewards stakeholders for steps taken toward the desired behavior change. Long-term change can be induced through repetitive, consistent, and informative use of various media if the rewards and social context are conducive to the changes sought.

12.7.4. Risk Communication: An Application of Informative–Educational and Persuasive Communication

Increasingly, wildlife managers and stakeholders are focusing on "risk" associated with human–wildlife co-existence; "risk" is now one of the fundamental lenses through which diverse

stakeholders view the world, leading wildlife managers to adapt the ways they go about resolving risks. Wildlife present an array of risks (negative impacts) to humans; these include economic, psychological, and health and safety threats. Communicating about these threats is important and becoming increasingly so in situations where risk perceptions of stakeholders are major factors in determining management objectives. In fact, managing negative impacts associated with human–wildlife interaction involves both informative and persuasive communication. Risk perceptions (i.e., intuitive judgments about risks) include some level of information and informative communication or education may modify risk perceptions, but, as we will see below, there is more to risk communication than just I and E because stakeholders may need an additional nudge to engage in behavioral change.

Concepts of risk communication—essentially a combination of application of informative and persuasive communication concepts—can serve managers as tools for effective wildlife management in situations where perceived risks associated with wildlife are key components of the management context. Risk communication theory provides insight about wildlife-related risk information needs and responses of different stakeholders to risk messages. We introduce you to some basic risk-perception and communication ideas relevant to wildlife management.

Risk includes a *technical* component that assesses the probability of occurrence and severity of consequence(s) as well as a *value-based* component that assesses the level of dread (and other emotions) associated with an event. Variations in stakeholders' risk perception can be understood by examining individual differences (e.g., demographics) or sociocultural influences (e.g., worldview) or a combination of the two. Considering public perceptions of risk is as important to risk assessment as determining exposure probabilities. It is essential to remember that public perceptions of wildlife-related risks do not always mirror the perceptions of wildlife managers; furthermore, stakeholders' perceptions are the reality that informs their beliefs, attitudes, and behaviors, including wildlife management preferences.

Media—from newspapers to television to blogs—are fundamental actors in the process of communicating risks to society. The media play a substantial role in disclosing risks to society and discussing management strategies to reduce risk. The effects of media on public perceptions of risk should also be considered; media may amplify or attenuate public perceptions about risk, such as those humans may experience from wildlife (e.g., bite or disease), through a process known as the social amplification of risk (Kasperson and Kasperson 1992).

Risk assessment efforts can be linked to wildlife decision making through communication. Risk communication facilitates decision making best by identifying and incorporating both public risk perceptions and expert assessments into the decision process. Managers might use a persuasive risk-communication approach to induce a change in attitudes, beliefs, or behaviors associated with decision alternatives, achieve acceptance of a preferred management strategy, or motivate action in response to a problem (e.g., wearing insect repellent to minimize infection with West Nile Virus). A participatory risk-communication approach might be used to facilitate decision-making processes, resolve stakeholder conflict through participation, or communicate how decisions could be, and are, made.

Gore and her colleagues reviewed a variety of studies applying risk communication concepts to wildlife management. In an early study, communication was identified as one tool available to wildlife managers to modify human behavior and reduce risks associated with human–black bear interactions. Gore et al. (2006, 2007) conducted interviews with campers and caretakers at seven Adirondack Park campgrounds in northern New York to explore nine factors potentially influencing stakeholder-perceived risks. The researchers identified eight constructs salient among campground users and managers: (1) volition, (2) environment, (3) responsiveness, (4) seriousness, (5) dread, (6) agents, (7) trust, and (8) frequency. Overall, perceived risk associated with human–bear conflict was low; caretakers had a higher risk perception associated with negative human–black bear interactions than campers. Results suggested that risk associated with negative human–bear interactions in campgrounds could be most effectively communicated interpersonally rather than using mass media.

Evaluating the impacts of risk communication programs on public perceptions of wildlife-related risk is part of long-term wildlife management. The operating environment, or the sphere of activity within which a campaign functions, can influence a campaign's ability to achieve outcomes, so it is important to understand that environment prior to launching risk-communication efforts. Gore and Knuth (2009) studied human–black bear conflict in southeastern New York and the wildlife-related risk-communication campaign called the New York NeighBEARhood Watch Program (NYNWP; see Fig. 12.1). Based on the social amplification of risk framework (Kasperson and Kasperson 1992), they explored whether mass media could affect the operating environment of the NYNWP and, if so, identify the magnitude and direction of the effect. The researchers used a mail survey of residents in four towns to analyze black bear–related risk perception and collected all black bear–related media coverage during the program implementation period (May–Sep). Only newspaper exposure (not the NYNWP itself) was found to have an effect on respondents' increase in perceived likelihood of bear-related risks; thus, mass media can influence the operating environment of a wildlife-related risk-communication campaign. For more information on a wildlife management campaign explicitly targeted to risk issues, see Box 12.6.

Overall, integrating risk considerations into wildlife management can advance managers' understanding of stakeholders and expand the knowledge base used for decision making (Gore et al. 2009). Specific knowledge of communities or groups of stakeholders, such as that from the studies presented above, is particularly useful for communication planning. In the absence of situation-specific knowledge, communication theory can be a wildlife manager's best source of insight and direction.

12.8. STAKEHOLDERS AND PUBLICS

Stakeholders have different interests and characteristics. It is important to take those differences into account when planning communication. Communication specialists have approached this by thinking of stakeholders as "publics." In communication terms, a public is a group of people who are aware of an issue and have mobilized at some level to address it. We can think of this as a special case of stakeholders.

For any issue one can describe three types of publics: (1) a non-public, (2) a latent-aware public, and (3) an active public (Grunig 1983). The non-public is simply unaware of the issue. The latent-aware public has gained knowledge of the issue; its members recognize the dimensions of the issue and may even be predisposed to take action, but for a variety of reasons they do not. The active public has mobilized around the issue; its members not only recognize that it exists but have taken steps to achieve solutions reflecting their goals.

The number of people in each public segment for most issues usually can be described in a pyramidal structure

Box 12.6 *WILDLIFE-RELATED PERSUASIVE RISK COMMUNICATION*

Persuasive risk communication can motivate responsible wildlife-related behavior change, help foster acceptance for wildlife management strategies, increase knowledge and understanding related to human–wildlife conflict, and be used to increase public participation. Researchers at Cornell University worked with biologists at the New York State Department of Environmental Conservation (NYSDEC) to develop, implement, and evaluate a pilot persuasive risk-communication program called the New York Neigh-BEARhood Watch Program (NYNWP). The NYNWP created a set of eight types of materials whose content focused on residential behaviors that could be changed to reduce the proximate risks associated with human–black bear conflict. Six human behaviors specifically were emphasized: (1) refraining from hanging bird feeders during warm-weather months, (2) feeding pets indoors rather than outdoors, (3) storing barbeque grills indoors when not in use, (4) curbing

garbage the morning of pick-up and storing it indoors at all other times, (5) keeping home compost contained and secure, and (6) picking up fruit dropped from fruit trees and harvesting fruit from trees before fruit falls. In all, 11,117 individual messages (billboards, bear-o-meters, brochures, magnets, posters, lawn signs, article reprints, fact sheets) were distributed. Persuasive risk messages included relative risk levels on the "bear-o-meter" (e.g., high, medium, low bear activity), motivation to comply with the lawn sign (i.e., "I'm a good NeighBEAR—I don't feed the bears"), and knowledge about risk on the billboard (i.e., "Bear problems? Remove the food and you'll remove the bear.") Analysis of pretest and posttest survey data of residents in the NYNWP implementation region revealed no short-term evidence of black bear–related behavior change after the NYNWP was implemented. However, public attitudes toward black bears changed as a result of the program.

Preparation of materials for New York's "NeighBEARhood Watch" education program was informed by social science research and materials were evaluated for effectiveness (photos courtesy Meredith Gore)

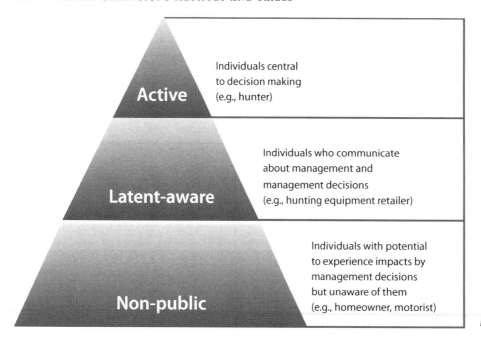

Figure 12.2. Types of publics

(Fig. 12.2). For example, though active hunters are a far smaller group than non-hunters, in many states there are tens of thousands or even hundreds of thousands of hunters who may mobilize and seek voice in a wildlife issue that affects their interests.

Communication planners design communication campaigns according to the type of public they want to reach (Table 12.2). When dealing with non-publics or people unlikely to see themselves as stakeholders for a given issue, informational communication is the order of the day. In general, being informed is a precondition to mobilization. The non-public is a diverse group typically lacking any specific, special channels to reach them, so wildlife managers use the mass media.

It is common for segments of the public to overlook the fact that they have a voice in a wildlife issue. In some cases, latent publics have become very active when faced with controversial management decisions such as those focused on suburban elk, bear, mountain lion, deer, or coyote management. Agencies can use education and persuasion to mobilize this public. Messages should deal with (1) perceived constraints (the factors that are preventing them from being more active), and (2) locus of control (so they understand that their actions can make a difference).

The greatest chance for short-term persuasion is with active publics, especially when the desired change is for such a public to take a particular action (in a referendum, for example, or toward a desired practice for managing wildlife within the community). Even with active publics, however, mass communication that encourages group participation is necessary to maintain a sense of involvement. Active publics may be enlisted to help disseminate a message.

Even if a public is opposed to an agency's position, the principles of communicating with the group are similar, though the messages might differ. An actively opposed public, for example, is best dealt with in person and intensively. A latent-aware public that leans toward opposition to an agency position might be dealt with effectively through mass-communicated information. A communication specialist should be able to analyze and recommend the reasonable approaches for various contexts.

12.9. DETERMINING EFFECTIVENESS

A common error in communication planning is neglecting to gauge the probable effectiveness of communication (usually expectations are too high given the expertise and other resources available to implement a communication campaign) and assess the reason for the result. An agency often will produce a Web page, conduct a meeting, or carry out a mass media–based public information campaign without actually determining whether it had the desired effect. Evaluation (in this case, "summative" evaluation) is paramount for subsequent communication planning.

Significant segments of the public often remain resistant to a change in wildlife policy and management even in the face of extensive and concerted stakeholder engagement and public communication on an issue. It is all too common to hear staff of wildlife agencies or NGOs say things such as "Our message is not getting out," "Stakeholders are misinformed," or "The information campaign didn't work." Evaluation research, one form of human dimensions research, can help answer several important questions:

- Was our message received? By whom?
- Was it interpreted as desired?
- Did it change attitudes and beliefs as desired? Why or why not?
- Did it change behavior as desired? Has the change endured over time?
- Were there any unanticipated side effects?

Table 12.2. Communication planning for types of publics

Type of public	Sender activities	Channel focus	Messages	Audiences
Non-public	Education	Media, publications	Define problem; are frequent	General public
Latent-aware public	Education, recruitment, conversion activities	Mixed	Focus on constraint recognition, locus of control	Targeted audiences
Active public	Meetings, consultation, demonstration projects, reward activities	Interpersonal, small groups, organizational	Involving; focus on data, solutions	Known audiences

A "no" answer to any of those questions (and that is often the case) indicates that the manager should take corrective action. In cases where the message simply was not received, media-planning issues are implicated. If it was not interpreted correctly, message design may be implicated. If it did not change attitudes and beliefs, it may simply be that not enough time has passed (and not enough repetitions have taken place). If the message did not change behavior, the manager should determine why, reformulate the planned communication accordingly, or even reassess the behavioral objective.

It is important to keep in mind what level of effectiveness is reasonable and should be expected and how media channels and outlets are perceived by stakeholders. A variety of studies have explored assumptions about the effectiveness of persuasive or informational communication, particularly when the mass media are involved.

In a study conducted at Roma Tre University (Carrus et al. 2009), researchers examined the role and effect of mass media communication associated with biodiversity conservation in protected areas before and after the designation of the Tuscan Archipelago National Park in Italy. The researchers found that the media emphasized both positive and negative aspects of the issue, and that local identity and sense of community drove community opposition to the park. The authors concluded that because media can drive local perceptions of issues it is imperative to develop alternate ways of communication, including media campaigns, public debates, and consultation initiatives.

Another study examined the role of "framing" (how information is presented) in media coverage of environmental policy (Hovardas and Korfiatis 2008). Researchers studied local newspapers during three periods of ecotourism development within the Dadia Forest Reserve in Greece to search for three main themes (ecotourism, forest management, environmental awareness). They found that the emphasis in the newspapers changed over time from an expert-led approach to a participatory approach when discussing environmental policy; therefore, "frames" can guide a community's interpretation of an issue. Framing is an important means by which media can affect environmental and wildlife management.

Most managers are familiar with the incendiary role that media can play in crisis situations such as a fatal wildlife attack or zoonotic disease outbreak. Crisis communication planning is just as much a part of communication planning as is planning for daily activities or scheduled programs. During crises, publics respond more positively to managers who provide in-

formation quickly and who are perceived as trustworthy and sensitive to stakeholders' concerns. Audiences often are just as concerned with the perceived fairness of the communication process as they are with the outcome. That is, audiences who perceive that their voice was heard and that there is a possibility that their voice will have an effect on management are potentially as satisfied with group communication processes as those who perceive that their desired position is being advocated. Thus, while a particular management decision may be determined largely by scientific or policy constraints, the communication dimension should not be ignored. Involvement of stakeholders throughout any process is one way to show that voices are being heard.

One tendency that is common among wildlife managers, but that should be avoided, is assuming that the media are more powerful than they really are. The Media and Society Research Group at Cornell University (Besley and Shanahan 2004) used secondary data from the General Social Survey data to explore Bjørn Lomborg's (2001) hypothesis that the media drive environmental "hysteria." The researchers did not find strong relationships between media use and concern about environmental issues. Although the media do obviously play an important role in all environmental issues—including wildlife management—managers should be aware that mass media effects in wildlife-related issues are typically small to moderate. This observation is not meant to dissuade you from actively engaging mass media in wildlife-related communication, but be realistic about the effect of your efforts and those of others using the media to communicate information and opinions about any particular issue. Clearly, it is potentially problematic if the public consistently receives only one view from mass media when that view is not consistent with current policy and program, so it is foolhardy not to plan a mass media component for any wildlife program.

12.10. PAYING ATTENTION TO FEEDBACK

Communication between managers and stakeholders should be two-way; listening to stakeholders' opinions about a management issue is as important as telling them your viewpoints. Building effective mechanisms for soliciting or capturing feedback is essential to two-way communication and for meaningful public involvement in decision making.

Feedback can be intentional or unintentional. Intentional feedback occurs when stakeholders communicate with managers to make certain managers know how stakeholders interpret and feel about the content of the message they received. Unintentional feedback is information the manager receives about stakeholders' reactions through observation or a second-hand source; it may be useful for assessing the need for additional communication efforts.

Managers can give unintentional feedback to stakeholders and sometimes that feedback can send the wrong message. As a manager, you will need to guard against being perceived as uncaring, aloof, or insincere in the way you deal with input from stakeholders. Remember our opening comment about traits for success in communication: sensitivity and awareness of others' perspectives are crucial. Regardless of your intended meanings, people's interpretations of your verbal and non-verbal feedback are what matter. Be inquisitive so that you learn how people are interpreting your messages.

Wildlife managers should not leave feedback to chance; it is too important. Purposeful solicitation of feedback has two distinct advantages. First, managers are more likely to obtain accurate, timely, and representative information if they develop the feedback mechanisms carefully (e.g., a survey of the birdwatching public). Second, seeking input has important implications for public image. By seeking input, managers are saying that they value stakeholders' opinions. Mechanisms to encourage feedback may be as simple as providing the address of an agency contact person on brochures and Web pages or as formal as conducting surveys and public hearings. Formal inquiry to obtain feedback from stakeholders is covered in detail in Chapter 11.

12.11. USING COMMUNICATION

The trend toward greater public participation in management decision making highlights the importance of effective communication between and among stakeholders and managers to ensure that all players are aware of each other's interests and needs regarding wildlife. Similarly, communication between managers and stakeholders can improve stakeholders' understanding of and support for management policies, objectives, and activities.

Without effective communication, wildlife management can be frustrating, unproductive, or ineffective for wildlife managers and stakeholders alike. Poor communication about management programs can result in confrontations between and among stakeholders and managers and, in worst-case situations, the pursuit of political solutions or litigation by the public. Uninformed or misinformed stakeholders may ultimately circumvent the agency's management process through non-participation, legal action, and non-compliance. The agency can greatly diminish the probability and severity of these unfortunate outcomes through carefully planned and executed communication efforts designed as public-involvement mechanisms.

Keep these points in mind:

- Contact does not mean communication; make sure your message is noticed and interpreted as you intended.

- Communication is a multi-way interchange leading to mutual understanding.
- The public is not monolithic; diverse individuals and groups of people have a stake in wildlife management.
- You must communicate with professionals inside your agency as well as with external stakeholders.
- You need to understand each group's characteristics to communicate effectively with them.
- You will need to make effective use of communication skills to realize the full benefit of public involvement in wildlife management.
- Be a persistent and patient communicator; do not always expect success in the short term.
- Understand how, why, when, and whether others received your message; seek and understand feedback.

SUMMARY

Wildlife professionals must be prepared for communication in a highly connected and rapidly changing environment. This chapter provided an introduction to communication theories, communication planning, and provided examples of effective communication with stakeholders.

The intent is a primer that helps wildlife managers think more deeply about planning and delivery of communication, and as a catalyst for learning more about communication skills, which often are identified as the most important skills in wildlife management.

- The opportunities for communication successes and failures, and therefore opportunities for management successes and failures, have never been greater, primarily because of the convergence of three factors: (1) more diverse stakeholders are interested in wildlife and its management; (2) more governmental and non-governmental entities are involved in governance of wildlife; and (3) the media available for communication are changing dramatically.
- A source–message–channel-receiver model of communication is guided by answers to five basic questions: (1) "Who? (2) Says what? (3) To whom? (4) Through what channel? (5) With what effect?" More sophisticated theories are available to guide strategic communication associated with stakeholder engagement, where multi-way exchanges of information are needed.
- The key to effective communication is good planning; this planning begins with an assessment of how communication will contribute to overall program objectives. Communication objectives can include information sharing, education, persuasion, dialogue, and behavioral change, among others.
- Media planning is a process for choosing the most effective channels for communication. In mass communication, media planning uses marketing and demographic data to reach targeted stakeholders. In an agency, however, media planning more commonly refers to the informal process of determining the best way to reach the intended audience.

- Communication is most likely to reach the intended audience when multiple channels (e.g., newspapers, Internet, TV) are used. It is advisable to use channels "in parallel" (several channels that reach the audience directly) rather than in series (one channel that reaches another channel that then reaches the audience).
- Persuasive communication campaigns—those intending to change beliefs or behaviors—are most effective if they follow fundamental tenets of persuasion to guide planning. Changes in beliefs or behaviors typically require long-term, intensive, and extensive efforts to influence attitudes or behaviors.
- Risk communication—a combination of informative and persuasive communication—can serve managers in situations where perceived risks associated with wildlife are key components of the management context.
- Feedback and monitoring of communication is critical to learning how to improve communication and has implications for public image. By seeking input, managers are sending an implicit message that they value stakeholders' opinions.

Suggested Readings

Corbett, J. B. 2008. Communicating nature: how we create and understand environmental messages. Island Press, Washington, D.C., USA.

Fazio, J. R., and D. L. Gilbert. 1986. Public relations and communications for natural resource managers. Second edition. Kendall/Hunt, Dubuque, Iowa, USA.

Hayes, R., and D. Grossman. 2006. A scientist's guide to talking with the media. Rutgers University Press, New Brunswick, New Jersey, USA.

Jacobson, S. 2009. Communication skills for conservation professionals. Island Press, Washington, D.C., USA.

Peter Sandman's Risk Communication Website: www.psandman.com

PART V · HUMAN DIMENSIONS APPLICATIONS

Part V examines various ways that human dimensions research and resulting insight are used in wildlife management. Chapters focus on three broad categories of management: (1) abundant wildlife, (2) scarce wildlife, and (3) uses and users of wildlife. Examples of how human dimensions insights and stakeholder engagement techniques are applied in management are presented in each chapter. The intent is to demonstrate how human dimensions research is advancing effectiveness—achieving desired types and desired levels of impacts—in wildlife management.

Chapter 13 draws attention to the challenges of managing abundant wildlife populations. When President Theodore Roosevelt took office, it was probably inconceivable that one day the focus would not be only on how to save wildlife but also on how to manage overabundant wildlife. This chapter takes on the issue of defining overabundance, which is not always directly related to high wildlife population densities; overabundance is more closely related to the frequency and quality of human–wildlife interactions. An issue today is to establish whether society's needs for wildlife are being met and what wildlife population size or density would yield desired net benefits (positive impacts exceed negative impacts) for society. With increased urbanization, suburbanization, and exurbanization occurring throughout much of the world, issues that require management of human–wildlife interactions relative to abundance are likely to magnify.

Some of the most contentious issues in wildlife management are about scarce wildlife. Various applications of human dimensions methods and insights in the context of managing species that are naturally rare or whose abundance is greatly reduced from exploitation or habitat degradation are discussed in Chapter 14. It is interesting to compare the management of abundant species and management of scarce species. Many of the same methods and information needs are apparent in both situations, yet the context often is radically different.

A long-standing area of human dimensions concern—management of wildlife uses and users—is addressed in Chapter 15. This is a broad topic area that spans traditional concerns related to recreational hunters and hunting, subsistence use of wildlife, and cultural uses of wildlife by Native Peoples. This chapter illustrates the influence wildlife use has had on the direction of wildlife management, including on development of knowledge about the human dimensions of management and how this knowledge is applied to identify and understand wildlife-related activity trends and patterns; sociodemographic changes that influence activity involvement; wildlife recreation as a social process; impacts associated with wildlife-use activities; public attitudes toward wildlife-use activities; and importantly, public support for existing and alternative sources of program funding.

13
HUMAN DIMENSIONS OF ABUNDANT WILDLIFE MANAGEMENT

MICHAEL R. CONOVER AND JONATHAN B. DINKINS

Abundant populations of some wildlife species provide valued opportunities for society to experience positive impacts from wildlife viewing and hunting. Abundant wildlife also presents the potential for negative impacts. Managing impacts associated with abundant wildlife, whether through game management (Bolen and Robinson 2003) or wildlife damage management (Conover 2002), has been a dominant feature of public wildlife management for much of its history and remains a focus of wildlife management worldwide.

Human dimensions research and its application have been directed at issues caused by wildlife abundance for decades. Early studies of impacts from abundant wildlife focused on describing and understanding farmers' and ranchers' tolerance of crop damage and livestock depredation. During the 1980s, some wildlife managers turned their attention to wildlife impacts in urban and suburban environments as expanding populations of wildlife spread into these environments. "Abundant" wildlife may not necessarily mean a large number of animals in some situations. Even one animal can sometimes be enough to cause undesirable and perhaps intolerable impacts (Box 13.1).

When a population of wildlife becomes abundant and causes serious negative impacts it is often referred to as "overabundant." Management of overabundant wildlife is often approached through the lens of "too many animals of species X at a particular location," with "too many" sometimes determined based on assessments of habitat carrying capacity (a biological or ecological assessment) but more often judged by problems that confront people (an assessment of human tolerance of impacts). Reducing a wildlife population is a means to an end and not the end in itself; that is, population reductions are pursued to achieve some larger fundamental objective.

13.1. DEFINING WILDLIFE AS ABUNDANT OR OVERABUNDANT

Abundant wildlife can be defined in ecological and sociological terms. Based on an ecological perspective, the term "abun-

dant wildlife" means a population that is at or near its long-term, sustainable biological carrying capacity. This definition implies that an overabundant wildlife population is one that is above the habitat's carrying capacity and is overusing the available resources in its habitat; it is ecologically unsustainable. Based on a human dimension perspective, "abundant wildlife" means a wildlife population that is large enough to satisfy human needs. Abundance of wildlife is a perception of population size relative to humans' desires for positive impacts produced by wildlife and the capacity of humans to tolerate the negative impacts experienced from wildlife. This is a sociological, social-psychological, and economic assessment. From a human dimensions perspective, "overabundant" wildlife denotes a population that occurs in such numbers or densities that people have lost their tolerance for that species because of its increasing threats to human health, safety, property, or enterprises. In such cases, the positive impacts provided by wildlife that are experienced by stakeholders are exceeded by the negative impacts, thus making the net human–wildlife relationship negative for the stakeholders involved. Conversely, scarce wildlife (the topic of Chapter 14) can be defined as a wildlife population that is too small to meet all human needs and interests, including a desire to avoid extinctions of wildlife.

Unlike the ecological assessments of abundance and overabundance, where wildlife population needs and environmental effects are compared with thresholds defined by scientific evaluation of habitat, sociological and social-psychological assessments can vary among stakeholder groups (people experiencing similar kinds of impacts) and among individuals within a group. This makes determining status a "messy" problem (Lachapelle et al. 2003). Essentially, a population of the same number of animals can be abundant and overabundant (and sometimes scarce) depending on the kinds, extent, and level of interactions and impacts people experience versus what they desire. In the case of negative impacts, overabundance may be largely a function of the capacity of those experiencing such impacts to tolerate them.

Box 13.1 *SMALL NUMBERS–LARGE IMPACTS*
Conserving Tigers in Nepal

There are now more than 100,000 protected areas covering approximately 11% of the world's land area. The main purpose of these areas is to protect wild biodiversity. These lofty purposes notwithstanding, human–wildlife conflicts are pronounced along the edges of protected areas and threaten effectiveness of such areas for conservation. Conservation of wildlife species usually depends on the capacity of local people to tolerate impacts created by increasing or expanding wildlife populations. In some of these situations, the wildlife species creating impacts may be endangered. Such is the case in Chitwan National Park (CNP) in Nepal. CNP is a globally important protected area for conservation of the endangered tiger, but increasing rates of human–tiger conflicts in and around CNP threaten the long-term coexistence of tigers and people.

We collected data on how human interactions with tigers

modify several psychological elements, which, in turn, influence local stakeholders' preferred future tiger-population size in the nearby forests (tiger acceptance capacity [TAC]). A survey of 499 individuals living less than 2 km from CNP and the nearby buffer zone was conducted to quantify TAC and factors that may influence TAC, including interactions with tigers, perceptions of tiger population trends, attitudes toward tigers, risk beliefs, risk perceptions, and risk tolerances. Our results indicated that multiple dimensions of risk belief, attitudes toward tigers, cognitive risk perception, and risk intolerance of threats to livelihoods and human health and safety were significantly associated with TAC. Stakeholders living nearest to tigers value the tigers and broadly support tiger conservation; yet stakeholders are intolerant of impacts tigers have on their livelihood and well-being. Important conservation gains may be achieved if we modify the psychological context in which people perceive impacts (positive and negative) from locally abundant wildlife (even if it is endangered globally), instead of focusing only on altering the physical environment in which wildlife and humans interact.

NEIL H. CARTER

Researchers set up a trail camera to collect information about movements of tigers in Chitwan National Park, Nepal. Ecological and social information is needed to inform tiger conservation programs. (photos courtesy Neil H. Carter)

Carrying capacity in wildlife management is most often viewed in the classical ecological sense of biological carrying capacity, which is the natural limit of a wildlife population. A human dimensions perspective on carrying capacity can help managers integrate biological limits with social, economic, institutional, administrative, cultural, and legal limits. It is based on the assumption that bounds exist on the impacts that stakeholders will accept—it acknowledges that they will tolerate negative impacts associated with wildlife only to a point, beyond which wildlife become "overabundant" for those whose tolerance has been exceeded (Box 13.2). Concepts such as wildlife acceptance capacity (Decker and Purdy 1988), cultural carrying capacity (Ellingwood and Spignesi 1986), social carrying capacity (Minnis and Peyton 1995), and wildlife-

Box 13.2 *DEER-RELATED IMPACTS PERCEIVED BY THREE STAKEHOLDER GROUPS*

A study on deer-related impacts in Michigan helped clarify relationships between wildlife populations, human values, and acceptance capacity for an abundant yet popular animal. Open-ended interviews and a mail-back questionnaire to rural residents of southern Michigan were used to (1) compare how three commonly identified stakeholder groups (hunters; farmers; and non-hunting, non-farming rural residents) perceive impacts created by deer; and (2) how those impacts affect acceptance capacities for white-tailed deer. Comparisons among stakeholder groups revealed commonalities and differences in perception of impacts resulting from specific interactions with deer. The findings indicate acceptance capacity is influenced by perceptions of impacts—the context-specific perceptions of the effects of interactions with a wildlife population—rather than simply the frequency of interactions. One of the most important impacts for non-hunting, non-farming residents arises from seeing deer, which residents indicated was a perceptual signal that they lived in a healthy environment. On the other hand, negative impacts arising from deer–vehicle collisions reduced acceptance capacity for deer, regardless of the stakeholder group to which residents belonged. Participation in hunting and farming were poor predictors of acceptance capacity for deer.

This study was subsequently applied statewide to explore how a better understanding of impact perceptions might lead to management actions that address desires of multiple stakeholder groups across varied and changing landscapes. Stakeholders concerned about deer in Michigan's Upper Peninsula, where interactions with deer were the least frequent and strongly tied to hunting satisfaction, were more willing to tolerate deer and wanted more deer

than stakeholders in the Lower Peninsula, where deer–vehicle collisions and other problems caused by deer were frequent. This research aided statewide planning for deer management in Michigan, where objectives take into consideration not only deer populations but also the impacts created by those populations. Managers recognized that actions focused exclusively on conventional manipulations of wildlife populations and habitats may not be sufficient to achieve all impacts desired by stakeholders.

STACY A. LISCHKA
Source: Lischka et al. (2008)

A Michigan study found that negative impacts arising from deer–vehicle collisions reduced acceptance capacity for deer, regardless of the stakeholder group with which residents identified themselves (courtesy Joe Paulin)

stakeholder acceptance capacity (Carpenter et al. 2000) were developed to consider human tolerance for impact levels.

Implicit in the term "overabundant wildlife" is the notion that reducing the population will solve problems caused by that population, but that is not always the best approach. Every situation needs careful consideration before launching into wildlife population management for two reasons: (1) all wildlife populations (even those that are so small that they are endangered) can create negative impacts and lead to human–wildlife conflicts; (2) many wildlife damage problems cannot be alleviated by reducing a wildlife population.

13.2. RELATIONSHIP BETWEEN WILDLIFE POPULATIONS AND THE IMPACTS ARISING FROM HUMAN–WILDLIFE INTERACTIONS

Any wildlife population can be thought of as a resource that provides a large array of values for humans, which arise from impacts that humans experience when interacting with

wildlife (Giles 1978). Many of these impacts are beneficial to people (e.g., health benefits, income from wildlife-associated employment, and an enhanced quality of life). Other impacts are detrimental to people (e.g., health and safety risks and economic losses) with the effect of decreasing a person's health, wealth, psychological well-being, or quality of life (Conover 1998, 2002). Negative values are produced when wildlife damage property, injure humans by attacks, increase frequency of human diseases, harm businesses, threaten public safety (e.g., deer–vehicle collisions), or cause a nuisance (Table 13.1). When all of the positive and negative values of wildlife populations are summed, however, the net benefits that wildlife populations provide to society greatly outweigh the costs (Conover 1998), a reality manifested in the public and private support that Americans and others provide for wildlife conservation. The extent to which this support is sustained depends, in large part, on how effectively impacts from abundant wildlife are managed.

Both positive and negative impacts associated with wildlife

Table 13.1. Annual costs of wildlife damage in the United States (including both property damage and expenditures to prevent the problem from occurring), 2001

Problem	U.S. dollars (in billions)
Deer–vehicle collisions	1.6
Bird–aircraft collisions	0.3
Wildlife damage to agriculture	4.5
Wildlife damage to timber industry	3.4
Wildlife damage to households	12.5

Source: Conover (2002)

populations vary as numbers of wildlife change. A description of a management scenario frequently starts with the statement "there are too many [species *X*]," but this is misleading because the actual problem is that the balance of impacts that stakeholders experience from wildlife has shifted from net positive to net negative. When an issue in wildlife management is described as "There are too many . . ." then a human dimensions approach leads to questions about the specific kind, extent, and level of impacts experienced (both positive and negative); who is experiencing them (stakeholder identification); and how to weigh the aggregate positive and negative impacts associated with the wildlife population.

Problems associated with many wildlife species increase in degree and frequency as their populations expand because there are more animals to cause problems and humans become less tolerant as negative impacts increase or cross a threshold of tolerance. One human dimensions specialist referred to this phenomenon as "the tragedy of becoming common" (Leong 2010). When tolerance of negative impacts is exceeded then the wildlife population of interest is "overabundant" from a human–wildlife interaction perspective. For instance, people are willing to tolerate some risk of car collisions or contracting Lyme disease because they enjoy having deer in their neighborhood; but, as the risk of collisions or disease increases for them and their families, some (perhaps many) people become so concerned that they start to view local deer unfavorably (Stout et al. 1993, Leong 2010).

The benefits of wildlife (i.e., positive impacts) also increase as a wildlife population increases, because there are more animals to view or hunt; but, for many species of wildlife, there are limits to positive impacts as the wildlife become more abundant. For example, the satisfactions (positive impacts) that a person receives from shooting his or her 10th deer of an annual hunting season is often less than he or she felt from shooting the first (Conover 1997*b*).

13.3. MANAGING WILDLIFE: BALANCING POSITIVE AND NEGATIVE IMPACTS

When wildlife populations in North America were recovering from overharvest and habitat degradation in the early 1900s, the focus of wildlife agencies was on species sought by hunters and trappers. The goal was to produce the greatest number of

animals that could be harvested by people, and the underlying fundamental objective was to have satisfied hunters and trappers, which at the time was taken to mean managing game populations at levels capable of producing the experiences and attendant satisfactions (benefits) desired by hunters and trappers.

As wildlife became more numerous during the latter half of the twentieth century, more people began to complain that some species were causing damage in excess of what they were willing to tolerate. For example, many people were thrilled when Canada geese first visited their local park, but when those few geese increased to hundreds of geese then people became concerned about goose feces fouling their park and wanted the geese removed. In response to numerous complaints about overabundant wildlife (and ongoing concerns about scarce wildlife), many wildlife agencies gradually shifted objectives toward maintaining wildlife populations within a range where they would be considered adequately abundant and simultaneously tolerable for stakeholders. This objective requires wildlife agencies to ascertain whether society's needs for wildlife are being met or whether a larger or smaller wildlife population would yield greater benefits for society. Making such an assessment is not easy because there are many kinds of stakeholders who have different perceptions of what constitutes the ideal wildlife population. For instance, a deer hunter may view the local deer population as scarce because he or she did not see any antlered adult male deer during the hunting season. An avid gardener may view the same deer population as overabundant because deer destroyed his or her garden.

The upper and lower limits of a wildlife population that define its range of tolerance can be ascertained by first determining the population size range associated with net positive impacts for various kinds of stakeholders. This is the population range within which most stakeholders would consider a particular wildlife population to be generally satisfactory. From a societal viewpoint, a wildlife population can be considered to be within its range of desired abundance when most (but not necessarily all) stakeholder groups perceive the population as being at an acceptable level.

Wildlife agencies that pursue this management approach rely upon a combination of human dimensions research and stakeholder engagement practices to assess the impacts produced by different populations of various wildlife. That is, wildlife agencies use multiple sources of insight to assess whether society would benefit more if a wildlife population increases, decreases, or stays the same. Wildlife agencies are most likely to err in estimating what is an acceptable density for a wildlife population when any of three conditions exist: (1) the wildlife population is either increasing or decreasing rapidly, which poses the risk that the policies of wildlife agencies may not reflect the current situation; (2) the wildlife population simultaneously provides great benefits and causes great problems for society; and (3) stakeholder groups are divided as to what represents an appropriate level of abundance, which happens when the wildlife is harming the interests of some stakeholders while benefiting others.

If a wildlife agency errs in estimating the appropriate density for a wildlife population, then elected officials may intervene to correct the error or dissatisfied stakeholders may pursue litigation. Elected officials can be expected to respond when impacts from a wildlife population are outside the socially acceptable range because these officials are charged with attending to the best interest of their constituents. If some stakeholders have captured the policy-making process such that others do not feel they have a voice with the decision makers, then the alienated set may believe their only recourse is litigation.

13.4. ISSUES OF SCALE IN MANAGING WILDLIFE IMPACTS: BALANCING LOCAL, STATE, AND NATIONAL INTERESTS

One ongoing political debate relevant to managing impacts of abundant or overabundant wildlife centers on which decisions should be made by local governments in communities experiencing impacts and which should be made by state and national governments. In the United States, management of wildlife is a responsibility of state government except in those cases where the actions of one state affect another state. Management of endangered species is under the jurisdiction of the federal government because citizens of all states share in the loss if a species becomes extinct; therefore, no state is allowed to take a management action that might cause a wildlife species to be become extinct, even if that action is supported by the state's citizens. Oversight for management of migratory birds also falls to the federal government because decisions made independently by one state in a flyway could lead to mismanagement of migratory birds with far-ranging effects. In addition, agreements about migratory birds are needed among countries and only the federal government can enter such agreements. Migratory bird management in North America is controlled by a treaty signed by the United States, Canada, and Mexico because so many birds spend part of the year in all three countries.

Wildlife management decisions often become contentious when the concerns and interests of local stakeholders differ from non-local stakeholders (Box 13.3). The potential for contention increases when managers deal with wildlife species that are overabundant and causing intolerable negative impacts in some locations but not in others. In these cases, local and non-local stakeholder perceptions of species abundance often diverge and the perceived need for population management differs in the extreme.

Multiple perspectives about abundance of a wildlife species is illustrated in the case of grizzly bear and wolf management in the Yellowstone ecosystem, where these carnivores present risks to both the safety of local citizens and the livelihoods of local ranchers. Because of these impacts, many local stakeholders perceive wolves and grizzly bears as "dangerous" and "destructive" predators and prefer that their population sizes be managed at low levels. Concomitantly, non-local residents across the United States enjoy knowing that wolves and bears

---

Box 13.3 *NATURE PRESERVES IN CAMEROON AND MOZAMBIQUE*
Balancing the Interests of Local, National, and International Stakeholders

Wildlife reserves have been established across Africa to conserve charismatic megafauna, such as elephants. Though classified as an endangered species, African forest elephants are locally abundant and can be a major source of crop damage in localities surrounding a reserve. Elephant conservation depends on the ability of managers to balance national and international interests in protecting elephants with local interests for relief from the crop damage by elephants.

In developing nations, the creation of a nature preserve is often opposed by local residents who fear restrictions on their historical use of natural resources within preserves, crop damage from herbivores venturing out of preserves, and loss of livestock and human lives due to an increase in local predator populations. Crop damage is often cited as the main reason why neighbors dislike nature reserves (Parry and Campbell 1992, Heinen 1993, Newmark et al. 1993). For example, elephants venturing out of Cameroon's Maza National Park caused considerable damage to local villagers. In addition to experiencing crop losses, local farmers were forced to guard their crops 24 hours a day; most of the guarding was performed by school children, who could not attend school while guarding crops. In a survey, 54% of the local villagers around the park believed the elephants provided no value to them and 75% reported that elephants posed a threat to their safety or livelihood (Tchamba 1996). Most local villagers (73%) thought that more elephants needed to be killed to protect crops.

In contrast to Maza National Park, most local villagers near Mozambique's Maputo Elephant Reserve had a higher opinion of that elephant reserve. Most (88%) reported that they liked and used the Maputo Reserve; 71% harvested plants from the reserve and 21% hunted there (De Boer and Baquete 1998). The ability of local villagers to continue using resources within the reserve helped improve their attitude toward it. The local residents reported that the biggest problem with the reserve was crop damage by marauding wildlife, and the opinion of local residents was correlated with their crop damage experiences. De Boer and Baquete (1998) believed that participation of the local communities was critical to the development of management strategies for the Maputo Elephant Reserve and that crop damage by reserve animals must be addressed.

exist in the Yellowstone ecosystem; many of these non-local stakeholders perceive wolves and grizzly bears as "magnificent" apex predators and prefer that their population sizes and range of distribution be expanded. Their interests are represented by powerful national non-governmental organizations (NGOs). Wolf numbers in the Yellowstone ecosystem, for example, are viewed locally as being out of control primarily because of impacts on livestock and big game, but they are viewed nationally as being uncomfortably close to extinction. Obviously, any decision about how wolves should be managed will be contentious and displease or even anger some number of people who have a stake in (i.e., desire certain impacts from) management decisions.

National interests trump local interests in situations that involve federally managed species (migratory species, endangered species), but the federal government does not have jurisdiction in other situations where abundant populations of wildlife are generating negative impacts. For example, management of coyote populations in Wyoming is determined by county boards that are locally selected. The federal government also has recently ceded management of some locally overabundant migratory birds to the states. In Mississippi, where large numbers of double-crested cormorants cause problems for the local aquaculture industry, the state can issue depredation permits that allow catfish farmers to kill these birds under certain conditions without having to seek federal permission on a case-by-case basis.

Stakeholder engagement can help wildlife managers determine the appropriate weight to give to the various interests of such multi-scale issues. The weighting of impacts among stakeholders, including those at different scales, is necessary to arrive at fundamental objectives for management. The wildlife manager is faced with the question, "How do I weigh the tangible and immediate concerns of local stakeholders against the intangible concerns of stakeholders in a far-off state, who may never visit the area?"

A common corollary to the question posed above is, "How do I weigh the concerns of local farmers and ranchers, who are providing wildlife habitat and incurring damage from wildlife inhabiting their land, against the concerns of community residents, who enjoy viewing or hunting those animals?" In this case, one set of stakeholders incur all the costs, and may enjoy some benefits, whereas the other set of stakeholders enjoy the benefits without incurring any costs. Obviously some trade-offs will be needed in the balance of impacts experienced by the stakeholders involved. Fortunately, there are two kinds of actions that can be taken to inform managers in this situation. First, human dimensions inquiry can shed light on what levels of impacts or what balance between positive and negative impacts is acceptable to the farmers (Brown et al. 1978, Decker and Brown 1982, Conover 1998). That is a key piece of information. Second, stakeholder engagement processes at the local level can lead to compromises that are satisfactory (if not perfect) to all parties (Lauber and Knuth 1997, Schusler and Decker 2002, Lauber and Brown 2006; Box 13.4).

13.5. METHODS FOR MANAGING THE IMPACTS OF OVERABUNDANT WILDLIFE

The negative impacts associated with wildlife abundance can be addressed by wildlife managers through a variety of management actions. Some of these have been alluded to earlier in this chapter. In this section, a few of the more common categories of actions are highlighted with emphasis on the human dimensions considerations involved with their use.

13.5.1. Hunting and Trapping

The usual approach for addressing problems caused by an overabundant wildlife population is direct reduction of the population. The conventional method to achieve this for many species of wildlife (e.g., ungulates, furbearing mammals, geese) is through the use of regulated hunting and trapping. Involving hunters and trappers in population reduction has the dual advantage of (1) creating opportunities for additional positive impacts from utilization of the resource, and (2) being cost-effective by engaging "volunteers" in management rather than hiring people to cull animals. For this method to be successful, however, there must be a sufficient number of hunters and trappers willing to harvest enough animals in aggregate to regulate the size of a wildlife population. This method is challenged by a multi-decade decline in hunter numbers (Brown et al. 2000a, b).

Three approaches are commonly used to influence effectiveness of hunting and trapping for controlling overabundant animals: increasing allowable take per hunter, increasing duration of hunting and trapping seasons to provide more opportunity for harvest success, and focusing harvest on females in deer, elk, and other polygynous species. The motivations of hunters and trappers, as well as the amount of time and effort they will commit, must be understood in order to design specific regulations to achieve increased harvest of wildlife. For example, each hunter has a personal upper bound on the number of game animals that he or she wishes to harvest during a year. This number is often limited by how many animals the hunter can consume or give away to friends. A hunter may be unresponsive to enhanced opportunities to harvest more animals when that number is reached.

13.5.2. Influencing Behavior of People Who Experience Negative Impacts

One way to reduce or avoid negative impacts caused by overabundant wildlife is to change the behavior of stakeholders who are experiencing the impacts so that these people are less vulnerable. Examples of steps to reduce wildlife damage include the following:

- Sheep ranchers may be able to reduce the number of lambs killed by coyotes by using herders, guard dogs, predator-proof fencing, outdoor lighting, or shed-lambing (Conover 2002).
- Farmers can reduce their losses to Canada goose herbivory by planting less palatable crops or by not planting palatable

Box 13.4 *USING CITIZEN TASK FORCES TO RESOLVE CONTROVERSIAL WILDLIFE PROBLEMS*

In the metropolitan area surrounding Rochester, New York, management of the local deer population had become controversial and divisive. Local citizens had formed three very vocal groups concerning deer: (1) "Deer Action Committee," composed of people concerned with deer problems, (2) "Alliance for Wildlife Protection," opposed to the use of lethal control of deer, and (3) "Save Our Deer," also opposed to the killing of deer.

To resolve the dispute, the New York Department of Environmental Conservation, Cornell Cooperative Extension, and the Human Dimensions Research Unit formed a Citizens Task Force (CTF) to give local citizens an opportunity to determine the optimal deer density for the area. CTF functioned by allowing people from diverse groups (farmers, sportsmen, foresters, conservationists, animal welfare supporters, animal rights activists, and members of the three citizen groups) to meet and discuss the costs and benefits of increasing, decreasing, or making no change to the deer population. After receiving input from wildlife biologists, health officials, and various stakeholders, CTF recommended that the optimal deer density was 8-10 deer/km². CTF conducted an aerial survey to count deer and found that deer densities in some areas were four times higher than the population goal. The CTF therefore recommended that the number of deer removed equal the number of deer killed in local deer–automobile collisions and in the following year that twice as many deer be culled if there was no decrease in the number of deer–automobile collisions. CTF recommended that hunters be used to reduce deer densities where feasible and that sharpshooters be used elsewhere. The state wildlife agency gained a better understanding of the concerns and attitudes of local residents about deer because of the CTF. The wildlife agency also gained local credibility by employing a bottom-up approach in deciding how to manage deer. The local community developed a sense of ownership in the deer management program (Curtis et al. 1993) when its community and stakeholder leaders became involved in these decisions.

A street sign warning motorists of a popular deer-crossing area in a residential neighborhood of Amherst, New York (courtesy Joe Paulin)

crops close to water bodies frequented by geese (Washburn et al. 2007, Radtke and Dieter 2010).

- Catfish farmers can reduce their losses to double-crested cormorants by using other fish species as buffer prey, constructing smaller ponds, placing dyes in the water (to reduce cormorant ability to locate fish), changing fish stocking times, locating fish of vulnerable sizes close to human infrastructure, and harassing cormorants more persistently (Glahn et al. 2000).
- Motorists can reduce frequency of deer–vehicle collisions by paying more attention to deer along a road's right-of-way, slowing down in deer-crossing areas, and practicing defensive driving (Grovenburg et al. 2008, Mastro et al. 2008, Ng et al. 2008).

Note that educational and informative communication efforts (informational guides, training, etc.) may be useful in encouraging or enabling the behaviors of stakeholders. Human di-

mensions studies can help wildlife agencies design and deliver these programs effectively.

13.5.3. Increasing Tolerance of Wildlife Impacts

Negative impacts of wildlife populations are diminished when the affected people become more tolerant of the negative impacts; essentially, tolerance increases when the balance of positive and negative effects experienced from one's interactions with wildlife shifts so the positive aspects of interactions exceeds the negative. This can be achieved in a variety of ways, and insight about the individuals involved can aid in selecting an optimal approach. Several studies have shown, for example, that farmers who enjoy hunting are more tolerant of wildlife damage on their farms than those who do not hunt (Gabrey et al. 1993), and they are more likely to favor an increase in the deer population in their area (Tanner and Dimmick 1983). How can that insight be put to use? Human dimensions studies have shown that becoming a hunter is largely a sociocultural process that is unlikely to occur in a short time frame; yet, in the long term, encouraging children from farm families to become hunters may be a reasonable element of a hunter recruitment program given that these children are likely to become farmers in the future. Furthermore, the involvement of youngsters in hunting may generate parental interest in wildlife (and perhaps tolerance of negative impacts of game species) in the near term. This approach has limitations. Farmers who lose enough crops to wildlife that their livelihoods are jeopardized are unlikely to find their losses tolerable even when they or their families also enjoy hunting.

Many landowners become more tolerant of economic losses caused by game species when they can offset these losses by generating income through leasing the right to hunt game on their property. Landowners generate more income from hunting leases than from raising livestock or growing crops in some parts of the United States and Canada (Sovoda 1980, Burger and Teer 1981, Messmer 2009). Rather than be-

ing annoyed by the negative impacts of game animals on their property, these landowners may encourage wildlife use of their land by building ponds, adding water guzzlers, or planting food plots (Box 13.5).

Most people want to be appreciated by others for their contributions to the greater good. This is also true for landowners that ensure wildlife abundance for the enjoyment of others. For example, a landowner's tolerance for wildlife damage can be increased by other people acknowledging that the landowner's tolerance is appreciated and recognizing landowners for their efforts to accommodate wildlife. Telephone calls, letters of thanks, personal visits, and offers of help all convey the message to landowners that other people appreciate their efforts and sympathize with their losses. Many wildlife agencies and NGOs have landowner appreciation awards and award events (dinners) where recipients are publicly recognized.

Society is often divided about the management of impacts resulting from overabundant wildlife populations because

Box 13.5 *ALIGNING LANDOWNER INTERESTS WITH PUBLIC INTERESTS THROUGH COOPERATIVE HUNTING UNITS*

Wildlife must have inherent values that are competitive with alternative land uses for wildlife to exist on private land (Berryman 1981). Unfortunately, private landowners have had little economic incentive to invest in wildlife (Wigley and Melchiors 1987); however, increasing demands for high-quality hunting areas and a decreasing supply of wildlife habitat (Doig 1986) have stimulated interest among landowners to lease the rights to hunt on their property (Wallace et al. 1989).

In response, Utah created the Cooperative Wildlife Management Unit (CWMU) program, wherein hunting is allowed by permit only. Some of these permits are distributed free to the public through a public drawing conducted by the Utah Division of Wildlife Resources; others are given to participating landowners, who can then sell them. Hunters like the CWMU program because they gain access to private lands, have greater chances of harvesting an animal, and get the opportunity to experience a high-quality hunt (Messmer and Dixon 1997). Landowners like the CWMU program because it provides them a new source of revenue and a mechanism to manage hunter access and control trespassing. CWMUs also increase landowner tolerance of wildlife damage because landowners are able to recoup some economic losses by selling hunting permits (Messmer and Schroeder 1996). For instance, one CWMU pays local farmers $200 (U.S. currency) for every acre of irrigated alfalfa they grow. The payment is designed to compensate farmers for any big game damage to their alfalfa, which has increased farmer tolerance of wildlife damage. Equally important, CWMUs motivate landowners to enhance wildlife habitat on their property.

The Utah Cooperative Wildlife Management Unit program creates economic incentives for private landowners to tolerate mule deer and elk, which may compete with livestock for forage (courtesy Michael Conover)

the benefits and costs of wildlife do not fall evenly upon all members of society. Hunters, outdoor enthusiasts, and wildlife viewers all benefit from wildlife and often want large populations of wildlife. Alternatively, many farmers and ranchers prefer smaller populations of wildlife because these people are often victimized when wild animals eat their crops or kill their livestock. Some farmers, ranchers, and forest landowners question why they should be expected to endure problems from wildlife when wild animals are owned by society. The next section focuses on these various types of landowners.

13.5.3.1. Focus on Farmers, Ranchers, Woodland Owners, and Other Landowners

Emphasis on mitigating the negative impacts associated with wildlife in North America has focused on various types of landowners, in part, because they control a large portion of the land base and habitat for wildlife and are important for access to wildlife for many Americans. Many farmers, ranchers, and woodland owners tolerate damage to their crops, predation on their livestock, or diminished regeneration of valuable tree species in their forests because they reap positive impacts from the intangible values of wildlife. These people may enjoy observing wildlife on their land or gain a sense of satisfaction from knowing that they are sharing their land with wildlife. By supporting wildlife on their land, many people gain a sense of satisfaction from knowing that they are doing a good job managing their property. Others enjoy the rewards of providing their family and friends with opportunities to hunt or view wildlife on their property. These people may believe that the trade-offs of negative and positive impacts experienced from wildlife using their lands are, on balance, positive. This positive balance can be enhanced when appreciation of wildlife is increased by creating a sense of stewardship of wildlife among landowners. This can be achieved by allowing and encouraging landowners to help with the planning and implementation of management plans for wildlife (Treves et al. 2006).

Usual mitigation methods for landowners are of the following types: permits that allow landowners to kill depredating animals, loan of equipment to repel or exclude animals, provision of special services by government to handle problems, infrastructure to exclude wild animals from high-value crops, and direct compensation for lost income. Most state agencies (and U.S. Department of Agriculture [USDA]–Animal and Plant Health Inspection Service–Wildlife Services [WS]) engage in these activities, which are commonly referred to as wildlife damage management.

Depredation permits. One way of reducing the negative impacts of wildlife is to issue depredation permits to landowners who incur wildlife damage; these permits allow them to kill animals in the act of causing damage on their property. From a human dimensions standpoint, these permits increase the level of control that individuals have over the damage they incur. Depredation permits may grant landowners or their agents the right to shoot game species out of season, to shoot more animals than would otherwise be permitted, or to kill animals that are otherwise protected by law. When depredation

permits are issued for big-game species, some states require that carcasses be turned over to conservation officers or other agents of the state who will process the meat for donation to a charity, food bank, or needy family. This requirement allays concerns that game animals might be "wasted."

Depredation permits seem like a simple solution to the depredation problems, but human dimensions studies have revealed that this method is not universally embraced by landowners for several sociological and social-psychological reasons. For example, farmers in western New York State (Brown et al. 1977, 1978) and Northeast Michigan (Dorn and Mertig 2005) reported that killing deer outside of hunting season carried with it a social stigma among neighbors; it was not an accepted behavior within the community. This made the use of depredation permits untenable for many farmers.

Loan of equipment. Many wildlife agencies help farmers who experience wildlife damage by loaning them equipment (e.g., propane cannons or other fear-provoking stimuli) to scare animals away from their crops. From a human dimensions perspective, this kind of program obviously alleviates concerns about having to purchase specialized equipment but it also increases farmers' sense of self-efficacy, a factor that studies have shown is important to reduce perceptions of risk.

Special services. Some wildlife agencies have their employees chase away or harass wild animals that are foraging in fields owned by farmers. The USDA WS helps ranchers with predator problems by killing the predator that is responsible. The rationale for having WS employees kill offending predators, rather than having ranchers do so, is based in the value society places on predators. Trained wildlife biologists can use their expertise to identify and kill only the offending predator and thereby minimize the number of predators that have to be killed to alleviate a problem. For the same reason, state wildlife agencies often deploy their own personnel to shoot damage-causing deer and elk rather than allowing landowners to do so. Another important collateral effect of this method is that the involvement of government specialists increases community perceptions of legitimacy of the activity and relieves landowners of some social pressure.

Infrastructure. When wildlife damage is likely to recur year after year, wildlife agencies may help pay for the cost of projects that minimize future damage, such as building a deer-proof fence around an apple orchard. In these cases, wildlife agencies often provide the fencing material if farmers will install and maintain the fence. In some instances, state and federal waterfowl refuges grow their own crops for the purpose of luring depredating birds to refuge fields and to stop them from damaging the fields of neighboring farmers. This method reduces the negative impacts experienced by farmers and also reinforces the perception that society values the habitat and other resources that farmers and other landowners contribute to wildlife conservation.

Compensation. Several states have compensation programs that address economic losses incurred by landowners from wildlife. Even NGOs have partnered with state or federal agencies to lessen the financial burden of some wildlife species

Box 13.6 *ADDRESSING LANDOWNER TOLERANCE FOR WILDLIFE THROUGH COMPENSATION*

An approach used in some situations to mitigate the wildlife damage that agricultural producers experience is to pay them compensation for their losses (Wagner et al. 1997). Payments usually are made only after monetary losses have been confirmed by trained wildlife personnel and payments are prorated based on the amount of money available for compensation and the extent of losses. These compensation programs are usually limited to damage caused by game species. The rationale for this limitation is that wildlife agencies are trying to maintain high densities of game species (i.e., abundant populations), and by doing so their management actions may be contributing to a higher level of damage. Compensation is therefore more likely for damage caused by deer, elk, pronghorn, wild turkeys, and pheasants than for damage caused by non-game species. Some wildlife agencies also compensate ranchers for livestock killed by bears, cougars, and wolves rather than allowing ranchers to kill them because other stakeholders do not want these large predators to be killed.

Compensation programs work best when farmers and ranchers who have endured wildlife damage are reimbursed for all or most of the losses. For example, a poorly funded compensation program did little to help farmers in India who were suffering from livestock losses due to predation by snow leopards and wolves. Both species were protected so local farmers could not kill the predators even when their losses exceeded half a family's income. Unfortunately, government compensation only covered 3% of losses (Mishra 1997). With this level of compensation, it is no surprise that Mishra (1997) found that local livestock owners preferred to kill predators rather than seek compensation.

Many people believe that compensation programs are an appealing alternative to lethal control but this is not always the case (Wagner et al. 1997). Compensation programs are expensive and wildlife agencies often have to pay for them using money obtained from the sale of hunting licenses or taxes on the sale of ammunition. This depletes money from the budgets of wildlife agencies that would otherwise be used for other purposes, such as habitat restoration. Not surprisingly, many hunters oppose the use of wildlife funds raised through their licenses and specific taxes on their equipment to pay for compensation programs. Many agricultural producers also do not consider compensation as a panacea; they note that compensation

programs usually only cover part of their actual loss and that the paperwork required for compensation is excessive. Ranchers also note that compensation is only offered when a dead lamb or calf is located and when the predator responsible for killing it can be identified. Most carcasses are never found when livestock on rangelands are killed by predators. For example, Brian Palmer and several technicians searched every day for dead lambs in pastures located on Cedar Mountain, Utah. Despite their intensive search efforts, they were only able to locate carcasses for 112 of the 898 lambs that disappeared from the herds during the study (Palmer et al. 2010).

Open grazing systems may expose livestock to predators. Farmer and rancher compensation programs may increase tolerance for carnivores among the stakeholders most likely to incur direct economic losses to predators. (© Dario-Fotolia.com)

(especially large predators, such as wolves) that consume livestock.

Direct compensation addresses the economic aspects of tolerance, but studies have shown that farmers and ranchers have a psychological attachment to their livestock and value their personal identity as people practicing good animal husbandry,

which includes protecting their animals from harm. This has sociocultural implications that financial compensation programs do not address; thus, what appears on the surface to be a straightforward economic impact with a financial solution may have much more complex human dimensions considerations (Box 13.6).

13.5.3.2. Focus on Urban and Suburban Residents

Although many negative impacts from wildlife historically occurred and still occur in rural, agricultural environments, the attention of wildlife management increasingly has turned to managing negative impacts of human–wildlife interactions in urban and suburban areas (Conover 1997a). Wildlife managers find diverse stakeholder perspectives in urban and suburban situations. Human dimensions research in these areas during the past two decades has helped managers understand this diversity and the advantages and disadvantages of various approaches to address urban and suburban wildlife management issues (stakeholder engagement approaches are described in Chapter 11).

Efforts at community-based management occur across the United States and Canada, often including stakeholder advisory committees that develop objectives and select actions acceptable to the community. Management in urban and suburban situations typically takes on political overtones. The ends are usually more easily identified and agreed upon than are the means for achieving them, especially when the decision is between lethal versus non-lethal actions. Almost always there are efforts to increase residents' tolerance of the problems they are experiencing by providing information and access to services to "fix" the problem (e.g., remove the offending animal or erect physical barriers). Although decades of national surveys conducted by or for the U.S. Fish and Wildlife Service have demonstrated that wildlife observation is popular in urban and suburban areas, other studies (Butler et al. 2003) are also revealing what might be the front edge of a growing negative reaction to the health (wildlife-associated diseases), safety (vehicle collisions and attacks from wildlife), and economic (property damage, including plantings and motor vehicles) risks that wildlife present. Managing tolerance of impacts in this situation requires sophisticated combinations of education, information, and technical assistance programs that, in aggregate, diminish negative impacts (including unnecessarily high perceptions of risks associated with wildlife).

13.5.4. Education and Outreach

Some impacts, especially those based in perceptions of risks associated with wildlife and those that can be mitigated through changing human behavior can be influenced through education. Many people have unfounded fears of wild animals and may forego outdoor activities because they fear encountering wildlife that might harm them. Other people may venture outside but not enjoy themselves to the fullest because they are concerned about the risk of being harmed by wildlife. These situations reduce the positive impacts of outdoor experiences generally and wildlife experiences specifically. When a person's perceptions of risk are exaggerated, educational programs about wildlife can allay unfounded fears by providing information about the actual risks and ways to avoid negative interactions.

Human–wildlife conflicts involving physical injury to humans often occur when people find themselves in unsafe situations after failing to recognize the danger that wildlife can pose to their safety. Many people approach wild animals too closely because they are motivated by the desire to get a closer look, take a better photograph, or feed a wild animal as if it were a pet. For example, 56 visitors to Yellowstone National Park were gored by bison between 1978 and 1992 (a much higher number of people than the number attacked by bears or cougars). Many of those gored had approached too closely in an attempt to photograph a bison (Conrad and Balison 1994). Other people are injured when they make physical contact with a wild animal, perhaps in an attempt to pet it or pick it up. Intentional contact is also a factor involved in cases of snakebite in the United States. For example, thousands of people are bitten by venomous snakes annually in the United States, and half of all bites occur when the victim attempts to catch or handle a snake (Minton 1987). These injuries could be prevented if people were better educated about such risks. Land and wildlife management agencies at all levels engage in outreach efforts to visitors and others to prevent them from engaging in risky behaviors. Surprisingly, few studies have evaluated the effectiveness of these outreach efforts. One intensive program and evaluation to reduce behaviors of residents that attracted black bears into residential areas in a community in New York yielded discouraging results (Gore et al. 2008).

Education programs, such as those offered by state land-grant universities and their statewide extension and outreach systems, can provide agricultural producers and homeowners with research-based information about future risks they face from wildlife damage to their crops, property, or gardens. This information can allow them to take proactive steps to prevent problems before they occur. Some possible solutions are technically simple. For example, damage to sunflowers by blackbirds could be prevented if farmers planted fields close to blackbird roosts with crops that are unpalatable to blackbirds (e.g., soybeans) rather than sunflowers. Equally valuable is information for farmers and homeowners to discourage use of techniques that do not work.

In addition to becoming habituated to the presence of humans, wildlife (such as the mountain goat pictured) can become conditioned to approach humans that offer food. Food conditioning may contribute to unsafe human–wildlife interactions. (courtesy Bret Muter)

SUMMARY

As some wildlife populations increased during the 1900s, wildlife agencies broadened their management objectives from increasing numbers of wildlife to maintaining populations within a range of abundance that yields impacts acceptable to most stakeholders. The latter objective is difficult because stakeholders vary widely in what positive and negative impacts they experience from wildlife and in their perceptions of the size of the wildlife population that produces desirable levels of such impacts. This chapter revealed how human dimensions insights are applied to management of abundant wildlife.

- Although assessment of wildlife being "abundant" and "overabundant" is commonly approached as a technical biological or ecological condition, in most cases the assessment is more accurately estimated from the perspective of human tolerance of impacts irrespective of wildlife population size.
- "Abundant wildlife" is defined as a wildlife population sufficiently large to satisfy human needs and interests. Abundance is a perception of population size relative to humans' desires for positive impacts experienced from wildlife and human capacity to tolerate the negative impacts experienced from wildlife. Abundance becomes a management issue when tolerance is exceeded.
- Wildlife populations are considered overabundant when humans experience such a high level of negative impacts from wildlife that tolerance is exceeded. When overabundance occurs, one solution is to reduce the population of wildlife, but this is not the only solution and sometimes not one that will produce the desired effect. Other options include influencing people's tolerance of wildlife impacts, changing the behavior of people so they can protect themselves, and compensating people for their economic losses.
- A chief wildlife issue today is to ascertain whether society's needs for wildlife are being met or whether a larger or smaller wildlife population would yield greater net benefits (that is, positive impacts exceed negative impacts) for society. Questions of scale often center on whether determinations of overabundance should be made by local governments in communities experiencing impacts or whether determinations should be made by state and national governments.
- Increased urbanization and suburbanization throughout much of the developed world likely will lead to these areas being a center for issues of overabundance. Management in these urban and suburban situations typically becomes political and occasionally hotly contested. The ends—the fundamental objectives—are more easily identified and agreed upon than the means—the management alternatives—for achieving them, especially when the alternatives are between lethal versus non-lethal actions.
- Information and education campaigns play a role in changing current situations where overabundance is occurring. Avoiding habituation of wildlife to humans can prevent many negative impacts and human dimensions insights are usually required to help people make informed choices that lead to tolerable levels of human–wildlife interactions.

Suggested Reading

Adams, C. E., K. J. Lindsey, and S. J. Ash. 2006. Urban wildlife management. Taylor & Francis, Boca Raton, Florida, USA.

Barton, D. 2004. The beast in the garden: a modern parable of man and nature. W. W. Norton & Company, New York, New York, USA.

Conover, M. R. 2002. Resolving human–wildlife conflicts: the science of wildlife damage management. Lewis Brothers, Boca Raton, Florida, USA.

Decker, D. J., T. B. Lauber, and W. F. Siemer. 2002. Human–wildlife conflict management: a practitioners' guide. Northeast Wildlife Damage Management Research and Outreach Cooperative, Ithaca, New York, USA.

14

HUMAN DIMENSIONS OF SCARCE WILDLIFE MANAGEMENT

JODY W. ENCK AND ALISTAIR J. BATH

Managing scarce wildlife species—whether to conserve existing populations or to restore species to places where they have been extirpated—is vital to mitigating worldwide anthropogenic changes that are leading to large-scale fragmentation of landscapes, habitat degradation, and imperilment of a growing number of species and ecological systems worldwide. The specifics of wildlife conservation and restoration necessarily differ across the globe, depending on a variety of contextual factors. These include stakeholder interest and acceptance at the local level; forms of local, national, and even pan-national governance; and the capacity of stakeholders, non-governmental organizations (NGOs), and government conservation groups to do the work. The range of human dimensions considerations and information needed to inform conservation and restoration decisions is therefore broad and diverse.

In this chapter, we describe some of the human dimensions considerations in managing scarce wildlife in different contexts, and we situate human dimensions information needs with respect to managing scarce wildlife in the management cycle detailed in Chapter 7. We first define what we mean by scarce wildlife, describe the concept of conservation, and identify actions used to accomplish species restoration. Then we explain management of scarce wildlife from an impacts perspective. We also provide examples in which human dimensions considerations, in the forms of social science concepts and theories as well as specific empirical data, have been applied to management of scarce wildlife. Throughout the chapter, we augment the application of human dimensions considerations with information about the general historical trajectories of wildlife conservation and restoration efforts in North America, Europe, and Central America to demonstrate the contextual importance of governance and capacity issues. We discuss how those differences may require different kinds of human dimensions insights and considerations.

14.1. WOLF RESTORATION: AN EXAMPLE OF SCARCE WILDLIFE MANAGEMENT

Some of the most contentious wildlife issues in the history of modern wildlife management have been about scarce wildlife. Social-psychological, sociocultural, and economic considerations abound in these issues. Managing scarce wildlife has the potential to call on all areas of human dimensions insight and skills discussed earlier in this book. A brief review of gray wolf conservation in the United States illustrates the range of human dimensions considerations underlying management of any scarce wildlife species. Throughout the chapter, we return to wolf management for examples of human dimensions considerations when managing scarce wildlife.

Gray wolves were extirpated from nearly all of the contiguous 48 U.S. states by the 1930s (USFWS 1987). Increasing levels of bounties provided economic incentives to kill wolves as they became scarcer (Nesslage et al. 2006). When the U.S. Federal Endangered Species Act (ESA) was passed in 1973, gray wolves were one of the original species receiving federal protection, and for which a recovery plan was mandated (USFWS 1987). Between the passage of the ESA in 1973 and the mid-1990s, the idea of restoring wolves to the United States, especially the northern Rocky Mountain region around Yellowstone National Park, was the focus of intense regional, national, and even international debate (e.g., McNaught 1987, Bangs and Fritts 1996). Few people were ambivalent about wolf restoration; and opponents passionately debated restoration in the Greater Yellowstone Ecosystem (GYE) and the idea of wolf conservation in general (Bath and Buchanan 1989).

Managers had a conundrum: the ESA mandated that wolves be restored but left many operational questions unaddressed. Where should restoration focus in this area of thousands of square miles? How many wolves (or packs) were required to be successful? Whose idea of the natural order of things and acceptable associated sets of positive and negative impacts from the presence of wolves should receive the greatest consider-

© Outdoorsman-Fotolia.com

ation? Managers faced a complex, value-laden problem in this coupled socioecological system (see Chapter 3 for a discussion of the characteristics of "wicked" management problems). Many of the circumstances of wolf management are common in management of other scarce wildlife.

As with most wildlife conservation efforts, part of the debate about wolf management developed around issues of scale—including geographic scale (how large an area would be affected), temporal scale (how long it will take to see a response), and stakeholder proximity (whose interests carry the most weight). Different stakeholder perspectives on these scale considerations contributed to the difficulty of management planning and implementation. Some people pointed out that wolves are not endangered on a global scale or even in North America; thousands live in Canada and Alaska. With the discovery of a wolf den near Glacier National Park in northwestern Montana in the late 1980s, some people thought that perhaps recolonization to the GYE was only a matter of time; they suggested that patience coupled with protection (rather than active restoration) might be all that was needed to achieve wolf recovery. Wolves were increasing in other locations as well. For example, a population of wolves was expanding into the upper Great Lakes states of Minnesota, Wisconsin, and Michigan from adjacent populations in Canada (Fuller et al. 1992). By the early 1990s, wolf restoration was even being discussed as a possibility for the Northeast United States (Enck and Brown 2002). These developments further complicated discussion about the importance of active restoration of wolves in GYE.

From a human dimensions perspective, a scale issue of interest in GYE and many other efforts to conserve or restore popular scarce species of wildlife is that of stakeholder proximity to the location targeted for management of the wildlife species and its habitat. In the GYE wolf example, people living adjacent to the area targeted for restoration are concerned about the effects they may experience if the wolves succeed in establishing a breeding population and expand their range. On the other hand, people from across the country feel they have a stake in the wildlife living in national parks and forests. Understanding the human dimensions of this issue requires, in part, the identification of stakeholders across scales and the revelation of their beliefs, attitudes, organizational affiliations, and political activity. In addition, managers in this type of situation work with policy makers to determine what weight in the decision-making process to give the various perspectives held by stakeholders from near and far. Although legal parameters may be established (e.g., ESA in the United States), in reality no formulaic approach to weighting stakeholder interests exists that is uniformly applied in even one country (let alone common among nations) because of varying governance structures. The human dimensions of scarce wildlife conservation and management play out in their own particular way in each case.

In part, this is because impacts that people associate with human–wildlife interactions may differ from case to case. In the GYE, Native American tribes were supportive of the idea of restoring an important cultural symbol, and environmentalists liked the possible ecosystem services that might be provided by the restoration of a top predator (see, for example, Scarce 1998). Across the GYE, ranchers were concerned about the negative effects on their economic livelihoods and way of life; hunters were equally concerned about potential impacts on elk herds and the quality of elk hunting opportunities (Thompson 1993). Visitors to Yellowstone National Park expressed both hope that a missing component of the ecosystem could be seen or heard in the park and concern about the possibility that they or their pets could be attacked by wolves (Bath 1991).

Another set of considerations in scarce species management focuses on issues associated with various kinds of capacity. Biologists believed that abundant populations of elk, bison, and other large herbivores in the GYE would be more than adequate as a prey base for a substantial wolf population. Because of the uncertainty about the impacts of greatest importance to various stakeholders, however, it was not clear whether the wildlife acceptance capacity (Decker and Purdy 1988) for wolves would be exceeded for some stakeholders, such as ranchers and hunters. In addition, some businesses and NGOs were touting the potential ecotourism benefits of wolf restoration, yet there was uncertainty about whether municipalities in the Yellowstone area had the capacity to capture those potential benefits and to mitigate any negative impacts from wolves. The full social and ecological ramifications of wolf restoration into the GYE are still unfolding as the landscape, the wolf population, and stakeholders continually change.

14.2. GENERAL APPROACHES TO MANAGEMENT OF SCARCE WILDLIFE

Scarce wildlife are of at least three basic types: (1) species that once were abundant but whose populations have become greatly diminished, threatened, or endangered with extinction throughout their range; (2) species that may be common globally, but have populations that have become scarce or even extirpated from portions of their historical range; and (3) species with ecological and behavioral traits that result in their occurring at very low population densities throughout their ranges (i.e., they are naturally scarce). Wildlife management framed as "restoration" is most relevant in the context of the first two kinds of scarce wildlife. Restoration is the management work of actively re-establishing a self-sustaining, free-ranging population of a species in a geographic area where that species was extirpated or reduced to an unviable, unsustainable, or otherwise undesirably low population level. This definition is particularly important given the confusing set of terms that have been associated with the idea of "bringing back wildlife" (e.g., restoration, introduction, reintroduction, recolonization). Sometimes confusion over definitions has serious consequences. This occurred in the GYE wolf case, where one of the most contentious elements in the social and legal debate surrounding wolf restoration focused on whether the management action was really a "restoration" or a "recolonization."

Restoration means re-establishing a previously naturally occurring species in an area from which it has been extirpated (e.g., putting gray wolves back into Yellowstone National Park). Restoration is one part of the broader concept of conservation; that is, it is management to ensure long-term viability of a wildlife species, which may include human uses of the species (e.g., harvest and viewing). Conservation of species such as American bison, bald eagle, or eastern timber rattlesnake may involve, or even require, restoration. On the other hand, conservation of species that are scarce throughout their ranges because of their ecology (e.g., northern goshawk, pileated woodpecker) may depend upon protection rather than

restoration efforts. Protection in this sense means legal restrictions on any manner of direct and indirect take of individuals of a species and usually protection of its habitat, backed up with law enforcement.

Definitions are important for understanding how stakeholders react to management of scarce wildlife. Some stakeholders may make a distinction between protecting an existing population, assisting recolonization of a population, and restoring a population. Stakeholder support of management may differ depending on which approach is proposed.

14.3. IMPACTS TO BE MANAGED WITH RESPECT TO SCARCE WILDLIFE

Recall that *impacts* are highly valued effects of human–wildlife interactions. The categories of impacts that might be managed in the context of scarce wildlife are the same as impacts associated with nearly any species: human health and safety impacts, economic impacts, psychological impacts, sociocultural impacts, and ecological impacts (Box 14.1). As is often the case, different stakeholders may perceive or experience different impacts from a wildlife population or actions to manage it. For example, landowners whose properties support endangered species may be concerned that legal requirements to avoid disturbing habitat would diminish their economic return from possible land uses, whereas commercial ecotourism operations may gain economically from offering opportunities for wildlife viewers to encounter scarce species.

Real and perceived levels of risks to human health (e.g., rabies, West Nile virus) and safety (e.g., bear, cougar, and coyote encounters) likely will be increasingly important impacts to manage, regardless of whether the wildlife species involved are scarce or common. Some of the most challenging impacts to manage could be sociocultural conflicts among stakeholders who value a species in quite different ways (e.g., ranchers are concerned that restoration of wolves will negatively affect their livelihood, whereas Native Americans and other rural residents may place a high cultural value on the presence of wolves).

One main difference between management of impacts associated with scarce species versus common species is the legal framework guiding management of threatened and endangered species (Box 14.2). The U.S. Federal ESA (1973), similar state-level statutes, the Mexican Federal standard NOM-059-SEMARNAT (2001), and the Canadian Species at Risk Act (2002) all provide legal protection for hundreds of species of scarce wildlife throughout North America. Despite this legal protection, no clear guidance exists about which species should receive management attention, which locations should receive high priority, or what types of management actions should be taken (e.g., habitat management, population restoration, or some other action). These "should" questions remain open to situation-specific determinations and they remain the center of attention for some ethicists (see Chapter 16).

A point emphasized throughout this book is that important wildlife management decisions invariably are made within

Box 14.1 *MANAGING IMPACTS OF CYPRUS MOUFLON ON AT-RISK COMMUNITIES*
Pafos Forest, Cyprus

The Pafos Forest is located in the northwestern part of the Troodos massif in the Mediterranean island of Cyprus. Due to its ecological importance and rich biological diversity, the Pafos Forest was included in the European Union (EU) network of protected areas "Natura 2000," which was established under the EU Habitats Directive. The forest provides habitat to the only existing population of the Cyprus mouflon, a sub-species of wild sheep, endemic to Cyprus. The Cyprus mouflon is listed in the International Union for Conservation of Nature Red Book of threatened species and it is strictly protected.

For centuries within the Pafos Forest, small village communities have existed that depend on the forest for their livelihoods and that sustain the forest by protecting it

The mouflon is an iconic species in Cyprus. This animal was photographed in the national zoo of Cyprus. (courtesy Marina Michaelidou)

from fires and helping with forest regeneration when fires occurred (Michaelidou and Decker 2003, 2005). Today, a lack of employment opportunities, limited social services, and the unprofitability of local agriculture have forced many people to abandon their villages, migrating to the larger cities (Michaelidou and Decker 2005).

Most remaining residents in these communities earn their primary income by cultivating vineyards and small-scale orchards on the steep forest slopes. Their work often brings them in direct conflict with the Cyprus mouflon. Frequent and prolonged droughts in Cyprus in recent years have made it increasingly difficult for the approximately 3,000 mouflon to find forage within the forest. Mouflon are drawn to agricultural land, where they cause extensive damage to trees and vines and put an additional strain on local livelihoods (Michaelidou and Decker 2002, 2005).

Mouflon have both positive and negative impacts on the people of Cyprus. To the larger Cypriot society, the mouflon is an iconic national symbol depicted in many aspects of local culture. An island-wide quantitative survey found that 90% of the 1,100 respondents agreed with the statement "It is important that Cyprus always has a viable population of Cyprus mouflon in the wild" (Michaelidou and Decker 2003). Other studies have shown that the people of the mountain communities in the Pafos Forest also value the Cyprus mouflon and have a sense of pride that such an important wildlife species lives in their area, but the local villagers, who experience negative impacts from mouflon browsing, resent the impacts these animals have and feel neglected and unprotected by EU and national policies (Michaelidou and Decker 2003, 2005).

Farmers recently were able to receive partial compensa-

a complex social, political, and cultural milieu. Despite this complexity, or perhaps because of it, many human dimensions considerations can be easily overlooked in discussions about management of scarce wildlife species. When this happens, the manager may inadvertently prevent public input into management decisions and diminish the chances of management producing acceptable outcomes. Further complicating the management challenge for species that cross national borders is a complex array of legal authorities, decision-making capacities, and cultures from country to country (Box 14.3). These factors greatly complicate management of scarce species that migrate across national borders (e.g., golden-cheeked warbler, walrus) or are range-restricted in border areas (e.g., polar bears, Le Conte's thrasher).

Wildlife management interventions can have tremendous effects on ecosystems and on human communities regionally, nationally, and even internationally (as in the case with migra-

tory species that are globally scarce). Impacts of management may be manifested in many ways, and may be both positive and negative, depending on one's perspective. Management can influence people's livelihoods, whether through the shutdown of a pulp and paper mill to protect the endangered northern spotted owl (Yaffee 1994) or the creation of a tourism industry built on the attraction of viewing polar bears in Churchill, Manitoba (Lemelin and Wiersma 2007).

14.4. THE HUMAN DIMENSIONS OF MANAGING SCARCE WILDLIFE SPECIES

In North America, the decision-making context usually is framed by governmental natural-resource agencies serving (or serving as) decision makers that are faced with questions about where and how to manage a scarce species or its habitat. Typically technical problem-solving approaches are

tion for the damages caused by mouflon; although the money was not sufficient to cover all of their losses, it was income farmers could rely on (Michaelidou 2002). After the accession of Cyprus to the EU in 2004, the compensation scheme was terminated because it was considered a form of state aid, which is not compliant with EU regulations. This policy change left farmers who owned few resources vulnerable to negative economic impacts from mouflon; this, in turn, put the mouflon at risk. A new integrated management plan for the Pafos Forest that is currently being prepared explores several options for the Cyprus mouflon, such as providing economic benefits to local people who

agree to sustain cultivations that were abandoned by their owners, *for the benefit of the Cyprus mouflon.*

The challenge for Cyprus is to enhance the conservation of the Cyprus mouflon while ensuring that local livelihoods are protected and mountain communities sustained. Policies must recognize the interdependence that exists between wildlife conservation and mouflon impacts on the people of the Pafos Forest. Through participation in decision making, local people can overcome feelings of neglect and share the knowledge they have accumulated over the course of centuries.

MARINA MICHAELIDOU
Source: Michaelidou and Decker (2005)

The Pafos Forest, Cyprus
(© M. Andreou)

taken (Primm 1996). Such approaches often prove inadequate because stakeholders' beliefs, attitudes, and capacities for action usually do not reflect the rational choice model of decision making (and the assumptions inherent to it) commonly used to guide management planning (Primm and Clark 1996). Rather than make assumptions about stakeholder traits, wildlife managers can benefit from working with human dimensions experts to apply appropriate concepts, theories, and methods for collecting and using data about stakeholders. Human dimensions considerations pertaining to the management of scarce wildlife species can be found in a variety of social science disciplines, including psychology, social psychology, sociology, and economics.

14.4.1. Values and Value Orientations

Human values are an important part of the context within which management of scarce wildlife occurs. As noted

in Chapter 4, values are deeply seated and the basis of attitudes and actions, but they are difficult to measure directly. To overcome this limitation, human dimensions researchers use value orientations as a measurable surrogate to values. Value orientations often are used to better understand human populations at relatively broad scales by categorizing people according to their fundamental beliefs about wildlife and how humans should interact with wildlife (Box 14.4).

Teel and Manfredo (2010) identified several categories of value-orientations among residents of states in the western United States. These orientations essentially reflected a continuum with domination (i.e., humans are more important than wildlife) and utilitarian (i.e., it is appropriate for humans to use wildlife) value orientations at one end of the continuum and a mutualism (i.e., humans and wildlife should co-exist because they benefit each other) orientation at the other end.

Box 14.2 *MANAGING SCARCE WILDLIFE IN THE UNITED STATES*
How Context Broadened during the Twentieth Century

Public interest in managing scarce wildlife species in North America emerged in response to an era of almost unfettered exploitation of wildlife through the end of the nineteenth century. Management actions at that time usually took the form of translocations in an attempt to restore extirpated or declining populations of various hunted species. Most of these management efforts were undertaken to address the interests of a very narrow set of stakeholders (e.g., local "game protective associations," and precursors to the Boone and Crockett Club, National Audubon Society, and Izaak Walton League of America) without much coordination, scientific basis, or human dimensions considerations, although these actors were motivated to conserve species based on their underlying value systems.

The first coordinated, scientific phase of management that focused on scarce wildlife in the United States did not occur until after the establishment of modern wildlife management during the 1930s. It was at this time that human values were explicitly codified into conservation laws. For example, federal legislation and state cooperation led to the Federal Aid in Wildlife Restoration law in 1937, which directed federal excise taxes on firearms and ammunition to state wildlife agencies and required states to use funds from hunting-license fees for wildlife management. Under that law, management attention was on restoration of game species, which reflected the funding source for the program and for state wildlife management agencies (taxes and fees imposed on hunters). The laws and management actions to support conservation of scarce wildlife during the mid-twentieth century still mainly highlighted the values, attitudes, and preferences of a narrow range of stakeholders, including hunters and anglers.

A second coordinated phase of wildlife restoration emerged from the environmental movement of the late 1960s and early 1970s, most notably through the Endangered Species Act (ESA). Federal interest in the management of scarce wildlife species emerged through the ESA, and conservation efforts broadened to include many non-hunted species. The set of stakeholders whose values, attitudes, and beliefs were incorporated into management considerations also broadened during that time. Starting in the 1970s, managers began sponsoring human dimensions research to gain information and insights about stakeholder interests and concerns with respect to management of scarce wildlife species and insights about public acceptance of various management options.

A third phase of public interest in management of scarce wildlife began in the 1990s, when several non-governmental organizations began public campaigns to "bring back" several extirpated and charismatic species of wildlife, including gray wolves, American bison, and grizzly bears. Human dimensions considerations can help managers tackle difficult decisions about why any of these species should be managed, which species should be managed, where management should focus geographically, and what interests and concerns (i.e., impacts) should be addressed through management.

The black-footed ferret recovery program is a multi-agency–multi-organization restoration program created and led by the U.S. Fish and Wildlife Service. Partners in the effort recognize that building and maintaining public support for restoration actions is critical to program success. (courtesy USFWS)

14.4.2. Beliefs and Attitudes

Wildlife managers use attitudinal information in several ways for management of scarce wildlife species. These include some basic needs: to understand the degree to which private landowners are likely to accept having endangered species on their properties (e.g., Brook et al. 2003), as a surrogate for understanding the social feasibility of restoring extirpated or endangered species (e.g., Morzillo et al. 2007), and to predict public responses to the idea of managing to conserve a scarce species (e.g., Enck and Brown 2000). Attitudinal research has also been used to improve communication messages to the public about conservation of scarce wildlife (e.g., Kaltenborn et al. 2006). One of the most important considerations for inquiry into stakeholders' attitudes about scarce wildlife is to get the referent right: attitude toward what? For example, is it most important to assess attitudes toward conservation of a particular scarce species, or is it more important to assess attitudes toward the species itself? Public attitudes data about conservation also can be used as an index of potential social feasibility for specific management actions, but this application has several limitations. Perhaps most important, attitudes can change substantially over time (e.g., Enck and Brown 2000), particularly when people gain first-hand experience with the wildlife species after it is restored.

Wildlife managers, working with partners and stakeholders, can interpret attitudinal data and use them to set management objectives and then to evaluate outcomes of management. Perhaps because human contact with scarce wildlife is infrequent, one needs to be mindful that beliefs and attitudes about scarce species may be based on vicarious or secondhand interactions, which can create the potential for misinformation or misperceptions about potential impacts. For example, concern about the relationship between particular populations (numbers and distribution) of wolves and livestock depredation, or the relationship between higher populations of a large herbivore, such as bison or moose, and additional vehicle accidents, may be based on beliefs that these negative impacts will be greater than will be experienced. Human dimensions inquiry can help sort out whether misperceptions are playing a role in public attitudes about a scarce species, and it can signal the need for an educational communication effort with the aim of improving public understanding of the species and the benefits and costs of living with it.

Such inquiry may also reveal that social or demographic characteristics of stakeholders influence attitudes toward scarce wildlife species or management actions directed at their conservation in a particular situation. For example, Kaltenborn et al. (1999) reported more negative attitudes toward large carnivores among Norwegian respondents who were older, female, less educated, and living in rural areas. Differences in attitudes toward management of scarce wildlife are reported in several studies of U.S. residents to be influenced by place of residence, such as urban–suburban environments, or other sociodemographic attributes (Pate et al. 1996). In the United States, women are more supportive of the ESA compared with men, are more likely than men to believe that conservation of non-human species has greater importance than human property rights, and give more importance to ecological reasons over economic reasons for conserving wildlife (Czech et al. 2001). National data do not necessarily reflect the local situation, however, and because management of scarce species is place-specific, one needs to be cautious about using broad human population data as a shortcut to characterizing local populations. For example, Pate et al. (1996) found no relationship between attitude toward restoration and age, education, or gender among Colorado residents.

Insights about public attitudes can be measured and interpreted most reliably within the context of a social science theory. The theory of reasoned action and the theory of planned behavior have served as theoretical underpinnings for understanding future potential behaviors and level of support for conservation of scarce wildlife species as different as endangered songbirds (e.g., Sorice and Conner 2010), large predators (e.g., Pate et al. 1996), and sea turtles (e.g., Dimopoulos et al. 2008).

Conservation of scarce wildlife may require positive con-servation behaviors or the elimination of negative human behaviors (e.g., disturbance, poaching, other forms of human-caused mortality). Attitudinal and socioeconomic data have been used to curtail poaching of endangered species, such as Saiga antelope in Russia (Kuhl et al. 2009). Similar kinds of human dimensions information have been used to design management actions to improve compliance of Florida boaters with speed restrictions in areas inhabited by manatees (Jett et al. 2009) and to reduce illegal killing of scarce wildlife species that are causing crop damage adjacent to a national park in India (Heinen and Shrivastava 2009).

Management of scarce wildlife species is sometimes confounded by conservation policies developed for large geographic scales because the wildlife species range widely (e.g., large land predators, marine mammals, Neotropical migratory birds) that need to be translated into specific management interventions informed by local human dimensions considerations. Throughout the world, community-based management efforts are being developed to address these kinds of challenges. Such efforts provide opportunities for engagement of local residents and incorporation of local knowledge, which are two aspects of community-based management that improve local attitudes toward conservation of scarce species (e.g., Low et al. 2009).

14.4.3. Motivations and Satisfactions

Sometimes wildlife managers need to understand how stakeholders will interact with the scarce wildlife species being managed. Is there a demand for wildlife viewing opportunities? Would stakeholders comply with requests to avoid den sites or nesting areas during critical periods in the species' life history? Would ranchers participate in compensation programs to mitigate depredation of their livestock by scarce predators? These questions indicate managers' need to understand stakeholders' motivations (the personal goals or reasons why people engage in particular behaviors) and the level of satisfaction or dissatisfaction (degree to which their motivations were fulfilled) that affect stakeholders' evaluations of their interactions with scarce wildlife. Knowledge of motivations and satisfactions–dissatisfactions are valuable in designing and delivering conservation education messages and wildlife viewing opportunities. For example, many participants in a "swim with the wild dolphins" program in New Zealand (Amante-Helweg 1996) were found to be motivated to participate by a desire to experience interspecies sociability, to exhibit altruistic behavior toward another mammalian species, and to have a nature-based "spiritual" experience. Participants were uninformed about dolphins and their behavior, however, which led to opportunities for communication as a management intervention.

Segmentation of stakeholders (whether by their value orientations, attitudes and beliefs, or motivations and satisfactions sought from wildlife-associated experiences) is common in wildlife management. Segmentation is done for various reasons, including market segmentation to better target communication messages to particular groups (see, for example,

Box 14.3 *MANAGING SCARCE WILDLIFE IN MEXICO*

The definition of scarce wildlife species needs to be understood in its ecological context. This is most evident in tropical countries with high biodiversity but also high endemism and many range-restricted species. For example, of Mexico's approximately 1,070 recorded species of birds, more than 200 are found *only* in Mexico and most have remarkably small ranges linked to specific habitats. These habitats result from the combined influences of temperature, elevation, and moisture—or lack of moisture—that greatly influence both the structure and diversity of vegetation in the country. Mexico's bird species also include several hundred species of North American passerines, raptors, and others that winter in Mexico or further south.

Some of the factors that make management of scarce bird species so challenging in Mexico include decentralized governmental oversight of environmental and conservation issues, limited or absent coordination of land-use planning, and lack of a coordinated approach to developing conservation priorities. Little land is privately owned in most areas of the country (which is quite different from the United States or Canada). Instead, much of the land is collectively farmed, forested, or used in other ways by residents of small communities (*ejidos*). Residents of each ejido decide on the land uses and resource extraction priorities for that local area. Although this increases opportunities for impacts of greatest importance to local stakeholders to be considered in the context of managing wildlife, it also makes it difficult to address impacts of importance to stakeholders elsewhere (who are interested in long-distance migratory birds or in

the sustainability of range-restricted species). Part of this challenge stems from a situation in which state governments and the federal Instituto Nacional de Ecologia have little regulatory or management influence at the local level.

Mexico does have 64 federally protected areas conserving some of the country's most spectacular geography and important Aztec and Zapotec ruins. Of these protected areas, approximately 20 are designated biosphere reserves because of their incredibly high endemic biodiversity. Most of these areas also receive additional protective status from state and local governmental units and receive a label as a United Nations–protected area, but designation as a protected area does not have the same meaning in Mexico as it does in the United States. Virtually all such areas have ejidos within their boundaries, with hundreds to thousands of local residents dependent on economic activity occurring within the boundaries of the protected areas.

Relatively little human dimensions research has been conducted in Mexico but the need is great. Given the weak national governance with respect to wildlife management issues and the ejido-based landownership structure, human dimensions considerations are likely to be most appropriate when framed in community-based management approaches rather than studies of particular user groups or other single-issue stakeholders. Human dimensions concepts of greatest use may include (1) understanding the attitudes and beliefs ejido members have about conservation and the potential for integrating ecotourism enterprises into the community economy; (2) community members'

Eubanks et al. 2004) and determination of specialization categories based on how and why people interact with wildlife and their capacity to engage in conservation activities (see, for example, Lee and Scott 2004). Segmentation can increase efficiency of communication and education as well as allow for specific management responses for impacts of concern to various segments.

14.4.4. Social Norms

Understanding social norms helps managers predict what people believe they ought to do in the context of interacting with scarce wildlife species or supporting management of these species. That we humans should conserve species and not allow them to go extinct is a normative assessment reflected in national policies and laws. Nevertheless, as indicated earlier, wildlife managers responsible for conserving scarce wildlife need human dimensions information to make the best possible decisions about where and how a species is to be managed. For example, restoration of the American bald eagle was a continent-wide effort that required public support,

which was not hard to mobilize but also resulted in a huge upwelling of interest in eagles and demand to observe our national symbol firsthand. An approach was developed to meet this demand while protecting the birds from excessive human disturbance—the placement of TV cameras in nests and placement of the monitors for remote observation of nestlings in visitor centers. The National Wildlife Refuge System used this technique, and visitors were enthusiastically receptive. This approach certainly seemed to become an accepted norm among visitors as a safe way to observe the young eagles and their parents.

14.4.5. Social Movements

Grounded in social norms and value orientations is the concept of social movements. Two social movements have emerged in the context of managing scarce wildlife species as being particularly important as a basis for understanding conflict and helping to identify interventions to ameliorate that conflict: the environmental movement and the wise-use movement. The wise-use movement is characterized by sev-

willingness to substitute wildlife-related interactions (e.g., guiding birdwatchers vs. capturing birds for the caged-bird trade); and (3) the degree to which range-restricted species play a role in community members' sense of place and their desire to sustain it. In such challenging management situations, managers cannot shy away from human dimensions

considerations but rather need to tailor those considerations to the situation.

Bird species endemic to Mexico include the maroon-fronted parrot (left) and thick-billed parrot (right). Endangered thick-billed parrots nest in loose aggregations in the mountains of the Sierra Madre Occidental of Mexico. They typically nest in old-growth pine trees, but also may utilize cavities in aspen trees (as seen in this photo). (© Steve Milpacher, all rights reserved [left], and © Ernesto C. Enkerlin Hoeflich [right])

eral premises: (1) that humans have dominance over wildlife and other resources that are managed for human benefits; (2) that private property rights supersede any conservation considerations; and (3) that rugged individualism is a symbolic characteristic of people who live in wild, open spaces where many scarce wildlife species occur. Conversely, the environmental movement is characterized by (1) a naturalistic orientation to human–wildlife interactions (i.e., humans are considered to be an integral part of nature, not apart from it); (2) a belief that although public lands are necessary as wildlife refugia, private landowners have a responsibility for stewardship of scarce wildlife on their properties; and (3) a belief that sustained existence of scarce wildlife is symbolic of a world in which humans have a much less dominating influence on landscapes.

Local conflict can occur when adherents to different social movements reside and pursue their livelihoods in the same community. The conflict can be particularly severe when these adherents to different worldviews have access to the same ecological data on which to base their arguments but interpret

them differently. Consider that the term "extinction" is likely to be recognized and defined similarly by either group, but that the term "endangered" is open to very different meanings and interpretations about when a wildlife species warrants conservation concern. Arguments supporting or opposing conservation of scarce wildlife species, which are grounded in opposing social movements, have been at the crux of many conflicts over endangered species management (e.g., Peterson et al. 2002).

For an example of clashing social movements we can turn again to the case of wolf restoration to the GYE (Wilson 1997). On one hand, a study found that many residents living and deriving their livelihoods around Yellowstone National Park held beliefs consistent with the wise-use movement. They tended to see wolf restoration as what they referred to as a "power-grab" by the federal government and the urbanized general public who were, through policies and actions aimed at restoring wolves to the GYE, endangering the way of life and livelihoods of local residents. Proponents of wolf restoration, however, espoused an "overriding commitment to increased

Box 14.4 *VALUE ORIENTATIONS AND CONSERVATION OF MASSASAUGA RATTLESNAKES*

Human–wildlife interactions affect stakeholder attitudes and impacts perceived from various wildlife species. In the case of snakes, encounters with humans often lead to death of the snake. The eastern massasauga rattlesnake and the timber rattlesnake are species under consideration for listing on the Endangered Species Act (ESA). Both snakes are listed as endangered, threatened, or as a species of special concern in most of the states and provinces in which they reside. Although their survival is tied to human behaviors that adversely affect these species, measures of human tolerance for living with these frequently overlooked taxa were lacking. Christoffel (2007) developed a mail survey instrument to gain information on factors affecting acceptance capacity for these two species of rattlesnakes among a sample of residents in Michigan and Minnesota.

Surprisingly, people who lived in the presence of either rattlesnake species expressed more positive attitudes toward snakes than people who did not. A rattlesnake stewardship score derived from a series of questions that determined feelings of obligation toward snakes explained the greatest degree of variation in acceptance capacity. Risk perceptions were the most important determinant of whether or not a respondent would kill snakes.

Results of this study indicated an acute need for education and communication efforts to provide people with positive human–snake interactions to align mistaken beliefs and reduce heightened risk perceptions about non-venomous snakes and rattlesnakes. The situation with eastern massasauga rattlesnakes is an example of the conservation challenges presented when the acceptance capacity for impacts perceived by stakeholders is well below the population levels needed to insure survival of the species.

Findings from this research provided input for development of experimental workshops to test the efficacy of various outreach methods in improving knowledge and attitudes about non-venomous snakes (Christoffel et al. 2010). First-hand experience with snakes yielded the most significant positive attitude change.

REBECCA A. CHRISTOFFEL

Graduate students from Michigan State University capture an eastern massasauga rattlesnake during a research project (left, courtesy Michael Shaffer; right, courtesy Ed Schools)

social control of land use through [national] environmental legislation" in support of their belief that wolf conservation and restoration of one of the last remaining large predators in the GYE must take precedence over extractive and utilitarian relations with the land (Wilson 1997:461). Thus, wildlife managers understand that the conflict over wolf restoration in GYE is rooted firmly in antagonistic social movements that have a history of disagreement about multiple issues. Even with these insights, however, wildlife managers have found it difficult to develop interventions to reduce the conflict.

14.4.6. Culture

Geographical variation in public support for conservation of some species often reflects cultural differences in how people think about and interact with scarce wildlife. Managers can sometimes achieve conservation outcomes by incorporating insights from the local culture, especially in situations where a scarce wildlife species needs protection or stakeholders need relief from negative effects of the species to maintain a viable population. Local cultures sometimes have strict taboos (a form of informal institution or social norm) against the use of particular species.

An example of this occurs in the eastern forests of Madagascar, an island nation off the southeastern coast of Africa that is home to more endemic species of wildlife than any other place of its size in the world (Jones et al. 2008). Even when species such as lemurs and fossa (Madagascar's largest mammalian carnivore) are being harvested, cultural taboos avert overexploitation by preventing subsequent sale of species of conservation concern and by limiting harvest to certain seasons or particular categories of animals within a species. These taboos seemed to emerge from social norms and disapproval expressed by important members of the community, rather than from norms of sustainable management of those species or from a perceived cultural penalty (such as supernatural retribution). When a nearby national park was established and conservation rules were imposed on the area by external entities, the social norms underlying conservation of lemurs and fossa started to break down. Knowledge of this unintentional effect of conservation regulations helped wildlife managers make changes that included social and cultural control of activities in addition to external management regulations (Jones et al. 2008). Integration of knowledge about cultural factors affecting human–wildlife interactions is particularly vital in places such as Madagascar where formal governmental institutions are weak and conservation depends on the maintenance of local, informal institutions.

14.5. CAPACITY CONCEPTS

Wildlife managers are familiar with the concept of "biological carrying capacity" with respect to wildlife populations. Managers also may be familiar with the concept of capacity as it relates to people through the term "social carrying capacity," which typically is used to describe some acceptable number of people in an area (e.g., national park visitors, wildlife viewers) based on people–people impacts (e.g., crowding) or people–wildlife impacts (e.g., harassment of or stress on wildlife). The following capacity concepts can be applied to help understand the human dimensions of managing scarce wildlife.

14.5.1. Wildlife-stakeholder Acceptance Capacity

Wildlife stakeholder acceptance capacity (WSAC), defined as "a mixture of tolerance of problems and desires for benefits from wildlife" (Carpenter et al. 2000:5), recognizes that the capacity of any stakeholder group to accept a wildlife species occurring in their area depends largely on their evaluation of the trade-offs of positive and negative impacts associated with that species. Knowledge of the WSACs for key sets of stakeholders can help managers pursue actions that achieve an

acceptable balance of these positive and negative impacts. This is a complicated task because different stakeholders within the same area will interact differently with a wildlife species and will associate various impacts with the wildlife population.

Understanding that stakeholders may have different WSACs for any given size of wildlife population in a particular area is critical in scarce wildlife management. It helps managers recognize that their efforts should include managing the various impacts that affect WSAC rather than only managing the size of the wildlife population (e.g., impacts of wolves may be more important to ranchers than actual number of wolves). In addition to identifying impacts of importance for various stakeholders and the minimum desirable levels of positive impacts and maximum desirable levels of negative impacts (Fig. 14.1), managers assessing WSAC need to be aware of several scale issues, including spatial, temporal, and human organizational. The spatial scale at which a scarce species exists (e.g., range-restricted vs. widespread but uncommon) and the behavior of the species can affect both the likelihood of interactions that produce impacts and the number of individuals or stakeholder groups involved. WSAC at any level of human organizational scale (e.g., individual, stakeholder group, community of place) can change over time as people experience various interactions with the scarce wildlife species and each other.

Human dimensions experts point out that WSAC for a population of people is likely to be influenced by many factors, including characteristics of the scarce wildlife species they are interacting with, past experiences, and various social-psychological variables, such as values, beliefs, and attitudes (Zinn et al. 2000). These factors all play a role in determining which impacts will be important to stakeholders and what levels of those impacts will be acceptable. For example, beliefs and attitudes related to the cognitive and affective components of risk perception have been found to influence WSAC for cougars in Montana (Riley and Decker 2000). Knowledge of factors that affect WSAC can help managers choose appropriate interventions for managing impacts.

14.5.2. Community Capacity

A second kind of capacity issue related to wildlife governance and community-based management critical to management success is the capacity of stakeholders, communities, and local institutions to participate in management (Raik et al. 2005). Participation can occur in establishing management objectives or in helping to conduct management activities. Capacity to participate in these ways differs among stakeholder groups within a community and spatially across the landscape from community to community.

Knowledge of community capacity can provide managers with insights about where restoration discussions might be raised most effectively with stakeholders. Sometimes wildlife agencies face the dilemma of needing to explain the idea of managing a scarce species (especially of endangered species) to a public that may generally understand conservation values, but that has concerns about secondary or collateral effects in

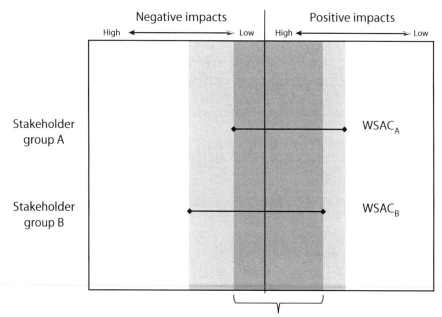

Figure 14.1. Theoretical overlap in wildlife stakeholder acceptance capacity (WSAC) for two stakeholder groups in the context of managing a scarce wildlife species

their local area. Making restoration decisions that are both ecologically sound and socially acceptable is more difficult for large carnivores or herbivores that have great symbolic meaning or potential for negatively affecting human livelihoods or lives (e.g., Scarce 1998, Naughton-Treves et al. 2003). Managers can use information about capacity to approach communities and stakeholder groups who are in the best position to take advantage of management-related benefits and overcome or mitigate any potential negative management-related impacts associated with restoring or protecting scarce wildlife.

One context in which consideration of community capacity is particularly important is in conservation of biodiversity "hot spots"—geographic areas that contain large numbers of endemic species (species that occur only within that region and nowhere else in the world). It has historically been common for wildlife managers to call for species inventories in those areas as a precursor to identifying conservation actions. More recently, managers are recognizing that an "inventory" of human dimensions considerations also is invaluable as a precursor to identifying, and especially carrying out, conservation actions. In areas with a high degree of species endemism or with rare or endangered species, the greatest conservation outcomes may be achieved by identifying and mapping localities where human and social capacity are sufficient to initiate and sustain actions (Cowling et al. 2009). This integration of ecological and human dimensions considerations undoubtedly will increase conservation efficiency and effectiveness, given that opportunities for implementing management actions (e.g., landowner willingness to participate) often do not correspond spatially to areas having the wildlife species with the highest conservation need (Knight and Cowling 2007, Polasky 2008).

14.6. ECONOMIC CONSIDERATIONS

Conservation of rare or endangered species historically was considered a "non-use value" because conservation outcomes are "not revealed through observable economic transactions and are not measurable through market data" (Kotchen and Reiling 2000:93). This perspective changed as economists during the late twentieth century applied economic theory and methods to conservation issues in an attempt to frame costs and benefits of conservation in monetary terms. Especially in the context of endangered species management in the United States, economic valuation of those species and their conservation has become standard practice. Contingent valuation (e.g., willingness to pay) is one of a few techniques applied to scarce species valuation (Chapter 6).

Information about stakeholders' hypothetical willingness to pay for conservation of scarce wildlife is useful in the context of prioritizing and planning conservation efforts. One caveat, however, is that many of the larger, more charismatic scarce species with which people may be familiar (e.g., wolves, polar bears, tigers, elephants) are valued nationally or globally, but the costs of living with these species are incurred locally, where people have to contend with depredation of crops and livestock or a higher risk of human injury or death (Nelson 2009). To overcome such "'market failures' underlying carnivore declines" (Nelson 2009:381), wildlife managers are developing programs called "payments for ecosystem services," in which local landowners or communities are compensated in return for measurable local increases in carnivore populations.

Perhaps the most recognizable form of payment for ecosystem services is compensation for depredation to crops and livestock. The general approach is for landowners to receive financial compensation for the estimated value of crops consumed by an herbivore or each livestock animal killed by a large predator. The assumption behind compensation is that local willingness to tolerate the presence of these wildlife species in the area will increase when economic impacts are mitigated with payments. This approach has been used in conjunction with conservation of many species of large predators (e.g., wolves, grizzly bears, jaguars) and large herbivores (e.g., elephants). In at least some of these cases, however, compensation payments had no effect on tolerance of predators compared with people who experienced depredation but received no compensation (e.g., Naughton-Treves et al. 2003).

Sometimes managers have modified the compensation approach by building in incentives to increase the prey base for the predators. For example, wildlife managers in Mexico developed a program aimed at jaguar conservation in Sonora, which supports the closest breeding population of jaguars to the United States and one that had been declining because of illegal predator-control activities by ranchers (Rosas-Rosas and Valdez 2010). An innovative aspect of the program was to pay participating landowners $1,500 (U.S. currency) for each deer-hunt permit sold on their properties, but only if they also stopped controlling jaguars and cougars. Landowners reduced the number of livestock grazing on their properties to improve deer habitat, thus increasing their economic return from their land while also increasing the natural prey base for large predators.

In addition to understanding whether compensation does (or could) affect tolerance with respect to living with potentially damaging or dangerous scarce wildlife species, wildlife managers can benefit from knowing about the factors that affect people's willingness to make voluntary contributions to compensation funds. For example, in Wisconsin, where wolves are classified as endangered but where their population is increasing substantially, voluntary contributors to a wolf depredation fund are supportive of continuing compensation payments to farmers who lose livestock even when the wolf population is officially recovered. When surveyed, contributors stated that they were opposed to lethal control of problem wolves and approximately one-quarter were strongly opposed to the idea of opening a hunting season for wolves after the population was recovered. These findings led wildlife managers to consider instituting sunset-clauses in compensation regulations and highlighted the need for making decisions within an adaptive framework (Treves et al. 2009).

Compensation schemes and payments for ecosystem services also are developed to increase interactions between humans and scarce wildlife species. This is evidenced by the increasing trend in ecotourism opportunities around the world. Ecotourism enterprises that specialize in helping people experience scarce wildlife species are emerging for wolves (howling tours) in the northern United States and Europe, birds (birdwatching in areas of high avian endemism in Central and South America), marine mammals and turtles in North America, Australia, and Israel, and large carnivores and other scarce mammals in Africa. A key aspect for managers who are interested in ecotourism and wildlife viewing as conservation interventions for scarce wildlife species is to ensure economic benefit to the local residents "living in landscapes of high conservation value [who] are trapped between their dependence on natural resources to meet their development aspirations and the [state, national, and] international pressure to conserve those resources" (van Vliet 2010:1). In the context of using ecotourism to both conserve scarce wildlife species and to contribute to local economic development, several key human-dimensions considerations include (1) motivations and satisfactions of potential ecotourists, (2) willingness of non-visitors to pay for conservation of scarce wildlife, (3) resident–wildlife and resident–visitor impacts to be managed, and (4) the capacity of local communities and residents to help manage the scarce species and benefit from ecotourists.

SUMMARY

Some of the most contentious issues in wildlife management are about scarce wildlife. With respect to the management cycle described in Chapter 7, however, human dimensions considerations for managing scarce species are much the same as with abundant species except for the unique situations that occur with scarce species. Managing scarce wildlife requires skillful application of knowledge from all areas of human dimensions discussed throughout this book.

- Scarce wildlife are defined as being of at least three types: (1) species once abundant but whose populations are greatly diminished—often classified as threatened or endangered—throughout their range; (2) species common globally, but where certain populations are scarce or extirpated from portions of their historical range; and (3) species that, for whatever reason, occur at low population densities throughout their ranges (i.e., they are naturally scarce).
- Specific contextual factors of scarce species differ across the globe and include (1) stakeholder interest and acceptance at the local level; (2) forms of local, national, and even pan-national governance; and (3) the capacity of stakeholders, non-governmental organizations (NGOs) and government conservation groups to do the work.
- Multiple approaches to management of scarce wildlife exist (including restoration, recolonization, and introduction), and these interventions usually create different sets of stakeholder-perceived impacts.
- One prominent difference between management of scarce species versus common species is the legal frameworks (such as the Endangered Species Act or state designation of "species of special concern"); however, such frameworks do not answer questions about which species should be managed or where they should be managed. Answers to those questions rely on identification of various positive and negative impacts associated with management of the species.
- Preventing or mitigating conflicts likely to arise in management of some scarce species (such as wolves, grizzly bears, or rattlesnakes) necessitate measuring (1) stakeholders'

motivations and satisfactions associated with activities made possible (or impeded) by management of scarce species, (2) beliefs about management outcomes, and (3) some indication of the capacity of stakeholders to either participate in or benefit from management.

Suggested Readings

Decker, S. E., A. J. Bath, A. Simms, U. Linder, and E. Reisinger. 2010. The return of the king or bringing snails to the garden? The human dimensions of a proposed restoration of European bison (*Bison bonasus*) in Germany. Restoration Ecology 18:41–51.

Fischer, H. 1995. Wolf wars: the remarkable inside story of the restoration of wolves to Yellowstone. Falcon Press, Helena, Montana, USA.

Musiani, M., L. Boitani, and P. Paquet. 2009. A new era for wolves and people: wolf recovery, human attitudes and policy. Calgary University Press, Calgary, Alberta, Canada.

Valdez, R., J. C. Guzman-Aranda, F. J. Abarca, L. A. Tarango-Arambula, and F. C. Sanchez. 2006. Wildlife conservation and management in Mexico. Wildlife Society Bulletin 34:270–282.

HUMAN DIMENSIONS OF WILDLIFE USE MANAGEMENT

TERRY A. MESSMER AND JODY W. ENCK

With a use permit from the National Park Service, a professional photographer works for 2 weeks in the back-country of Glacier National Park. Sale of rights to publish the photos that she takes in national parks will be a big part of her annual income.

In northern Maine, an unemployed logger takes beaver and coyote pelts to a local fur buyer. Money he receives from selling the fur from harvested animals will help keep his family budget out of the red until he picks up more work in the timber industry.

Like generations before him, a tribal elder skillfully sorts bald eagle feathers he will use to create a ceremonial headdress. Possession of bald eagle feathers without a permit from the U.S. Fish and Wildlife Service (USFWS) is generally illegal, but federal law allows use of certain wildlife parts and products of traditional cultural importance to many Native American groups.

A retiree in North Carolina has her morning coffee on her back porch while watching hummingbirds that have been attracted to her flower gardens and feeders. For the price of a few feeders and gardening supplies, she has created an affordable recreation activity that she can enjoy every day from the comfort of her own backyard.

These are a few of the many ways that people purposefully use wildlife; they reflect diverse values of wildlife and distinguish these wildlife uses from other kinds of human–wildlife interactions described elsewhere in this book. This large class of human–wildlife interactions creates a host of sometimes conflicting expectations for wildlife management.

The purpose of this chapter is to illustrate how human dimensions contribute to managers' understanding of wildlife uses, users, and management of related impacts. We begin by discussing the role wildlife use has played in the direction of wildlife management in the United States, including development of research on the human dimensions of management. We then illustrate how human dimensions inquiry, based on foundations in social psychology, sociology, and economics, is applied to identify and understand

- wildlife-related activity trends and patterns,
- sociodemographic changes that influence activity involvement,
- wildlife recreation as a social process,
- impacts associated with wildlife-use activities,
- public attitudes toward wildlife-use activities, and
- public support for existing and alternative sources of program funding.

15.1. WILDLIFE USES AND THE EMERGENCE OF HUMAN DIMENSIONS INQUIRY

Conditions for wildlife that prevailed in North America for perhaps 100 centuries changed quickly with the influx of Europeans. Populations of many species of American wildlife declined markedly between the mid-seventeenth and late nineteenth centuries. Following the creation of the United States, the demands of an expanding nation for food, fiber, energy, land, and water resulted in staggering overexploitation of wildlife and degradation of habitat during the 1800s; this overwhelmed the natural processes that had previously allowed wildlife to prosper.

Wildlife populations in the United States first plummeted in the eastern half of the nation, where the combined effects of subsistence use (hunting, gathering eggs, trapping), market hunting and commercial trapping, deforestation, and land conversion for agriculture took their toll. Many species (e.g., elk, caribou, mountain lion, gray wolf, moose, beaver, white-tailed deer, black bear, wild turkey) were in severe decline or extirpated in extensive areas of the East and Midwest by the mid-nineteenth century. During the mid- and late 1800s, the discovery of gold in California and the prospect of obtaining farm- or ranch-land fueled a massive westward migration of Americans, which created population pressures that adversely affected wildlife in the Midwest and West. The plights of species such as American bison and pronghorn antelope are well-known examples of exploitation depressing populations to near extinction.

A remnant herd of bison at the National Bison Range, Montana (courtesy Steve Hillebrand, USFWS)

Wildlife management agencies in the United States were established primarily to manage impacts associated with wildlife use. The U.S. wildlife management institution (Jacobson and Decker 2006), as we know it today, is a direct product of the conservation era—a social movement during the late nineteenth and early twentieth centuries that focused, in part, on stopping unregulated exploitation of wildlife. By the late 1800s, a body of basic protective laws was in place in most states to restrict harvest and harassment of the scarce wildlife that remained in many places. Habitat protection and improvement also became imperatives for wildlife conservation. Those actions were soon followed by propagation and stocking programs. Protection and restoration efforts reflected the dominant values of the times and were typically focused on species popular for hunting, many of which had been badly affected by overconsumption and habitat loss.

The earliest era of modern wildlife management (1890s–1930s) had a clear focus—prohibiting or tightly restricting all wildlife uses and propagating game species. In time, some wildlife populations were restored and use restrictions were eased under a renewable-resource management regimen. Although the genesis of wildlife management focused on replacing unregulated exploitation with wise use (i.e., sustainable use of a renewable resource), the practice eventually transitioned into an era of allocation. Decisions about allocation required information on rates of harvest of game species. Managers conducted simple inquiries of hunters and trappers about species harvest and made harvest reporting a requirement for some species to inform these allocation decisions.

During the early decades of the twentieth century, the work of naturalists—observation and description—evolved to include more experimentation with habitats and wildlife populations. Scientists began to develop the biological foundation for wildlife management. Those who wanted to manipulate wildlife populations or improve habitat for specific purposes questioned the relationships of such populations with their environments. The emerging field of ecology was a source of insight for them. The fields of wildlife science and wildlife ecology grew out of those pursuits.

Aldo Leopold, ca. 1940 (courtesy of the Aldo Leopold Foundation; www.aldoleopold.org)

In 1933, Aldo Leopold provided a foundation for the emerging art and science of wildlife management with his landmark book *Game Management*. The concepts he articulated prevailed for decades. Leopold's definition of game management—"the art of making land produce sustained annual crops of wild game for recreational use" (Leopold 1933:3)—was a reflection of the time. His use of the words *game, crops,* and *use* plainly indicated the utilitarian and agrarian focus of wildlife management. That focus on management of game and exploited species persists as a key feature of wildlife management to this day.

An extensive biological research initiative was launched in the United States during the 1930s with the establishment of a federal system of cooperative wildlife research units at many land-grant universities, followed by the creation of USFWS research centers coupled with the formation of research groups within many state wildlife agencies; funding for these research initiatives was supported by the Federal Aid in Wildlife Restoration Act of 1937 (Pittman–Robertson Act). The goal of the research initiative was to meet the information needs of wildlife conservation and management. At the time, the primary management goal was producing game species for use and harvest. Research focused on the biological science needs of game management. Regulation of location and amount of human-induced mortality on wildlife populations was a central concern.

The process of managing uses and users became more complex over time. New types of uses developed as significant recreational pursuits and the number and diversity of stakeholders considered in decisions about wildlife use increased markedly. It is easy to visualize conflicts developing between consumptive user groups and new non-consumptive groups. Early on in such cases, the wildlife agency usually allied with the consumptive user groups with whom they had interacted since the inception of the profession. Agency managers eventually were perplexed to find themselves dealing with conflicts between segments of their longtime consumptive stakeholders, mainly due to resource and opportunity allocation issues across different kinds of hunting methods preferences (e.g., use of primitive vs. modern implements, such as bows, muzzleloaders, and modern rifles for big game). Wildlife managers began with a relatively simple set of stakeholders, which was mainly developed along game species lines. For example, stakeholder sets might include big game, small game, upland birds, and waterfowl hunters, all of which were relatively easy to satisfy; therefore, intergroup conflict was kept low. Conflict developed when segments within these groups wanted different sets of regulations or special considerations based on methods used. Allying with one side or the other would result in public wildlife managers alienating a portion of their stakeholders.

By the late 1970s, some wildlife management agencies began to recognize that judgment about users (their motivations, factors affecting participation, etc.) based on observation and basic harvest monitoring was insufficient. Managers sought more sophisticated and representative human dimensions information to make informed decisions. Demand emerged for social science research aimed at understanding the human dimensions of management. Initially, understanding consumptive users of wildlife became imperative and seeded interest for human dimensions inquiry. Today, many innovations in human dimensions research are employed in the service of managing wildlife uses and users. They are an institutionalized part of management and are applied across a wide array of use types and user-management issues. It was human dimensions information explaining concepts such as multiple satisfactions, developmental stages, and specialization among wildlife-dependent users that provided managers with the understanding of the complexity of stakeholder management and paved the way for better public involvement tools for dealing with increasingly complex issues. Successes in these areas have fixed a place for human dimensions as an essential component of wildlife management.

15.2. CATEGORIES OF USE

By the 1950s, state and federal management agencies were monitoring wildlife use in three broad categories: consumptive, non-consumptive, and subsistence use. Activities within those categories give rise to a mix of impacts and management considerations, some of which we highlight in the following sub-sections. We focus on this classification system because it reflects actual practice (these terms have been in standard use in North American wildlife management for more than half a century). The reader should note that other use classifications are available that can be helpful to wildlife managers in conceptualizing impacts associated with different classes of wildlife use (see Decker and Goff [1987] for examples).

15.2.1. Consumptive Use

Consumptive use involves activities that permanently remove animals from the wild. This broad category encompasses a range of activities from recreational to commercial and from legal to illegal. Consumptive uses include regulated hunting, regulated trapping of furbearers, and legal and illegal commercial hunting and trapping (removal of animals for trade in wild meat, horn, ivory, exotic pets, etc.). Other consumptive uses include wildlife removal to manage nuisance, damage, or threats to human safety; regulated fee-hunting on private preserves; scientific research; removal of individuals for propagation of endangered species; and educational display of wildlife. Hunters, trappers, wildlife damage-control agents, game ranchers, wildlife researchers, zoo professionals, wildlife rehabilitators, and others seek special permits to legally engage in consumptive use activities.

15.2.2. Non-consumptive Use

Non-consumptive use involves activities in which people benefit from wildlife without permanently removing animals from the wild. Non-consumptive uses include wildlife viewing, wildlife photography, wildlife feeding, and wildlife-related education or interpretation. The non-consumptive label can be misleading, because one might assume that if wildlife are not removed from the wild directly then non-consumptive use has only positive impacts. Ecological research has revealed that both consumptive and non-consumptive uses can disturb wildlife in ways that cause physiological stress and disrupt wildlife feeding, habitat use, and breeding behavior, among other

Consumptive wildlife uses include legal and illegal commerce in all manner of wildlife products, including trade in live animals (courtesy USFWS)

things (i.e., significant ecological impacts can occur even when no wildlife are killed directly). Moreover, human dimensions research has shown how conflicts among non-consumptive users (and between them and consumptive users) can create a range of negative psychological, sociocultural, and economic impacts on the people involved.

15.2.3. Subsistence Use

Wildlife resources have been used for subsistence purposes wherever wild creatures and resource-dependent human populations have come into contact. Wildlife are harvested to meet various needs ranging from human survival (e.g., to obtain food and medicinal products) to maintenance of traditional culture (e.g., for social, aesthetic, or spiritual purposes). Subsistence hunting and trapping activities are common worldwide, including in North America.

In developing countries, most forms of edible wild animals are consumed. The list of wildlife that are hunted, trapped,

or gathered includes insects, amphibians and reptiles, birds, and marine and land mammals. Many species are harvested for subsistence purposes globally; large cervids, such as moose and caribou, are objects of subsistence hunts in North America.

Wildlife management authorities around the globe often distinguish between levels of subsistence use. For example, subsistence rights have been given legal definition in Alaska, with different regulations governing subsistence use by Native Americans and subsistence use by other Alaska residents. These kinds of regulatory structures recognize that subsistence use occurs at two levels.

For some people, subsistence equates to involvement in a few wildlife harvest activities (e.g., harvesting moose for personal consumption in Alaska, selling small amounts of bush meat to local farm workers in rural villages of Brazil) to supplement one's diet or income. In much of the developing world and in many remote areas of North America, people continue to rely on subsistence activities to provide food and other products for their survival and well-being. Participation in and regulation of these activities can impact people's nutrition and health, their income, and their sense of self-reliance.

Subsistence has a culturally expansive meaning for other people, including some in enclaves within advanced industrial societies. For this group, customary and traditional uses of wildlife and plants are part of a "way of life" that maintains kinship groups, ethnic cultures, and other social relationships. This type of subsistence enables a resource-dependent lifestyle that includes some combination of hunting, trapping, fishing, gathering, herding, cultivating, trading, bartering, resource sharing, and household- or village-level consumption.

Subsistence-based households are not necessarily primitive or socially isolated and most have adapted to economic, social,

Non-consumptive wildlife uses vary widely, from local bird-watching to whale watching and international ecotourism travel (*top*, courtesy Tammie Sanders, Cornell Laboratory of Ornithology; *bottom*, courtesy Julie Whittaker)

Subsistence uses, such as the egg collecting by this Native Alaskan, can be part of a way of life with social, cultural, and spiritual meanings (courtesy Rick J. Sinnott, Alaska Department of Fish and Game)

and technological developments. Modern rifles and traps are used for harvesting wildlife. Electric or propane freezers and modern food-preservation methods have replaced traditional food preservation techniques (Muth et al. 1987, Glass and Muth 1989). Nevertheless, these subsistence users still rely on wildlife to sustain their socioeconomic well-being and traditional culture (Glass and Muth 1989, Dick 1996). Participation in and regulation of subsistence lifestyles can impact people's cultural identity and survival, as well as their immediate health and well-being.

15.3. HUMAN DIMENSIONS INFORMATION NEEDS AND APPLICATIONS

Fundamental objectives for wildlife population management often relate to both wildlife use and wildlife conservation. Sustainable use becomes more difficult as human populations grow and pressures to use natural resources increase around the globe. In a range of contexts, from managing game animals in the American Midwest to conserving biodiversity in the tropics, it is becoming essential that decisions about wildlife uses and users are grounded in an integrated knowledge

base that includes human dimensions insights based in experience and research.

In the remainder of this chapter, we describe how concepts from social science disciplines such as social psychology, sociology, and economics are applied to understand six categories of human dimensions information related to managing wildlife uses: (1) documenting activity trends and patterns, (2) understanding demographic influences on wildlife recreation, (3) understanding wildlife recreation as a process, (4) identifying and managing impacts, (5) clarifying public attitudes toward and support for use activities, and (6) understanding and building support for new sources of wildlife program funding. The examples used are illustrative rather than exhaustive. They provide an introduction to the categories of human dimensions inquiry that have created a knowledge base on wildlife-related activity involvement and associated impacts.

15.3.1. Documenting Activity Trends and Patterns

Knowledge of trends in participation in wildlife-use activities is used by state and federal wildlife agencies for opportunity-allocation decisions and revenue projections. Wildlife managers continually need answers to a range of basic questions about wildlife-related use activities, questions such as the following:

- How many people are engaged in the activity?
- Where and when does the activity occur?
- Is the number of activity participants changing?

Monitoring involvement in hunting, trapping, wildlife viewing, and wildlife feeding has been a consistent priority in the United States and Canada. Periodic assessments of specific kinds of wildlife use and general outdoor recreation trends have allowed natural-resource management agencies to identify emerging patterns of activity involvement and anticipate effects on programs. This type of information is valuable for planning; wildlife agencies and public land-management agencies use it to evaluate policies, regulations, program plans and budgets, and if necessary, to revise them in ways that respond to changing recreation demands and pressures on wildlife.

Substantial shifts in wildlife-related activity involvement have wide-ranging implications for wildlife management and conservation. In the United States, state wildlife agencies receive a substantial amount of their operating budgets directly and indirectly from wildlife users. All state wildlife agencies in the United States receive funding from consumptive users in the form of license fees and redistributed federal excise taxes on specified hunting equipment (Organ and Fritzell 2000, Jacobson et al. 2010), so agencies have a strategic interest in sustaining participation (i.e., through recruitment and retention) in wildlife-related activities. All state wildlife agencies also depend on political support from stakeholders who utilize wildlife in some way or want agencies to manage the various impacts associated with uses of all kinds. In some situations, agencies depend upon hunters and trappers to control populations of wildlife that cause nuisance and economic damage.

State wildlife agencies have been monitoring involvement in hunting and trapping since 1937, when the states were first required to document annually the number of persons buying hunting licenses so they could provide data needed for distribution of federal funds (excise tax) to the states (Kallman 1987). State wildlife agencies also are required to provide the USFWS with the number of persons graduating from hunter-education courses in their states.

The USFWS has conducted a national survey of hunting and fishing approximately every 5 years since 1955. Data initially were collected for estimating national and regional participation and expenditures, but in 1980 the survey was expanded to allow estimates for each state. At the same time, participation in non-consumptive wildlife activities (such as observation and photography around the home and on trips) was added to the survey. The Canadian Wildlife Service conducted four national surveys: in 1981, 1987, 1991, and 1996. They were patterned after the U.S. national surveys and used for similar purposes.

State or regional analyses of wildlife-related activity involvement are typically more useful to decision makers than generalizations at a national level, because they identify regional and local variation masked within analyses of change at a larger scale. For example, a nationwide analysis indicated overall decline in hunting participation between 1991 and 1996. A regional analysis of the same data found that hunting participation rates were relatively stable in most regions of the United States during that time frame, while the number of hunters in the Middle Atlantic states declined by nearly 17% (Brown et al. 2000b). Subsequent national survey data have documented continued declines in hunting participation in most U.S. states (Responsive Management / National Shooting Sports Foundation [RM / NSSF] 2008). Regional analysis of national survey data found that hunter initiation rates (defined as the percent of children residing at home who have ever participated in hunting) declined in every region of the United States, and retention rates (defined as the percentage of people who have hunted at some time in their life and also had participated in hunting at least once in the previous 3 years) declined in several regions, between 1990 and 2005 (Leonard 2007).

The U.S. Forest Service has conducted the National Survey on Recreation and the Environment periodically since 1960 (every 4 years, on average). Originally named the National Recreation Survey, it focused on participation in a small number of seasonal outdoor-recreation activities, primarily on U.S. public lands. It was later expanded to provide data on 50 types of outdoor recreation. A core set of participation and demographic questions included in each of these surveys has made possible comparison of recreation trends relative to demographic changes in the U.S. population. National recreation surveys have been conducted in northern European countries (Denmark, Norway, Finland, and Sweden), but the contents and extent of those surveys vary considerably, which makes comparisons between countries more difficult.

Collecting information about wildlife activity involvement at a national level is costly and labor-intensive, even for developed nations. Canada, for example, has not conducted its national survey of recreation since 1996 due, in part, to cost considerations. Many nations lack both the resources and infrastructure to conduct wildlife-use assessments on a national scale but have been able to conduct useful recreation assessments on smaller scales (e.g., Robinson and Redford 1991).

Contemporary wildlife-use trend data demonstrate that many North Americans place a high value on wildlife, wildlife conservation, and outdoor recreation directly and indirectly involving wildlife. As the human populations and wildlife habitats change, however, participation in specific wildlife-related activities shifts as well.

National survey data for the United States suggest increased participation in birdwatching, photography, and other forms of wildlife viewing in recent decades (Bowker et al. 1999, Cordell 2008, Cordell et al. 2008, USFWS 2007). In 2006, 87.5 million out of 275 million U.S. residents 16 years old and older participated in wildlife-related recreation. Compared with 2001, 6% more people participated in wildlife-related recreation in 2006. This increase in participation was not experienced in all forms of wildlife-recreation use. Between 2001 and 2006, the number of people engaged in wildlife viewing increased by 8%. The number of hunters dropped by 4% (from 13.0 million in 2001 to 12.5 million in 2006) during the same time period. National survey data and hunting-license sales data suggest a long-term trend of decline in hunting and trapping participation (Enck et al. 2000, USFWS 2007, RM / NSSF 2008).

15.3.2. Understanding Demographic Influences on Wildlife Recreation

Wildlife management takes place within the larger context of human population changes because these changes affect wildlife recreation demand. When the factors influencing these trends can be identified and understood, public agencies can better plan and implement policies and programs to balance supply with demand, mitigate potential negative impacts to the resources, and reconcile conflicting uses.

Wildlife agencies and organizations have used human dimensions inquiry to identify and understand the implications of sociodemographic change for participation in activities such as hunting, ecotourism, and subsistence use. Valuable insights are derived from primary research, secondary data analysis, literature review, and synthesis of data collected by others (e.g., demographers, economists, geographers) to understand human population change. In this section, we outline several change processes that are anticipated to have profound effects on wildlife recreation and other uses of wildlife in the United States during the twenty-first century (Schuett et al. 2009).

15.3.2.1. Urbanization, Affluence, and Rural Gentrification

The combined 2010 human population of the United States (308 million) and Canada (34 million) exceeded 340 million. The combined population of the two countries is projected to top 400 million by 2030. Population growth will not be uniformly distributed and demographics will change dramatically. The fastest growth rates will occur in urban areas in the South and West in the United States and the Atlantic region of Canada.

Land development affects wildlife significantly but is difficult for government and non-governmental organizations to influence (courtesy Florida Fish and Wildlife Conservation Commission)

Approximately 80% of U.S. and Canadian residents (and approx. 50% of all people across the globe) live in towns or cities.

Urban growth is generated by several factors (e.g., transportation infrastructure, a growing human population), including economic development. Increased urbanization has given rise to a "convenience culture" where the newly empowered consumers have more choices and, because of changing work patterns, increased flexibility in how they spend their leisure time and discretionary funds. As economies mature and higher average incomes prevail, people who already possess tangible consumer goods may seek intangible products, services, and experiences. These products often include new recreational experiences. Increases in affluence have been correlated with increased participation in wildlife viewing, feeding, and other forms of outdoor recreation, which is important information for wildlife managers.

Within the larger trend toward greater urbanization in the United States and Canada is a coupled pattern of residential land use called exurban development (i.e., development of single-family dwellings on 5- to 40-acre [2–16-ha] parcels). As urbanization increases, a portion of the affluent residents move away from urban centers to rural landscapes where the quality of life is perceived to be better. In the United States, exurban development consumes 10 times more land than suburban and urban development combined (USDA 2003). Exurbanization is associated with habitat fragmentation, wildlife

habituation, declines in hunting participation, reduced support for hunting, and increases in human–wildlife interactions (Kretser et al. 2008, Glennon and Kretser 2005, Conover 2002, Organ and Ellingwood 2000, Stout et al. 1993). Access to rural lands for wildlife recreation can change quickly with exurban development, which divides rural landscapes into smaller parcels, increases recreational demand, and creates conditions for culture clash as newer residents with differing beliefs, attitudes, and behaviors regarding wildlife and their management populate and transform rural communities and landscapes.

The habitat loss that accompanies human population growth, urbanization, and exurban development is a far larger and more complex issue than simply causing an increase in problematic human–wildlife interactions. There are many indirect ways for the loss of wildlife to occur, such as construction of new roads, an increase in the demand for agriculture crops (e.g., ethanol production increases demand for corn, which causes an increase in the amount of grassland habitats being converted to cropland), and migration of people into the country (exurban development). These types of issues are especially difficult because state wildlife agencies do not have control over most land uses and, at best, only have a seat at the table along with many other interests. Wildlife managers have the knowledge and skills to manage wildlife habitat but have legal management authority over relatively little land. When managers lack authority over a basic element of the wildlife management triad (i.e., habitat), then a specialized set of skills is required for them to be effective in influencing desired outcomes for the benefit of wildlife. These skills are in the realm of communication and public engagement, which are discussed elsewhere in this book.

Human dimensions research has identified wildlife-use and recreation patterns related to urbanization, affluence, and rural gentrification. Hunting is probably the wildlife use that has been most affected by these factors. As society has become more urbanized participation in hunting has declined. Hunting participation has traditionally been highest in rural areas. People raised in rural areas or in hunting families (typically in rural areas) and introduced to hunting at an early age are more likely to be lifelong hunters. People raised in urban areas and introduced to hunting later in life are generally less likely to become hunters, and if they do take up hunting they tend to be less committed to it because they typically lack the familial support that reinforces hunting behavior. Hunting participation is, therefore, more related to the proportion of people living in rural areas than to access to hunting opportunities.

15.3.2.2. Aging
National census data and other demographic research document a long-term aging trend in the U.S. population, commonly referred to as "the graying of America" (Kausler and Kausler 2001). Demographers project that, by 2030, approximately 65 million Americans will be age 65 or older (approx. 20% of the U.S. population; Kausler and Kausler 2001). Aging of the North American population will affect participation trends in both consumptive and non-consumptive use activities.

Aging trends are expected to contribute to declining participation in consumptive uses of wildlife. Human dimensions studies have shown that attrition from hunting participation in the field (not from the social aspects of hunting) is associated with aging. After the age of 25, hunting participants drop out of the activity at a relatively steady rate as they age. By age 75, approximately 90% of hunting participants will have discontinued participation (Leonard 2007).

Conversely, aging trends are expected to result in increased participation in non-consumptive uses of wildlife. One model (Bowker et al. 1999) projected a 61% increase in birdwatching, photography, and other wildlife viewing between 2000 and 2050, driven largely by increasing age of the U.S. population. Large increases in wildlife viewing are expected as people aged 55 and older collectively engage in more multi-purpose recreation trips on which wildlife viewing is a complementary or incidental activity. Although it is good to know that interest in wildlife may shift rather than dissolve, the expectation that wildlife will be an "incidental" part of people's outdoor experience in the future has significant implications for program planning by the wildlife management community.

15.3.2.3. Ethnicity and Race
Immigration is projected to play an increasingly important role in population change in the United States in coming de-

Aging of the North American population will affect participation trends in both consumptive and non-consumptive uses of wildlife (courtesy Susan Steiner Spear)

cades. Demographers project that immigration and higher birth rates among non-white and Hispanic residents will create greater racial and ethnic diversity in the United States over the next 50 years. The U.S. Census Bureau projects that by 2042 non-Hispanic whites will no longer make up the majority of the population.

This projection has important implications for wildlife management because wildlife-use and recreation participation rates vary across ethnic and racial groups (USFWS 2007). Participation rates, particularly for hunting, are lower in minority groups. Because minorities are the fastest growing segments of the U.S. population, the percentage of the population that hunts is expected to decline. Additionally, the number of non-traditional households (i.e., households headed by single parents) is expected to increase, which has been shown to affect hunting recruitment negatively because youth from such households engage in hunting at lower rates than other youth.

15.3.3. Understanding Wildlife Recreation as a Process
Research on wildlife use expanded over time beyond efforts to merely track numbers of wildlife-related activity participants. The wildlife management community is keen to understand and influence activity involvement; thus, human dimensions

research includes efforts to increase understanding of participants' traits, motivational orientations, and activity-related satisfactions. In essence, the earliest human dimensions studies investigated questions of "where, when, and how much." Inquiry then went deeper and addressed a range of "how" and "why" questions related to activity involvement with questions such as the following:

- Why do people become engaged in wildlife-use activities?
- Why do they continue to participate or stop participating?
- How can participation be influenced?

Specific human dimensions information needs vary by context and activity. For a range of legal uses (e.g., trapping furbearers, wildlife feeding, collecting seabird eggs for subsistence use), managers often want to understand how to allow satisfying use activities that do not compromise wildlife conservation objectives. In the case of activities that directly or indirectly fund management and conservation (e.g., licensed recreational hunting, wildlife watching, or wildlife-related ecotourism), managers are interested in understanding how to recruit and retain activity participants. In cases where an activity threatens species survival (e.g., illegal trade in wild meat), managers seek information that helps them design interventions to reduce or eliminate the activity.

In the following sub-sections, we review some of the research contributing to understanding of wildlife activity participants and factors that influence processes of activity involvement.

15.3.3.1. Motivations and Satisfactions

Understanding stakeholder participation in various wildlife-use activities and forecasting changes in participation rates has been enhanced by human dimensions research focused on two key questions: (1) Why do people participate in activities related to wildlife use (and how do they get started)? and (2) Why do people continue (or cease) participating? The first question is related to the concept of motivations, which are cognitive forces that drive people to achieve particular goals. The second question pertains to the concept of satisfactions, which are the outcomes an individual receives from an experience and their evaluations of those outcomes.

Application of a multiple satisfactions approach (Hendee 1974) helps managers understand more about reasons for involvement in a range of wildlife-related activities, including recreational hunting (Decker et al. 1980, 1987) and trapping (Siemer et al. 1994), competitive birdwatching (Scott et al. 1999), subsistence hunting (Condon et al. 1995), wildlife viewing (Cole and Scott 1999, Hvenegaard 2002, Sali et al. 2008), and wildlife poaching (Forsyth et al. 1998, Muth and Bowe 1998, Eliason 1999).

Behavioral classifications have been developed to characterize motivations and satisfactions of wildlife recreationists (Table 15.1). For example, Decker et al. (1984) reported three main classes of motivations for hunters: (1) appreciative, (2) affiliative, and (3) achievement. Similar categories of motivations have been confirmed among non-consumptive users, particularly birdwatchers (McFarlane 1994). Recreationists with an appreciative motivation seek peace in the outdoors, want to feel immersed in nature, and want to feel "far away" from their everyday obligations and concerns. Those with affiliative motivations participate in wildlife-use activities because they enjoy those activities as a means to be with others and affirm or strengthen interpersonal relationships through shared activities.

Recreationists with achievement motivations are driven to meet self-identified standards such as shooting an animal with certain characteristics or a legal limit of game animals or adding a rare bird sighting to their species "life list." Any person may be motivated to use wildlife for all of these reasons but one reason usually is dominant. Both consumptive and non-consumptive recreationists can be segmented into meaningful specialization sub-groups based on the mix of motivations they seek (Box 15.1), which sometimes helps a manager create regulations that respond to the motivations of various segments and provide opportunities to experience satisfactions they value.

As we pointed out in Chapter 4, research on motivations and satisfactions identifies the benefits that recreationists seek and receive from wildlife-use activity, which has enabled managers to (1) identify the roots of conflict among recreationists and ways to reduce those conflicts; (2) develop distinct recreational experiences for different segments of a user group; and

Table 15.1. Typologies developed to describe goals, motivations, and satisfactions of wildlife recreationists

Goals that drive motivations and satisfactions	Wildlife recreation group	Source of motivation typology
Affiliation, achievement, and appreciation	Recreational hunters	Decker et al. 1984, 1987; Decker and Connelly 1989
Affiliation, achievement, appreciation, and wildlife conservation	Birdwatchers	McFarlane 1994
Appreciating birds, sharing knowledge, exploration, companionship, and spiritual experience	Birdwatchers	Sali et al. 2008
Competition, enjoyment, conservation, and sociability	Competitive birdwatchers	Scott et al. 1999
Nature appreciation, personal fitness, escape and/or relaxation, affiliation, wildlife damage management, and personal achievement	Trappers	Siemer et al. 1994
Commercial gain, household consumption, recreational satisfactions, trophy poaching, thrill killing, protection of self and property, poaching as rebellion, poaching as a traditional right, disagreement with specific regulations, and gamesmanship	Poachers	Muth and Bowe 1998

Box 15.1 *MOTIVATIONS OF BIRDWATCHERS IN ALBERTA*

Wildlife managers are often interested in why people participate in certain forms of wildlife recreation and, more specifically, why they seek certain types of experiences. An example from Canada is illuminating.

McFarlane (1994) examined the motivations of Alberta birdwatchers, using a mail survey conducted with a sample of 1,014 people in the province of Alberta who belonged to birdwatching-related organizations or purchased birdwatching supplies. She assessed motivations using a 25-item scale based on motivation research by Decker et al. (1987) and Beard and Ragheb (1983). Principal component factor analysis revealed the presence of four birdwatching motivations: (1) affiliation, (2) achievement, (3) appreciation, and (4) conservation.

McFarlane measured birdwatching specialization using a set of 11 items related to birdwatching experience, commitment, and centrality to lifestyle. She identified four specialization types by using cluster analysis, which she labeled casual, novice, intermediate, and advanced birdwatchers.

Motivational differences were found between specialization types. Advanced birdwatchers placed more importance on birdwatching as a source of personal achievement (e.g., expanding knowledge, improving skills). Casual birdwatchers placed more importance on appreciative aspects (e.g., enjoying and experiencing the outdoors). Conservation (e.g., the sense that one is helping wildlife or contributing to the conservation of birds) tended to be the primary motivation for novice and intermediate birdwatchers.

Recognizing differences in motivations can help mana-

gers address the varying expectations of birdwatchers at differing levels of specialization. McFarlane recommended that birdwatching programs strive to provide a range of opportunities tailored to meet the motivations and expectations of specific groups of casual to advanced birdwatchers. As government and non-government organizations consider the development of new wildlife-viewing opportunities, they should be aware of the participant groups most likely to be attracted to the new service, product, or viewing site. The potential to give birdwatchers opportunities to engage in conservation activities, for example, may have long-term conservation benefits.

Courtesy Donna Salko, Cornell Laboratory of Ornithology

(3) identify substitute activities that can offer similar benefits to recreationists. Research into motivations and satisfactions of wildlife recreationists helps wildlife managers deliver service and benefits to stakeholders.

15.3.3.2. Recruitment and Retention

Processes of wildlife-dependent activity recruitment and retention are of considerable concern to wildlife management agencies and non-governmental organizations (NGOs) because increases and declines in those processes have implications for wildlife program demand and funding. Long-term participation studies—even those focusing on geographically confined areas—are rare. Though it is a topic of considerable research, more inquiry is needed to understand recruitment and retention in wildlife-dependent activities well enough to influence those processes. Wildlife managers are particularly interested in obtaining practical understanding of factors leading to continued participation in activities linked to wildlife use (Enck et al. 2000). Assessments of factors that an agency or NGO could expect to influence, and their decisions about

potential actions, would be affected by knowing that activity initiation and continuation among people who have some propensity for hunting is determined largely by their social influences and support, their motivational orientations, and the satisfaction they experience.

The Human Dimensions Research Unit in the Department of Natural Resources at Cornell University has been studying hunting participation and hunters since the late 1970s. The Cornell studies have focused on why people hunt (Decker et al. 1984), how they become interested in hunting (Brown et al. 1981, Purdy et al. 1985, Purdy and Decker 1986), motivations for continued participation after initiation (Purdy et al. 1989, Enck and Decker 1991, Enck et al. 1993), and the satisfactions participants derive from hunting involvement (Decker et al. 1980; Enck and Decker 1991, 1994; Table 15.2). That research program also identified reasons people quit hunting (Purdy et al. 1989, Enck et al. 1993). In the 1990s, Cornell researchers began to examine the cultural dimensions of hunting. Those studies identified factors important to producing one's identity as a hunter (Enck 1996) and described a social network

Table 15.2. Conceptual framework to examine temporal and motivational dimensions of deer hunting satisfactions

Motivational dimensions	Temporal dimensions			
	Preseason	During season	Postseason	Multiple
Appreciative	• Seeing other wildlife while scouting for deer • Gaining a feeling of relaxation while scouting for deer	• Taking advantage of early season opportunities that offer less crowded conditions • Seeing other wildlife while hunting deer • Gaining a sense of belonging to the environment	• Skiing to favorite deer stand in the off-season • Personal reflection after the season	• Year-round observation of deer outside of a hunting context
Affiliative	• Sharing stories before the hunt • Talking over deer-hunting strategies with family and friends	• Hunting with family members • Hunting with friends • Helping others in the woods	• Sharing venison with others • Reminiscing about the hunt with friends	• Year-round story-telling • Passing the tradition of hunters on to others
Achievement	• Buying hunting equipment • Sighting-in bows or firearms	• Seeing deer • Shooting at deer • Bagging a deer	• Butchering deer • Eating venison • Tanning deer hides	• Reading "how-to" information • Year-round scouting
Multiple	• Getting ready for the hunting season • Anticipating the hunting season	• Getting exercise • Seeing other hunters bag a deer	• Cleaning or repairing firearms	• Photographing deer • Writing about hunting

Source: Enck and Decker (1994)

surrounding hunting that includes non-hunters who receive significant non-economic benefits through their association with hunting (Stedman 1993, Stedman and Decker 1993). By describing the social world of hunting that encompasses family units and social circles, those early studies demonstrated how decisions about hunting regulations affect people in addition to those who venture afield with firearms (Stedman and Decker 1996).

Human dimensions studies reveal the importance of social networks in cultivating and sustaining involvement in hunting and other wildlife-related activities (RM / NSSF 2008). The degree to which social support exists and recreation motivations are satisfied will have a great impact on the future of wildlife recreation. Wildlife agencies and NGOs cannot influence the sociodemographic trends discussed in the previous sub-section, but they can influence some aspects of recreation satisfaction and encourage social support networks, especially those essential to activity recruitment (Box 15.2).

15.3.4. Identifying and Managing Impacts

Managing impacts associated with wildlife use and users in the United States flows naturally from the mandates and statutory responsibilities states have assumed for wildlife governance. State wildlife-management agencies have several functions with respect to use. They regulate legal uses to conserve populations, to distribute opportunity among consumptive users equitably, and to allocate opportunity acceptably between recreational consumptive, non-consumptive, and subsistence users. In some cases, wildlife management agencies regulate opportunities for consumptive use to manage negative impacts associated with wildlife species (e.g., they rely on legal harvest to reduce negative impacts associated with white-tailed deer and snow geese).

Wildlife uses yield a range of positive economic, recreational, psychological, and social impacts for users. For example, positive impacts include personal enjoyment and increased income from participating in or offering wildlife-based recreation. Examples of specific ventures that yield such impacts include hunting or ecotourism, the culinary use of wildlife, selling products that are used for wildlife-related recreation, or simply observing wildlife. Negative impacts associated with wildlife use are also varied and may include personal injury, illness, economic loss, elevated fear or anxiety related to use, or unanticipated environmental consequences (e.g., endangerment of endemic species when exotic pets are intentionally or unintentionally released to new areas). Consider the trade-offs inherent in a multi-faceted question:

How are these wildlife-use activities affecting the following: wildlife mortality, participants, nonparticipants, the economy, social relationships, social support for conservation, and the wildlife management agency?

In the following sub-sections, we outline human dimensions insight used by managers for decisions about management of a sample of impacts associated with wildlife use.

15.3.4.1. Assessing Biological–Ecological Impacts

One of the earliest concerns of wildlife managers was assessing and regulating the biological impacts of consumptive use on game animal populations. Managers began seeking harvest data from hunters and trappers in the 1930s, not to understand the people involved, but to understand the magnitude of hunting-caused mortality on populations of game species. The kind of information collection that was typically done at game check stations and through hunter and trapper surveys (e.g., Where and when did you hunt? What game

Box 15.2 *FAMILIES AFIELD PROGRAM IN PENNSYLVANIA*

The Pennsylvania Game Commission has undertaken several steps in recent years to encourage the trial experiences and social support networks necessary to retain youth hunters. The agency hired a dedicated recruitment and retention coordinator, developed and published a "Youth Hunting Guide," and began connecting Pennsylvania students to the National Archery in the Schools Program, which teaches archery in physical education classes for 4th through 12th graders.

The Game Commission also supported passage of legislation to reduce barriers to youth participation in hunting. Pennsylvania's Mentored Youth Hunting Program was authorized when Governor Ed Rendell signed HB 1690 on 22 December 2005. This program authorized an apprentice hunting license that allows a licensed (experienced) hunter to take a youth hunter afield prior to his or her completion of a hunter training course. This mentor–apprentice

approach allows for safe first-year hunting experiences within a supportive social structure. According to the state Game Commission, Pennsylvania sold more than 100,000 mentored-youth hunting permits between the 2005-2006 and 2009-2010 hunting seasons. Many apprentice hunters go on to complete their hunter training certification in the following year.

Pennsylvania's Youth Hunting Program was part of the Families Afield campaign, a nationwide joint effort developed and promoted by a coalition of hunting and shooting sports organizations (i.e., the U.S. Sportsmen's Alliance, the National Shooting Sports Foundation, and the National Wild Turkey Federation). Since its beginning in 2004, more than 30 state legislatures have enacted legislation that creates a mentored youth hunting permit. Human dimensions studies will be needed to evaluate how such programs influence hunter recruitment and retention over time.

Family and friends sight in their rifles at a family deer camp in Pennsylvania (courtesy Jody Enck)

animals did you harvest? How many animals did you take? How many days did you set traps?) has been called surrogate biology (Mattfeld et al. 1984) because its purpose was to contribute to the biological information base (e.g., mortality, age and sex of harvested animals, and harvest per unit ef-

fort). Basic information about harvest pressure on game species remains important to managers and surrogate biology surveys occur annually in every state. This information is the foundation for setting season harvest quotas for many game species.

15.3.4.2. Assessing Economic Impacts

Information on the economic impact of wildlife recreation often plays a pivotal role in wildlife management decisions. Expenditures on wildlife-related recreation have traditionally been used to demonstrate the high value participants place on opportunities to pursue those activities. Economic studies are used to demonstrate how public-sector investments (e.g., land purchases or conservation easements) create positive economic impacts for the private sector (e.g., job-creation-associated additional expenditures on wildlife-related recreation). Economic studies also are used to inform decisions about mitigation of human-induced resource damages (e.g., loss of wildlife recreation opportunity due to oil spills).

Economic analyses are invaluable to wildlife agency directors when they are asked to discuss program budgets with their executive leaders and state legislatures. To the degree that those decision makers wish to base wildlife management budgets on economic considerations, they need answers to questions such as the following: How much will each dollar spent on wildlife programs generate in public benefits? What groups will benefit if more money is put into wildlife management? What will the value of the benefit be?

Two fundamentally different types of economic measures have been used with respect to assessing recreation in wildlife management: (1) the economic impact of wildlife recreation, and (2) the net value of wildlife uses. Those two measures have had very different applications, as illustrated below.

Economic impact of recreation. Economists have used expenditures of wildlife recreationists to understand the economic impact of wildlife-related activities on a geographic area or an economic sector within a geographic area. The USFWS has been collecting data on expenditures associated with consumptive and non-consumptive wildlife recreation every 5 years since the 1950s. These data have documented the positive impacts that money spent on wildlife-related activities has on national, state, and local economies in terms of income and jobs. Americans spent more than $80 billion (all dollar values in U.S. currency) in 2006 on wildlife-related recreation (including $45.7 billion related to wildlife watching, $22.9 related to hunting, and $11.7 billion on items used for both hunting and fishing (USFWS 2007).

In another example, economic expenditures research has documented the contributions an emerging nature-tourism industry makes to the U.S. and Canadian economies. Survey data show the close relationship between sustaining wildlife resources and public socioeconomic benefits. This relationship is of strategic importance because it provides stakeholders with a powerful argument to influence national, state, provincial, and local decision makers about how the loss of wildlife resources results in lost benefits to people and increased economic and social costs to communities.

Net value. When one wants to examine the value of a wildlife resource to society, or the value of wildlife-observation experiences to groups of recreationists, the appropriate economic measure is not expenditures (which are measures of direct economic impact) but the net value over and above the expenditures, also known as the net willingness to pay or the consumer surplus. It is the total value of the benefits minus the price actually paid. Economists have devised two useful techniques that are applied to understand the full value of wildlife use: the contingent valuation approach and the travel cost approach. Both techniques have been used extensively to inform wildlife management decisions.

Contingent valuation studies have been used to understand the economic value of wildlife uses as diverse as bird-watching during shorebird migrations in North America (Aiken 2009, Myers et al. 2010), ecotourism in Australia (Higginbottom 2004), and trophy hunting in various contexts across Africa (Lindsey et al. 2006, 2007). Such studies also have helped reveal the total value of wildlife to society by documenting the benefits people obtain from knowing wildlife resources exist even though they may not directly use them. For example, the 1996 survey on the Importance of Wildlife to Canadians reported that 13 million Canadians (over 50% of the adult population) would be willing to pay higher prices for selected goods and services to conserve wildlife populations and their habitats. In the case of the 1996 survey data, if passive values were considered then the total economic value of wildlife-related activities to the Canadian economy would have exceeded $3 billion, which is 300% larger than reported.

15.3.4.3. Assessing Sociocultural Impacts

Involvement in wildlife-related activities generates tangible and intangible benefits for individuals, groups, and communities. Human interactions related to wildlife use also can contribute to or disrupt community well-being. Studies of the social costs of human–wildlife and human–human interactions (i.e., studies focused on conflict between stakeholder groups) have been more common.

Clarifying conflicts between stakeholder groups. Many positive social impacts are associated with wildlife use (e.g., sociocultural values of wildlife use, cultural and community benefits that subsistence users gain from their use traditions), but negative social impacts that capture a great deal of managers' time and attention are produced as well (e.g., opportunity allocation conflicts arise between different groups using the same resource, conflicts between stakeholder groups emerge that are rooted in different beliefs about appropriate uses of wildlife or management policies related to wildlife use). Social science research, combined with better public-involvement processes, has helped managers understand and manage use conflicts (Messmer 2009). Perhaps the most comprehensive examples of such work developed in the context of suburban deer management in the United States (Box 15.3), but similar approaches have been applied to situations as varied as conflict between commercial hunting groups and bird conservation groups in the United Kingdom (Redpath et al. 2004) and wildlife conservation conflicts between local communities and international conservation organizations in Tanzania (Hausser et al. 2009).

Box 15.3 *BUILDING BLOCKS FOR SUCCESS*
Key Dimensions of Successful Community-based Deer Management

Suburban deer-management issues often involve conflicts over consumptive uses of deer (i.e., lethal removal of deer through recreational hunting or culling). The Human Dimensions Research Unit (HDRU) at Cornell University has, for many years, conducted studies of community-based deer management in suburban contexts across the northeastern United States. In 2002, HDRU recruited a panel of 10 experienced deer managers to develop descriptions of successful deer-management cases with which they had been involved. After working with managers to develop their case studies, HDRU staff convened the panel in a workshop setting to analyze their cases of suburban deer management. In combination with a synthesis of HDRU studies, this collaborative effort resulted in identification of a set of key dimensions of community-based approaches that might be considered building blocks for successful community-based deer management (see Decker et al. 2004*b*). These dimensions are restricted to the *process* of decision making in community-based deer management; that is, they are elements that are important in the process of getting to the point at which management actions can be implemented.

The investigators placed key dimensions into two categories: enabling conditions and intervention thrusts.

Enabling conditions. Community-based deer management can be enhanced by the existence or development of certain enabling conditions. These conditions represent characteristics of stakeholder groups or process conveners

A researcher kneels over an immobilized deer in Cayuga Heights, New York, one of the community-based deer-management efforts profiled in Decker et al. 2004b (courtesy Paul Curtis)

15.3.5. Understanding Public Support for Use Activities

Stakeholder attitudes toward use activities have been investigated for decades because those attitudes lie at the heart of many wildlife management conflicts. Human dimensions inquiries have helped managers understand topics as diverse as attitudes toward prairie dog hunting in the American West (Reading et al. 1999, Zinn and Andelt 1999, Lybecker et al. 2002), local acceptance of professional hunting concessions in African nations (Lewis et al. 1990, Weladji et al. 2003), and public opinion about commercial whaling (Freeman and Kellert 1992, Hamazaki and Tanno 2001). Collectively, these types of studies help managers understand how attitudes differ within and among stakeholder groups, especially those groups that can play a key role in generating or resolving wildlife management controversies. Perhaps more importantly, research has revealed whether conflicts over wildlife-use activities in particular situations are based in general concerns about ideology or ethics, with specific concerns ranging across topics such as animal treatment, behavior of activity participants, local control of decision processes, and equitable distribution of benefits.

Much of the research conducted to date simply documents level of support for uses or management actions. These studies

have described management conflicts that arise between factions of user groups (e.g., deer hunters who hunt in different seasons, with different implements), which often occur over allocation of opportunity or conflicts afield while engaged in incompatible activities. Studies also have explained the bases of conflicts between very different user groups (e.g., birdwatchers and waterfowl hunters, local and statewide residents, hunters and non-hunters), which may occur for similar behavioral interaction reasons or because of fundamental differences in ideas about what constitutes acceptable wildlife uses. This kind of research informs discourse, debate, or decisions about specific issues such as opening a new hunting season, legalizing new implements for hunting (e.g., crossbows), or policies about fur trapping on public lands. Many examples of such studies have focused on issues related to hunting carnivores, hunting ungulates, and trapping furbearers.

Some of the most useful research in this area has gone deeper than mere description to provide insight on why stakeholders differ in their attitudes about particular wildlife uses and management actions. For example, social science research has identified factors that explain support or opposition to regulated hunting. Differences in attitudes have been correlated with demographic traits (e.g., age, education, ethnicity, gender, income, urban vs. rural background), wildlife species

that contribute to community readiness for collaborative decision making. They can be encouraged by interventions, and they often result in improving the effectiveness and efficiency of a community's involvement in decision-making processes. The five enabling conditions identified are

1. adequate knowledge (among stakeholders and managers),
2. essential working relationships (partnerships and informal networks),
3. effective local leadership,
4. sufficient credibility, and
5. agency–community commitment to common purpose.

Intervention thrusts to achieve enabling conditions. Also identified were five important means for achieving the enabling conditions. These intervention thrusts are directed either toward stakeholders or toward the wildlife agency. The five intervention thrusts identified are

1. stakeholder involvement,
2. education and learning,
3. informative communication,
4. wildlife agency flexibility, and
5. inventory–assessment.

A radiocollared, ear-tagged deer walks among the homes in the village of Cayuga Heights, New York. With no significant mortality caused by predators and no hunting allowed within village limits, deer populations are limited mainly by deer–car collisions. (courtesy Paul Curtis)

(e.g., ungulate vs. carnivore, common vs. uncommon), value orientations, personal experience with wildlife, involvement in wildlife-related activities, economic interests, concerns about animal treatment, concerns about behavior of activity participants, and other variables. Some studies have found evidence that attitudes toward hunting may change over time if stakeholders come to experience additional or more severe negative interactions with a species. For example, research in New Jersey (Applegate 1984) documented that public support for white-tailed deer hunting increased over time concurrent with increases in the deer population and problem interactions with deer.

One relatively consistent finding is that public support for hunting is linked to the assumed motivations of hunters (Heberlein and Willebrand 1998). Surveys conducted over the past 30 years have indicated that most Americans approve of subsistence hunting or hunting to obtain meat but not hunting for recreation or to obtain a trophy (Kellert 1979, Duda et al. 1998, Duda and Jones 2009). For example, a national survey conducted in 2006 (Responsive Management 2006) found that more than 80% of Americans approved of hunting to obtain food, protect humans from harm, or to control an animal population, but fewer approved of hunting for sport (53%), for challenge (40%), or to obtain a trophy (28%).

15.3.6. Understanding and Building Support for New Sources of Wildlife Program Funding

The extent and nature of wildlife management and conservation actions are a function of program funding—conservation success in the twenty-first century will depend upon broad-based support from both consumptive and non-consumptive wildlife user groups. Decision makers (commissions, commissioners, legislatures) benefit from research on past wildlife-use patterns or projected activity involvement when considering proposals to expand traditional sources of program funding or to create new mechanisms to finance wildlife management and conservation. Management agencies also use economic research to support proposals to increase fees for existing licenses or to create license and fee programs for new use activities (e.g., creating a new license fee for a restored species such as moose, or creating a new opportunity to hunt antlerless deer or elk). Agencies have routinely used economic studies to estimate "elasticity" of demand—how changes in license fees are likely to affect license purchasing and associated revenue (Sandrey et al. 1983, Anderson et al. 1985, Brown and Connelly 1994)—which translates into activity participation effects. For example, economists in Canada used a consumer demand study to show provincial legislators how increases in hunting license fees for residents and non-residents were likely to affect short-term and long-term revenue to the province (Sun et al. 2005). Their study suggested that raising the resident license fee would be revenue-neutral in the long term, but raising license fees for non-residents was unlikely to reduce license sales demand and thus could be expected to provide long-term revenue increases (Sun et al. 2005). Human dimensions research also has been used to understand willingness to pay for proposed program expansion and support for alternative funding mechanisms (e.g., voluntary tax check-offs, dedicated sales taxes, fees from sales of specialty license plates, conservation trust funds, state lotteries, royalties, real-estate transfer taxes, surcharges on oil and gas leases).

Social science, public involvement, and coalition building have all played crucial roles in establishing and maintaining alternative funding streams. For example, during the 1980s, several human dimensions studies characterized contributors to newly established tax check-off programs to fund non-game management that would benefit non-consumptive users of wildlife (Mangun and Shaw 1984, Brown et al. 1986, Manfredo and Haight 1986, Moss et al. 1986). States such as Arkansas, Minnesota, and Iowa more recently have engaged in studies to determine citizens' attitudes about and support for proposed dedicated sales taxes to support environmental and wildlife conservation (Fairbank et al. 2006) and to assess the social and economic benefits expected from proposed conservation funding initiatives (Otto et al. 2007).

State wildlife agencies also have used human dimensions research to characterize stakeholder interests (largely related to wildlife-use activities) in programs funded by new alternative funding sources. Efforts by the Missouri Department of Conservation (MDC) illustrate this point. In 1976, Missouri citizens passed a voter initiative to create a dedicated sales-tax program to fund wildlife conservation (Brohn 1978). Between 1976 and 1983, MDC "completed nearly 40 social surveys to assist resource managers in understanding what people know about conservation and what they expect of Missouri's conservation programs" (Witter and Sherriff 1983:42). These and many subsequent studies have been used by MDC to guide program development and delivery and to maintain support for the dedicated sales tax in the face of multiple challenges to redirect the funding to other purposes.

SUMMARY

The process of managing use and users of wildlife has become more complex through time, as new types of consumptive and non-consumptive uses developed and the number and diversity of stakeholders considered in decisions about wildlife use increased markedly. Managers are seeking more sophisticated and representative human dimensions information to make informed decisions. Demand subsequently has increased for social science to better inform decisions related to the human dimensions of wildlife management. This chapter examined the history, evolution, and application of human dimensions insights into management of wildlife uses and users.

- Contemporary wildlife agencies in the United States were established primarily to manage impacts associated with consumptive, non-consumptive, and subsistence uses of wildlife. The earliest era of contemporary management concentrated on stopping unregulated exploitation of wildlife. With time, many wildlife populations were restored and more liberal use was possible under a renewable-resource management regime. The focus of management shifted to production of game species for sustainable use.

- Social scientists help identify sociodemographic changes and their potential influence on involvement of humans in wildlife-related activity. Agencies and organizations use these inquiries as input when designing policies or programs that balance supply with demand, mitigate potential negative impacts to the resources, or reconcile conflicting uses.

- Consumptive use involves activities that permanently remove animals from the wild, including uses that are recreational and commercial or legal and illegal. Examples of consumptive uses are (1) hunting and trapping; (2) wildlife removal to manage nuisance, damage, or threats to human safety; (3) regulated fee-hunting on private preserves; (4) scientific research; (5) removal of individuals for propagation of endangered species; and (6) removal of animals for educational display.

- Non-consumptive use involves activities in which people benefit from wildlife without permanently removing animals from the wild. Examples of non-consumptive uses include (1) wildlife viewing, (2) wildlife photography, (3) wildlife feeding, and (4) wildlife-related education or interpretation. The non-consumptive label can be misleading because there is an assumption that if wildlife species are not removed from the wild directly then non-consumptive use has only positive impacts.

- Research indicates both consumptive and non-consumptive uses can disturb wildlife in ways that affect (sometimes negatively) wildlife. Human dimensions research also has shown how conflicts among non-consumptive users (and between them and consumptive users) can create a range of negative psychological, social, and economic impacts on the people involved.

- Wildlife resources provide subsistence uses wherever wildlife and resource-dependent human populations come into contact. Wildlife is harvested to meet needs ranging from human food and medicinal products to maintenance of cultural traditions. In many cultures, customary and traditional uses of wildlife are part of a "way of life" that maintains kinship groups, ethnic cultures, and other social relationships. Not all subsistence-based households are primitive, socially isolated enclaves. Sustainable use becomes more difficult as human populations grow and pressures to use wildlife increase throughout the world.

- Contemporary human dimensions research continues to increase understanding of some key factors (e.g., participants' traits, motivational orientations, and activity-related satisfactions) that influence the process of wildlife-related activity involvement. Human dimensions inquiries help managers understand why people become engaged in wildlife-use activities, why they continue or cease to participate, and how participation can be influenced through policy or program.

- Stakeholder attitudes toward use and users receive substantial research attention because differences in those attitudes are the cause of many conflicts in wildlife management. Collectively, studies are helping managers understand how attitudes differ within and among stakeholder groups, especially groups who can play a key role in generating or resolving controversies focused on wildlife use. Human dimensions inquiries can reveal whether conflicts over wildlife-use activities in particular situations are based on concerns about animal treatment, behavior of activity participants, local control of decision processes, competition for scarce resources, equitable distribution of benefits, or other factors.

- Decision makers benefit from information about wildlife-use patterns, trends, and projections when considering proposals to expand traditional sources or create new sources of program funding. For example, managers use economic research to estimate benefits from existing programs, support proposals to increase fees for existing wildlife-use licenses, or to create license and fee programs for new use activities. Human dimensions inquiry also informs design of public involvement processes, coalition building, and program planning necessary to gain broad-based support for alternative funding proposals among consumptive and non-consumptive wildlife-user groups.

Suggested Readings

Gartner, W. C., and D. W. Lime. 2000. Trends in outdoor recreation, leisure and tourism. CABI, New York, New York, USA.

Knight, R. L., and K. L. Gutzwiller. 1995. Wildlife and recreationists: coexistence through management and research. Island Press, Washington, D.C., USA.

Manfredo, M. J., editor. 2002. Wildlife viewing: a management handbook. Oregon State University Press, Corvallis, USA.

PART VI • PROFESSIONAL CONSIDERATIONS FOR THE FUTURE

This part contains two chapters in which we discuss considerations that make wildlife professionals more effective: ethical analysis and education. We hope it is apparent from earlier chapters that wildlife management decisions are "informed" by science; natural and social science theory and scientific data do not "make" a decision. Choices made in wildlife management include high levels of scientific uncertainty about the management system; yet, even if professionals had perfect information, a method is needed to make trade-offs based on societal values and ethics. Chapter 16 suggests that ethics and ethical discourse provide a mechanism for thoughtful deliberation about the messy issues typical of wildlife management. Inclusion of ethical analysis and discourse does not necessarily make the job of a manager any easier but should aid in making management decisions more durable.

It is often said in the private-sector world of business that the only way to remain competitive is through continual learning—learning about the environment, learning about customers (stakeholders in public-sector wildlife management), and learning about yourself. In Chapter 17, we emphasize the kinds of education and experiential learning opportunities that should be part of anyone's preparation for employment and continued practice in wildlife management. We encourage you to view learning as a lifelong endeavor aimed at personal and professional growth.

The story of the countless roles for human dimensions in wildlife management is not finished with this book or at this point in time. Our final chapter is less a traditional summary of ideas presented in this book (look back to Chapter 1 for a summary) than our view of the future directions, challenges, and needs in the human dimensions of wildlife management. Understanding and application of human dimensions insights are rapidly improving and expanding as practicing wildlife professionals—in research and management—contribute to the growing body of knowledge and methodology. Rapid social and environmental change is affecting wildlife management directly and indirectly, which indicates the necessity of taking an adaptive approach to your work in wildlife management and of using an adaptive strategy to engage in transformative change of organizations in which you work. Future human dimensions issues are anticipated to center on continual improvement in stakeholder engagement, conservation organization structure and function in the face of changing stakeholder needs, and sources of financial support to conduct effective wildlife management.

16

ENVIRONMENTAL ETHICS FOR WILDLIFE MANAGEMENT

MICHAEL P. NELSON AND JOHN A. VUCETICH

Ethical issues influence nearly every aspect of wildlife management from its broad principles to specific decisions. The influence of ethics on the broad principles of wildlife management is illustrated by The Wildlife Society's (TWS) vision statement:

> TWS seeks a world where people and wildlife co-exist, where biological diversity is maintained, and decisions affecting the management, use, and conservation of wildlife and their habitats are made after careful consideration of relevant scientific information and with the engagement and support of an informed and caring citizenry.

How can this vision be realized without addressing issues such as: Why should people care about wildlife and biodiversity? Is the need to care only because wildlife and biodiversity are useful to humans or also because they are valuable in their own right? What exactly does it mean to conserve wildlife and their habitat? For example, both Aldo Leopold and Gifford Pinchot wrote about conservation, but their ideas about the meaning of conservation differ profoundly. An appreciation of modern environmental ethics helps one to address issues such as these intelligently and therefore fully understand the TWS vision statement.

Ethics also influences the details of many specific situations. Consider this example: Isle Royale National Park is a federally designated wilderness area, and home to a small wolf population that is isolated from other wolf populations and that shows signs of inbreeding depression. If you value healthy wildlife populations, you might consider the feasibility of genetic rescue, which entails introducing unrelated individuals to alleviate the negative consequences of genetic deterioration; however, you might think this a bad idea if you value designating a few places on the planet where humans intervene as little as possible. The idea is based on letting nature "run its course." This, you might suggest, is the purpose of federally designated wilderness areas in the United States.

On the other hand, you might think attempting genetic rescue represents a promising, but largely untested, conservation tool that could help conserve many other populations.

Isle Royale wolves might represent a model system for testing this tool, but how would intervention affect the health of the Isle Royale ecosystem? Because the effects of winter, ticks, and climate change on Isle Royale moose seem to be increasing, a more resilient wolf population could be importantly detrimental to the interactions among wolves, moose, and the forest. What about the welfare of the individual wolves? Evidence suggests that some of the bone deformities that Isle Royale wolves exhibit may also be painful to individual wolves—pain that might be mitigated in subsequent wolves by intervention. Isle Royale is but one example of a common challenge, the challenge of knowing how to balance values that may conflict when decisions are made about how to manage wildlife populations. Environmental ethics and conservation ethics are fields whose purpose is to help us handle these challenges.

16.1. WHAT IS ETHICS?

The social sciences (including social psychology, sociology, and economics) represent disciplines that can help to describe how humans value wildlife. Ethics is the discipline whose focus is formal and rigorous analysis of ethical propositions. The fundamental distinction between ethics and the social sciences you have read about in this book is that social science is primarily concerned with the analysis of descriptive propositions about human values, whereas ethics is concerned with the analysis of prescriptive propositions about human values. Descriptive propositions describe the nature of the world around us, and prescriptive (ethical) propositions are claims about how we ought to behave, value, or relate to the world around us. For example, a sociologist might work to describe what value or social norm stakeholders hold, and to understand why stakeholders hold a particular value. The purview of ethics, however, is to assess whether and why one *ought* to hold some value.

Ethical propositions are easily identified in that they can typically be expressed using words such as "ought" and "should." Do you ever think, "There ought to be fewer deer on the landscape" or, "The wolf population should be allowed to increase in abundance"? These are examples of ethical propositions. Ethics may also be defined as the analysis of propositions that assess what is good or what is right. For example, when Aldo Leopold said, "A thing is right when it tends to preserve the integrity, stability and beauty of the biotic community. It is wrong when it tends otherwise," he was making an ethical proposition that humans should relate to nature in ways that tend to preserve nature's integrity, stability, and beauty.

Insomuch as wildlife conservation involves propositions such as, "We ought to behave in this way (toward some aspect of the natural world) . . ." wildlife conservation can be considered ethics in action. Environmental policies and laws also reflect ethical commitments. For instance, the Endangered Species Act and the Wilderness Act of 1964 seem to reflect the ethical proposition that aspects of nature (in these cases, species and wilderness areas) deserve protection. These laws obligate us to protect species and ecosystems, not only because they benefit us somehow—physically, emotionally, or psychologically—but because they are also valuable for their own sake.

The ubiquity of ethics in wildlife management is also reflected in meanings of sustainability. Many consider (quite uncontroversially) sustainability to mean, "meeting human needs in a socially just manner without depriving ecosystems of their health"; but what is meant by "human needs," and what is a "healthy ecosystem"? Depending on how these terms are defined, sustainability could mean anything from "exploit as much as desired without infringing on future ability to exploit as much as desired" to "exploit as little as necessary to maintain a meaningful life." These two attitudes would seem to represent dramatically different worlds, and yet either could be considered sustainable depending on the meaning of ethical concepts that define sustainability (Vucetich and Nelson 2010).

Nearly all goals in wildlife management embody an ethical attitude about how society ought to relate to nature (Decker et al. 1991, Shrader-Frechette and McCoy 1994). A great deal of wildlife management, for example, is concerned with managing populations that are overabundant, too rare, or in need of restoration. Each of these cases represents an ethical attitude about how the world ought to be. Any claim that some wildlife management goal or action is inappropriate also reflects an ethical attitude. In this sense, ethical issues are not only ubiquitous but they are also inescapable; ignoring the ethical dimension of an issue does not make it go away. For this reason, it is wise to be adept at identifying and analyzing ethical issues in wildlife management, as illustrated by the examples involving sustainability, Isle Royale wolves, and The Wildlife Society vision statement.

Ethics is not merely asserting what is right or how we ought to behave. Ethics, as the academic tradition has been practiced for more than 2,500 years in the West, is also about understanding methods that reveal the most rational answers to these questions about how we ought to act. Much of this chapter is an introduction to these methods.

16.2. WHAT IS ENVIRONMENTAL ETHICS?

Environmental ethics is a relatively new field of study. In the early 1970s, a small group of philosophers realized that much of the controversy associated with natural resource management rises from unsettled ethical issues about how humans ought to relate to nature. At first they were interested in these sorts of questions: In what way or in what sense, if any, are humans really separate from the rest of nature? Does nature have intrinsic value and, if so, what does that mean? Though these questions remain important, they are better understood today than 40 years ago, and newer and different questions have emerged. The formal application of environmental ethics for the purpose of better understanding the human dimensions of wildlife is relatively uncommon (Box 16.1).

After four decades of development, the discipline of environmental ethics has given rise to distinct schools of thought that distinguish themselves primarily by the rational arguments they develop to support the type of value they conclude that nature possesses. These schools of thought also differ from one another by being more or less inclusive. For example, some argue that only humans are members of the moral community, whereas others argue that all living things should be included; some argue that species and ecosystems matter ethically. Before exploring different schools of thought in environmental ethics further, it is useful to describe a few of the most basic ethical theories (Box 16.2).

16.3. THEORIES OF ETHICS

One of the most important ethical theories is *consequentialism,* which asserts that the rightness of an action is determined by the consequences of an action. *Utilitarianism,* an important form of consequentialism (a form of which dominated American conservation in the twentieth century), presumes that we ought to act in ways that produce the most utility, happiness, or pleasure for the most people. Typically "people" has been equated with "human being," though not by everyone. *Pragmatism* is sometimes viewed as another school of consequentialist thought that claims truth or meaning ought to be judged by practical consequences. A pragmatic ethic is judged, therefore, by its ability to solve ethical problems, as we perceive those problems. Although pragmatism may seem commonsensical, it has long been deeply controversial among ethicists.

Deontology contrasts with consequentialism and judges an action's rightness by the intention or motivation for action rather than by the results of an action. Examples of deontological perspectives include treating others as you would want to be treated (e.g., the Principle of Ethical Consistency), respecting the rights of things that possess rights, performing an action out of a sense of duty, following certain pre-established rules, and only performing actions you would be willing to make into universal law. The Endangered Species Act seems to manifest a deontological perspective because it grants a basic

Box 16.1 *INSTRUMENTAL AND INTRINSIC VALUE IN ENVIRONMENTAL ETHICS*

Are wildlife and the rest of the non-human world valuable merely because they satisfy a variety of human needs and desires, or does the non-human world possess value that transcends "use value"? Although some people believe nature is valuable only as a means to serve human goals and objectives (Instrumental Values), others believe nature is valuable beyond its instrumental value (Intrinsic Value). This debate over the type(s) of value(s) attributed to nature is at the center of both wildlife management and environmental ethics and historically is illustrated by philosophical differences between Gifford Pinchot (1865–1946) and Aldo Leopold (1887–1948). Pinchot (1947:325–326) believed that "There are just two things on this material earth—people and natural resources." This assumption about the nature of reality (or metaphysic) served as the foundation for Pinchot's ethic, which suggested right actions could be prescribed as "the use of the natural resources for the greatest good for the greatest number for the longest time." Rejecting Pinchot's metaphysic of human distinction and instead arguing that humans were "plain member and citizen of" a "biotic community," Leopold (1949:204, 224–25) believed actions were right if they tended "to preserve the integrity, stability, and beauty of the biotic community," and wrong if they tended otherwise.

Gifford Pinchot, first Director of the U.S. Forest Service

right (i.e., the right to exist) to most species apart from their economic value.

Natural Law Theory and *Divine Command Theory* are similar, and presuppose that what is natural or divinely commanded is moral; while that which is unnatural or divinely forbidden is immoral. For example, if one were to expect the biophilia hypothesis (i.e., an innate or *natural* tendency to love life) to deliver specific moral mandates, then the biophilia hypothesis would exemplify Natural Law Theory (Box 16.3). Similarly, ethics developed explicitly from Christian ideals (e.g., an ethic of stewardship as a directive from God) or in reference to any divinity represent Divine Command Theory.

Virtue Theory holds that right actions arise from people who are manifestly virtuous, and that moral education ought to focus on identifying precisely which virtues ought to be manifest (e.g., generosity, respect, humility, courage) and how to cultivate such virtues in a person. A challenge for virtue ethics is to understand precisely which virtues are most important (e.g., justice or equality, modesty or magnanimity, and so on).

The Theory of Moral Sentiments stresses that reason and emotion are both critical for judging the rightness of an action. For example, in some cases, reason is necessary for indicating circumstances where moral attention is required, and emotional sentiments (such as compassion) motivate one to

manifest moral attention. Darwin's view on ethics (chapter 3 of *Descent of Man*) and Leopold's Land Ethic are both related to the Theory of Moral Sentiments, developed philosophically by David Hume (1739) and Adam Smith (1759) and as discussed in Callicott (1982).

Two other important terms in ethics are moral agent and moral patient. A moral agent is someone capable of extending moral consideration to others. Nearly all humans are moral agents. The extent to which some non-humans (e.g., chimpanzees and wolves) exhibit a very primitive form of moral agency is actively debated. A *moral patient* is anything that should receive moral consideration. Although moral agents also tend to be moral patients (e.g., a normal adult human), a moral patient is not always a moral agent (e.g., a 1-day-old human). Scholars in environmental ethics actively debate whether many non-human forms of life should be considered moral patients. Different theories answer environmental ethics questions in different ways (Box 16.4), each speaking to different values that people apply to environmental issues.

16.4. THEORIES OF ENVIRONMENTAL ETHICS

Although there are many ways to categorize the field of environmental ethics, it is centrally concerned with two entwined

Box 16.2 *ETHICS*
Misconceptions and Obstacles

Many misperceptions about the nature of ethics interfere with the effective application of ethics to issues in wildlife management. Below are some common objections [O] to ethics and responses [R] to those objections.

O1. Ethical problems are intractable and ethical attitudes change very slowly. For these reasons, ethical disagreements are inevitable and attempts at resolving them are not worth much attention.

R1. In this way, ethics is more like science than we often appreciate; that is, both require high degrees of rigorous thought and progress is often painfully slow. As discussed later in this chapter, ethical consensus, much like scientific consensus, is possible given the process of ethical discourse.

O2. Ethics is just non-rational and subjective, whereas only science is rational and objective; therefore, progress can be made with the latter but not with the former.

R2. In practice, science is not always as rational and objective as we sometimes think. More importantly, genuine ethical discourse relies on the formulation and assessment of rational and objective arguments.

O3. Ethics cannot be universally true; notions of right and wrong are only true from the point of view of a given culture or even a given individual. The belief that a wetland is better than a parking lot has no universal truth-value; we use the values we hold to decide which is better.

R3. Given that all humans and all human cultures have certain attributes in common or have common interests, then this simply might not be true. The values that ethical positions depend upon might, in fact, be as universal as many of the empirical premises that scientific positions rest upon.

O4. Ethics just seems to be a way to tell others what to do (i.e., another way to infringe upon freedoms and liberties).

R4. Fundamentally, ethics is about the understanding of what it is that we ought to do. In this sense, ethics is primarily a bottom-up, rather than top-down, exercise aimed at understanding the best way to live in the world.

questions: (1) the question of *moral considerability;* that is, what sorts of entities deserve membership in the moral community and what justifies that membership? (i.e., which entities are moral patients, and why?), and (2) the question of *moral significance;* that is, how humans ought to behave in an inclusive moral community (e.g., one that includes humans and non-humans) and how do humans sort out competing moral claims after we have established a moral community? Approaches to these questions begin with the application of standard ethical theories (rights theory, utilitarianism) to these questions and are therefore referred to as "extensionist" theories of environmental ethics because they work to "extend" those traditional ethical theories beyond the traditional bounds of moral inclusion (that is, beyond humans).

Anthropocentrism or Human Welfare Ethics. This perspective focuses on justifying how it is appropriate to believe that only humans are worthy of moral consideration and the consequences of that belief. Anthropocentrists care for non-humans—such as species, ecosystems, or non-human animals—only when human well-being depends on non-human well-being. For the anthropocentrist, only humans possess intrinsic value; all else is merely instrumentally valuable. Anthropocentrists agree with the famous eighteenth century philosopher Immanuel Kant (1930:241), who asserted, "all duties towards animals, towards immaterial beings and towards inanimate objects are aimed indirectly at our duties towards mankind." Anthropocentrists, therefore, believe we ought to conserve wildlife only be-cause their loss might negatively impact human beings in some manner.

Zoocentrism or Animal Welfare Ethics. This perspective is associated with the idea that, in addition to humans, certain non-human animals possess intrinsic value and deserve direct moral standing. The basis for this thinking begins with the observation that humans do not possess moral standing "just because"; instead, we have moral value because we have certain properties (e.g., consciousness or the ability to feel pain). Logical consistency, a zoocentrist then would argue, forces us to grant moral standing to anything possessing a morally relevant property. Hence, if certain non-human animals possess those morally relevant properties, then they too are intrinsically valuable and deserving of direct moral consideration. A zoocentrist would be supportive of, for example, efforts to conserve the habitat of morally relevant wildlife species (because these species would be made up of morally relevant specimens). However, he/she would also tend to oppose the killing of morally relevant animals that represent exotic species. An ethical dilemma is created in these situations when exotic, but morally relevant, animals harm native species.

Biocentrism. This perspective expands the moral community of zoocentrists by arguing that being alive is the morally relevant trait; that is, all individual living things deserve direct moral consideration. Albert Schweitzer (1923:254), perhaps the most popularly recognized biocentrist, summarizes the position thusly:

Box 16.3 *A COMMENT ON NATURAL LAW THEORY*

Natural law theory arguably serves as the foundation of many wildlife management ideas. Some people, for example, argue that hunting is a natural and, thus, ethical relationship to wildlife. Others defend the morality of wildlife restoration efforts on the basis that such efforts re-establish a "natural balance" to the world; therefore, the following "naturalistic" objection to environmental ethics might be expected: "Can't we dispense with all of this talk of ethics and just do that which is natural?" A couple of serious challenges, however, arise immediately when considering natural law theory. First, the theory assumes that what is natural can be discovered and defined objectively. Second, the theory assumes that what is natural is also what is good; this is an assumption philosophers sometimes call the "naturalistic fallacy." Third, even if what is natural can be defined, and even if what is natural is also what is good, it is not always clear what ought to be done. Appealing to naturalness, for instance, as a way to determine what should be done in the case of the inbred wolves of the Isle Royale will not necessarily establish clear guidance, because extinction (which would presumably suggest non-interference as the ethical course of action) and predation (which would presumably suggest genetic rescue as the ethical course of action) are both arguably natural. In the end, you should not be surprised when two wildlife conservationists, both committed to doing what is natural or "letting nature take its course," do not agree on a specific course of action.

Located in Lake Superior, Isle Royale is the location of a long-running study of wolf–moose interactions (photos courtesy Isle Royale Wolf–Moose Project)

Box 16.4 *HOW DIFFERENT THEORIES ANSWER ENVIRONMENTAL ETHICS QUESTIONS*

Should people be allowed to use snowmobiles in Yellowstone National Park (YNP), which might disrupt wildlife? Should YNP be treated differently from other places where snowmobiles are allowed? Should corporations, communities, or government entities be allowed to destroy a wetland important to waterfowl if they agree to create a wetland of equal size and wildlife value elsewhere? Different people evoke different ethical theories; therefore, different ethical theories will approach questions like these in different ways. It is important to note also that even those people who are employing what they think is the same ethical theory might not agree on a given course of action.

Consequentialism. A utilitarian would be obligated to try to account for the overall good versus the overall harm done by allowing some entity to trade one wetland for another, or by allowing a few people to benefit at the potential expense of others.

Rights. A rights theorist would consider how certain actions—here wetlands trading and snowmobiling in natural areas—might impact the rights of all those things that might be said to possess rights (human and non-human alike).

Virtue. Someone motivated by acting virtuously would consider whether wetlands trading or even certain types of recreation, such as snowmobiling in natural areas, are activities a virtuous person or society (a person or society who is respectful, caring, humble) would engage in or allow.

Natural Law or Divine Command. A person concerned with adhering to the laws of nature or the dictates of a particular divinity would work to discover which course of action (e.g., preserve an existing wetland or create a new one) most closely adheres to the laws of nature or the commands of that divinity.

A caravan of snowmobilers begins a trail ride into Yellowstone National Park (courtesy USFWS)

Ethics thus consists in this, that I experience the necessity of practicing the same reverence for life toward all with a will-to-live, as toward my own. Therein I have already the needed fundamental principle of morality. It is *good* to maintain and cherish life; it is *evil* to destroy and check life.

For the biocentrist, concern for, or policy regarding, the degradation of wildlife populations is motivated and justified by the effects such degradation might have on *all* individual living things: we ought to be concerned about the loss of wildlife, for instance, because of the effect it has on individual humans, fish, and trees (Box 16.5).

Some environmental ethicists argue that the extensionist approaches discussed above are flawed. The flaws arise from an exclusive focus on the moral consideration of individuals and do not accommodate the moral consideration of ecological collectives such as species, ecosystems, biotic communities, watersheds, or other things that seem important from an environmental perspective. Several theories give reasons why and how ecological collectives ought to have moral value. They include *ecocentrism,* which is related to Leopold's Land Ethic; *extended individualism,* which has ties to James Lovelock's Gaia Hypothesis; and *Deep Ecology,* which originated with philosopher Arne Naess. For people subscribing to these types

Box 16.5 *INCLUSIVE ENVIRONMENTAL ETHICS IN THE REAL WORLD*

Though there may be a temptation to ridicule or mock more inclusive moral theories, it is important to pause and appreciate how these theories actually appear and have force in the real world. In 2008, the government of Switzerland amended their constitution in a radically biocentric fashion. In the Federal Ethics Committee on Non-Human Biotechnology paper, "The Dignity of Living Beings With Regard to Plants," which explained the decision, the committee stated, "The Federal Constitution has three forms of protection for plants: the protection of biodiversity, species protection, and the duty to take the dignity of living beings into consideration when handling plants. The constitutional term 'living beings' encompasses animals, plants and other organisms" (Willemsen 2008:3).

In 2008, the government of Ecuador followed suit and forwarded an arguably ecocentric position, suggesting that nature "has the right to exist, persist, maintain and regenerate its vital cycles, structure, functions and its processes in evolution," and that "Persons and people have the fundamental rights guaranteed in this Constitution and in the international human rights instruments. Nature is subject to those rights given by this Constitution and Law."

Biocentric gestures even find their way into advertising. On 2 June 1998, the biotech company Monsanto Corporation proclaimed in a full-page ad in the *New York Times* that, "We believe in equal opportunity regardless of race, creed, gender, kingdom, phylum, class, order, family, genus, or species. All of life is interconnected . . . without a supporting cast of millions of species, human survival is far from guaranteed" (Rasmussen 2001:205).

of theories, the loss of wildlife populations is also a matter of concern because the health of species as well as specimens, watersheds as well as rivers, and forest ecosystems as well as individual trees matter and are negatively impacted by biological impoverishment.

Most of these theories, extensionist and non-extensionist, are controversial (i.e., contested) and active areas of scholarship. One of the greatest conflicts in recent years is between ecocentrism and forms of extensionism, such as animal welfare ethics. Are humans, for example, morally justified in killing many individual brown-headed cowbirds in order to preserve Kirtland's warblers? Resolving this conflict is one of the great ethical challenges of our day (Box 16.6). Most theories of ethics are also focused on the first of the great questions that define environmental ethics—moral considerability.

Originally, environmental ethicists focused on assessing questions of moral considerability. The future of environmental ethics will focus increasingly on the question of moral significance. The pursuit of this question will be much more applied in nature, and likely will provide ideas for how to solve conflicts such as those that exist between animal welfare ethics and ecocentrism. In principle, a solution to these conflicts begins by appreciating that many things have value; the challenge is in the detail of how to sort out competing moral claims in a world full of them.

Environmental justice, which focuses mainly on the distribution of environmental goods and harms, is one area of inquiry that is more applied and more focused on the question of moral significance. *Ecofeminism,* which draws important parallels between systems of oppression that harm nature as well as certain members of the human community such as women, is another area of inquiry that aims to apply environmental ethics more effectively, rather than to theorize about them. Other non-traditional or lay approaches to envi-

Wildlife management decisions and actions often raise ethical questions. For example, one might ask, "Is it ethical to remove brown-headed cowbirds to save the endangered Kirtland's warbler pictured?" (courtesy USFWS)

ronmental ethics have been powerfully articulated in popular forms accessible for the general population (Box 16.7).

16.5. WHAT IS ARGUMENT ANALYSIS?

Understanding ethics and environmental ethics (and therefore issues in wildlife management) begins with understanding arguments. When you peel back the layers of rhetoric, emotional manipulation, scare tactics, and assertions that are presented loudly or repeatedly, core arguments emerge. While irrational behaviors are common and normal, mentally healthy

Box 16.6 *TREE RING ANALOGY*

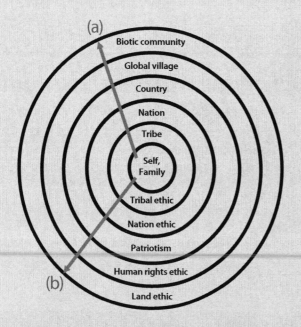

Tree ring analogy

Aldo Leopold (1949:203) points out an important commonality among all ethical systems: "All ethics so far evolved rest upon a single premise: that the individual is a member of a community of interdependent parts." The above "tree ring" analogy has been used in various forms to illustrate Leopold's point and make sense of our many nested, operative, and competing moral commitments. In the diagram, (a) is the development, driven by our rational faculties, of a sense of community over time, and (b) is the corresponding ethical attachment that must also develop in order for the level of community in (a) to continue to flourish. As Leopold (1949:203) summarizes this point, "The extension of ethics to [the land] is . . . an evolutionary possibility and an ecological necessity." Charles Darwin (1871, 1981:93), from whom Leopold takes many moral cues, likewise grounded this dual development of society and ethics in biology: "No tribe could hold together if murder, robbery, treachery, etc., were common; consequently such crimes within the limits of the same tribe 'are branded with everlasting infamy.'"

Leopold refers to these layers of ethical obligation as "accretions" or additions to, not replacements for, our previously existing social and moral commitments. As accretions, these additional layers are never smoothly adapted but are always challenging. Hindsight, however, tells us it is both possible and obvious to become more morally inclusive. Eventually such accretions become conventional, and we should at least entertain the possibility that the same is true with accretions of environmental ethics that we have just seen. This image of ethics allows us to address a common reaction to (against?) ethical accretions: the suggestion that we cannot actually live in a world where so many things matter ethically. We should, however, be aware that this same criticism has been offered every time an expansion of ethical commitments has been suggested (e.g., to end slavery, to give women the vote); hence, there is nothing unique about environmental ethics in this regard.

---

humans also desire and rely on appeals to reason, logic, and argument. Mentally healthy humans express themselves, forge their commitments, and attempt to persuade one another by the rules of reason expressed through argumentation.

Below is a brief sketch of the nature of argumentation. We briefly describe how one goes about laying out an argument and suggest a relatively simple method to think critically about arguments in general. In all fairness, however, facility with presenting and critiquing arguments demands much practice.

Argument Analysis. Ethical claims of various sorts are presented as arguments; therefore, it is most effective if they are also addressed as arguments. An argument is a systematic and coherent series of statements aimed at persuasion; it asserts both a claim, or a "conclusion," and supporting evidence, or "premises." Although we often experience ethical arguments informally in writing and speaking, there are rigorous methods to develop these statements into formal arguments and there is a way to formally analyze them as well. Although arguments come in a wide variety of "shapes and sizes," the basic form of an argument is as follows:

P1. Premise
P2. Premise
C1. Conclusion

Some examples might be instructive:

EXAMPLE 1
P1. All mammals are warm-blooded.
P2. Raccoons are mammals.
C1. Therefore, raccoons are warm-blooded.

EXAMPLE 2
P1. Mountain lions live either in Japan, New Zealand, or the United States.
P2. Mountain lions do not live in Japan.
P3. Mountain lions do not live in New Zealand.
C1. Therefore, mountain lions live in the United States.

If you are attempting to address an argument against the need to conserve wildlife habitat (or any argument for that matter) the first thing you should attempt is to lay out the argument in the above fashion. Quite often, the problem(s) with

Box 16.7 *NON-TRADITIONAL ENVIRONMENTAL ETHICS*
Narrative Non-fiction, Aldo Leopold, and the Ethic of Care

Ideas about how humans ought to live with wildlife and the rest of nature are sometimes articulated in forms other than traditional philosophical prose. The poems of Mary Oliver and Robinson Jeffers and the fiction of Wendell Berry and Edward Abbey also offer important environmental ethical insights. A particularly poignant example of interpreting the work of Aldo Leopold in a more literary fashion comes from philosopher and nature writer Kathleen Dean Moore. Moore (2004:60–67) argues that "Human beings are creatures who are drawn to one another . . . born into networks of dependencies and complications, hidden connections, memories and yearnings, births and rebirths, fierce, mysterious love—a web of relationships." "These are the facts," she writes, "of the greatest moral importance: If we value caring relations, then it makes sense that we commit ourselves to act in ways that strengthen and reweave and sustain the webs of relationships we value." This "ethic of care" also extends to the land or what Aldo Leopold refers to as the "biotic community," since we are also born into and dependent upon "relationships, not just with human beings, but with the land . . . the beautiful, complicated web of sustaining connections."

Moore writes, "I think the ethic of care has it right: The care we feel for people is the ground of our moral responsibilities toward them. And I think Aldo Leopold has it right: Our moral responsibility to care for the land grows from our love for the land and from the intricate, life-giving relationships between people and their places. Then does

not this follow?—that our moral calling must be to reknit and cherish healthy webs of connection not only to people, or not only to land, but also to families, human communities, landscapes, and biotic communities—all our relations. What we need next is a new ethic—call it an 'ecological ethic of care,' call it a 'moral ecology.' It's an ethic built on caring for people *and* caring for places, and on the intricate and beautiful ways that love for places and love for people nurture each other and sustain us all."

Aldo Leopold, photographed by son Starker Leopold while writing at the shack (courtesy of the Aldo Leopold Foundation; www.aldoleopold.org)

an argument become immediately clear when it is laid out in such a fashion, and quite often it is discovered that there is no argument at all—a fact that can be masked when the claim is not presented as a formal argument. Remember, arguments *must* have both premises ("evidence") and conclusions ("position") to be arguments; do not be fooled by conclusions just cleverly restated as premises. If any claim lacks supporting evidence, there is no argument. To ascertain whether or not an argument is present, it is helpful to run what can be called the "test of opposition." If you encounter a claim and are wondering whether or not there is an argument present, simply ask yourself, "Can I assert exactly the opposite claim and have no more reason to believe one over the other?" If there is no more reason to believe the opposite of a claim then there is to believe the claim itself, then no argument exists at present and you are advised to ask the person forwarding the original claim to provide support for that claim.

If an argument actually is present, however, then the method for critiquing an argument is fairly straightforward. An argument can only be wrong in one of two ways (theoretically bad arguments can be wrong on both accounts, but

likely that would mean the argument is barely intelligible). A critique of an argument would point out that there is either

1. a mistake in a premise (a factual mistake of sorts), or
2. a mistake in inference (a mistake in going from premises to conclusion; assuming that the conclusion follows from the given premises, when in fact it does not).

In sum, it cannot be the case that all of the premises in an argument are true, and that the conclusion follows from the premises, but that the conclusion then is false. If the conclusion of an argument is believed to be false, then your obligation is to demonstrate why the argument is faulty by employing one of the two methods above. Again, a couple of examples might prove helpful. Consider the following arguments:

EXAMPLE 1

P1. If something is not occurring then we should not worry about it.
P2. Anthropogenic global climate change is not occurring.
C1. Therefore, we should not worry about anthropogenic global climate change.

In this argument no problem exists with inference. The conclusion most certainly follows from the premises. If you are suspicious of this argument, it must be because of doubt in the truth of one (or more) of the claims made in the premises. If you wish to reject the conclusion of this argument, you must do so on the basis of a mistaken premise. In fact, in this argument, you might challenge the truth of Premise 1 on the basis that it is not clear that you should never worry about something that is not occurring, and you would then provide counter-examples to demonstrate how it might be wise to be concerned about and attempt to address potential threats to your health or to the safety of your home even if you are not currently ill or your home is not currently on fire or being robbed. You might also challenge Premise 2, both on the basis that it is far from clear that anthropogenic global climate change is not occurring or even with the claim that Premise 2 is patently false—you would then go on to make the opposite claim to refute such an argument.

EXAMPLE 2

P1. Crocuses bloom in the spring.
P2. The month of April is a spring month.
C1. Therefore, Highway 52 runs north-to-south through Michigan.

This is clearly a bad argument, but not because either of the premises is mistaken. Both premises are true. It is a bad argument because the conclusion simply does not follow from the premises provided. If the conclusion stated "Therefore, crocuses may bloom in April," then a sound argument exists. Notice, however, that because there are other spring months, if you concluded "Therefore, crocuses will bloom in April" you would again be making a bad argument on the basis of a mistaken inference. Of course most (but unfortunately not all) arguments that have a mistaken inference are not so blatantly wrong as the example provided; they are often more like the modified conclusion that "Crocuses *will* bloom in April."

In short, when attempting to address arguments, the method to follow is to

1. lay out the argument(s),
2. assess the premises for their truth-value,
3. assess the connection or inference between premises and conclusion, and
4. lay out the counter-argument using 2 and 3.

A final note on assessing and evaluating arguments: you greatly improve your chances of presenting good arguments and either avoiding critique or being able to fend off critique if you learn to become your own toughest critic. Examine your own arguments rigorously and question your premises as you would those of others, and forever attempt to consider the response of a would-be dissenter. Always hold in your mind as an ideal the following general rule regarding the strength of arguments:

> An argument's strength is *not* measured by the fact that it is persuasive to someone who already believes the conclusion of the argument. Rather, the strength of an argument

is measured by the force that it has over those who *dissent* from the conclusion of the argument.

16.6. APPLYING ETHICS TO WILDLIFE MANAGEMENT

Wildlife management is essentially environmental ethics in action, insomuch as both focus on assessing propositions about how humans should (or should not) interact with the natural world. Moreover, wildlife management is only justified when it is supported by reasonable ethical arguments. The application of environmental ethics to wildlife management involves extending argument analysis to public discourse associated with wildlife management.

16.6.1. A Preliminary Principle

Handling the ethical dimensions of wildlife management requires understanding how ethical knowledge is one of several distinct kinds of knowledge necessary for wildlife management to be ethical and effective. Clearly, wildlife management depends on knowledge from scientific fields such as ecology, sociology, and economics; but wildlife management also depends on, or presupposes, ethical arguments that always consist of two kinds of premises: ethical premises and scientific (or descriptive) premises intended to describe how the world is. Assessing the truth-value of descriptive premises requires scientific knowledge from various scientific fields (both biological and social), and assessing the appropriateness of ethical premises requires ethical knowledge.

There is value in comparing and contrasting contributions of each kind of knowledge. Obviously, ecology contributes knowledge about the environment. As critical as such knowledge may be, it is insufficient on its own for determining what management action *should* be taken. Consider this ecological knowledge: (1) Prior to persecution by humans, wolves inhabited much of the northeastern United States, and (2) today, coyotes seem to fill the approximate ecological niche once filled by wolves in that region. This ecological knowledge cannot determine whether it would be right to reintroduce wolves into the Adirondack region of New York. No amount of ecological knowledge, by itself, could determine the appropriateness of such a reintroduction. To make such a decision, wildlife managers also need to be informed by sociocultural knowledge, such as knowing the stakeholder acceptability of reintroducing wolves to the region. Economic knowledge is also valuable—knowledge about the effect that wolf reintroduction might have on portions of the economy associated with tourism or deer hunting (e.g., license sales, retail purchases, and travel). Science is primarily concerned with evaluating propositions about how the world was in the past (e.g., wolves used to live in the Adirondacks), how it is today (e.g., many people oppose wolf reintroduction), or how it will be in the future (e.g., wolf reintroduction might reduce the number of people who spend money on activities related to deer hunting, but increase ecotourism as people seek opportunities to see or hear wolves).

Though critically important, descriptions of how the

world was, is, or will be—no matter how abundant or reliable such knowledge is—cannot by themselves determine whether a wildlife management action is ethically appropriate. One must also identify and evaluate the appropriateness of ethical premises that underlie arguments for or against some wildlife management action. For example, consider this argument:

P1.   Wolves—through their predation on deer—once performed a vital ecological service in the Adirondacks, and human exploitation caused wolves to go extinct in this region.
P2.   Coyotes and human hunting on deer now perform the vital ecological service that wolves once provided.
P3.   Many people oppose wolf reintroduction.
P4.   Wolf reintroduction would likely harm aspects of the local rural economy in the Adirondacks.
P5.   The primary value of a species is its ecological function.
C1.   Therefore, we should not reintroduce wolves, because doing so comes at a cost (social and economic) but does not offer much benefit (because coyotes and hunters already do that for which wolves would be valued).

For the moment, take for granted that the premises are all true or appropriate. The conclusion is not determined by Premises 1 through 4, which represent scientific descriptions of the world. The conclusion is determined by Premise 5, which is the ethical premise. To better see how, evaluate the argument more carefully. First, consider that Premise 5 may be inadequate and might be replaced with this alternative:

P5a.   Species are intrinsically valuable. They are valuable not because of services they perform but just because they are an important manifestation of life.

If Premise 5 is replaced with Premise 5a, the conclusion is not justified by the argument. (Note: The apparent failure of this argument is not sufficient to show that wolves should be reintroduced. More work would be required to assess that conclusion.) The argument above is also likely missing premises along these lines:

P6.   It is wrong to enact policies that diminish local, rural economies.
P7.   It is wrong to enact policies that oppose public opinion.

Recall, a conclusion cannot introduce ideas that are not already entailed by the premises. Because the conclusion refers to the social and economic costs of wolf reintroduction, the argument must then contain premises speaking to these issues. The introduction of Premise 6 and Premise 7 raises the question, are they reasonable premises? They probably are not. If Premise 6 were true, then it might be wrong to criminalize prostitution or marketing tobacco products to young people. Premise 6 requires revision to account for the fact that we do not unconditionally promote local, rural economies. Premise 7 is also false. Great failures in management have occurred by allowing policies that were unwise but widely supported by stakeholders (e.g., overfishing of Atlantic cod). Sometimes leadership is required to promote policies that are ethically sound but unpopular. (Note: what has been outlined here is one approach to argument analysis. For an even more detailed treatment of argument analysis see Copi and Cohen [2005]. For a detailed treatment in the context of natural resource management see www.conservationethics.org.)

Even this simple evaluation shows how the appropriateness of wildlife management hinges on a careful understanding of the underlying ethical premises. Table 16.1 shows how several of the most important issues in conservation require complex ethical consideration.

Ethics and sociology. These domains of knowledge are similar in that both focus on values. They differ, however, in that sociology is more focused on understanding the values and attitudes various groups of people hold, the behaviors they exhibit, and how their values, attitudes, and behaviors

Table 16.1. Issues in conservation that require complex ethical consideration

Conservation concern	Related ethical issues
Ecosystem restoration	• Is the goal to restore a particular state of an ecosystem (e.g., old growth) or the processes that lead to "natural states"?
Removal of exotic species	• How do we know when the moral cost of killing individuals exceeds the moral benefit of removing an exotic species?
	• Are we removing the exotic species for our benefit or the benefit of the ecosystem?
	• Is it wrong to incur the moral cost of removing exotic species when so little is done to prevent their arrival in the first place?
Population viability	• Is genetic diversity important only insomuch as it affects population viability, or is it also valuable for its own sake, as another manifestation of biodiversity?
	• How much risk is too much risk?
Hunting	• Does our need for the resource outweigh the cost of affecting nature?
	• Is a hunt ethical if population health is unaffected, but there is no chance of achieving the management goal (e.g., controlling abundance)?
	• Where should the burden of proof lie: hunt unless there is reason not to, or do not hunt unless there is reason to do so?
Sustainable utilization: "meeting human needs in a socially just manner without depriving ecosystems of their health"	• What do we mean by "human need"?
	• What is a "healthy ecosystem"?
	• What counts as the socially just use of resources?
	• Do we care about ecosystem health only because of the benefits to humans, or also because ecosystems are intrinsically valuable?

may change over time. Ethics is more focused on using argument analysis to assess what values, attitudes, and behaviors people *ought* to exhibit. Holding some value or exhibiting some behavior does not mean you ought to do so. This distinction between "is" and "ought" raises some difficulties. On one hand, equating "how we are" with "how we ought to be" introduces all of the problems associated with naturalism. On the other hand, it cannot be ethical to expect behavior that is impossible to exhibit. To put it pithily: although ought to implies can, can certainly does not imply ought to. Although a policy prescribing wolf hunting would be contingent upon the ability of the wolf population to sustainably withstand a hunt, the fact that the population can withstand a hunt does not imply that we ought to hunt wolves.

Ethics and Economics. The relationship between ethics and economics is not simple. Our inspection of Premise 6 in the above argument suggests that ethical values often override concerns for economic growth. Many environmental protection laws entail some kind of curtailment of economic activity (e.g., U.S. Endangered Species Act and the U.S. Clean Air and Water Acts). Remember from Chapter 6 that economics is a social science that describes how society meets competing demands in the face of limited resources.

Economics strives to understand how economies work and how economic agents interact, where economies are systems involving the production, distribution, and consumption of goods and services, and economic agents are the people and institutions involved in those economies. Economics aims to describe how economies work, not how they ought to work. It is appropriate to ask, "How ought an economy to work?" However, that question requires careful treatment of ethical knowledge. The purpose of economics as a science is more or less limited to making reliable statements of this nature: *If we interact with these limited resources in these ways, then our economies very likely will respond in these ways.* It is more a matter of environmental ethics than economics to understand what counts as a right or wrong way to interact with wildlife and whether the projected economic outcomes are good or bad, in the ethical sense.

Ethics, Laws, and Politics. Although political–legal knowledge is also necessary to understand legal and political feasibility of wildlife management policies, this knowledge cannot by itself determine whether decisions made in wildlife management are right or good. A complex relationship between ethics, politics, and law rises from a few uncontroversial principles:

1. In a democratic society, laws and politics generally tend to arise from that society's ethical dispositions, rather than vice versa.
2. Following the law, however, does not adequately ensure that one is behaving ethically. That is, not all cases of rule-breaking represent unethical behavior (e.g., the civil disobedience of Rosa Parks, Martin Luther King, Jr., and Gandhi), and not all cases of rule-following are ethical (e.g., the defense or deflection of war crimes charges by referring to following orders from one's superiors).

3. Winning a political battle is not the same as being right. Many political battles are won because the winning side had more financial resources or political power.

These principles raise difficult and unavoidable questions about how a wildlife professional ought to act when there is a conflict between ethics and the laws or politics of wildlife management.

16.6.2. Ethical Discourse

Understanding the distinctions between various kinds of knowledge is critical for making wildlife management decisions, as is argument analysis for synthesizing knowledge. Implementing these principles in a process involving some kind of public discourse is called ethical discourse, which is best understood by describing some operational steps in the process and by comparing the process to political discourse. Many of the chapters in this book examine the role of stakeholder participation in wildlife management. The ethical discourse described here is another example of engaging in thoughtful deliberations as part of the management process.

16.6.2.1. Step 1: Catalogue and Inventory Reasons for or against a Particular Management Intervention

That is, identify all the reasons that each kind of stakeholder has for or against the management intervention (objective and related action) under consideration. Prejudging the reasonableness of various reasons should be held to a minimum. The most important mistake that can be made at this stage is overlooking a reason that is held important to a stakeholder. Beware: stakeholders sometimes offer one or more stated reasons but are motivated by other unstated reasons. Stated and unstated reasons are both important. It is also useful to categorize the reasons into sets of related reasons either supporting or opposing the management intervention. (See Nelson and Vucetich [2009] as an example of this kind of categorization as it applies to the debate about whether scientists should be advocates.)

16.6.2.2. Step 2: Argument Construction and Assessment

Now treat each reason as the conclusion to an argument that has not yet been articulated. The process can begin by articulating, in brainstorm-like fashion, facts and ideas that seem related to the conclusion. Then arrange these facts and ideas as premises for the conclusion. Upon being roughly constructed, the argument should then be assessed for missing premises. Identification of all the missing premises is important because doing so often frees an argument of mistakes in inference or at least draws attention to mistaken inferences. These steps look much like the steps we took as we began developing the wolf reintroduction argument above. The detection of missing premises is often difficult and may require a person with experience in argument analysis.

From this point, argument assessment can follow one of three strategies. Strategy 1 holds the conclusion fixed and judges whether the premises necessary for supporting the conclusion are valid. Strategy 2 revises invalid premises to make them valid, and judges whether the conclusion is supported by the revised premises. Strategy 3 revises invalid premises to make them valid, and then revises the conclusion to the extent necessary to keep the argument sound and valid. Strategies 1 and 2 are useful for assessing whether a reason for or against some management proposal is justified (in the sense that it is supported by a sound and valid argument). Strategy 3 is useful for discovering appropriate arguments when Strategies 1 and 2 seem to be revealing only inappropriate arguments.

Strategy 3, which involves revising conclusions to match revised premises, very often leads to surprising outcomes. An argument that, at first glance, seems to support some management intervention can often end up offering good reason to oppose it. For this reason, and because argument analysis is technically difficult (that is, it is easy to think you are doing it correctly when, in fact, you are not), it is important that those engaged in ethical discourse are willing to change their minds (perhaps by 180°) about the appropriateness of a management intervention (this willingness is known as intellectual honesty). That is, the purpose of ethical discourse is not merely to confirm what you already believe but rather to discover what you should believe. In this way, we assert that ethical discourse is similar to the scientific process.

Assessing scientific premises. When all the missing premises seem to have been identified and articulated, the appropriateness of each premise should be assessed. Presuming the argument contains only premises necessary for supporting the conclusion, the discovery of even a single inappropriate premise is often enough to determine that the argument is inappropriate.

Begin assessing premises by identifying the kind of knowledge each premise represents, and the kind of person qualified to pass judgment on that premise's validity. An ecologist, for example, would tend to be most qualified to judge the validity of ecological premises.

As described in Chapter 8 on decision making, treatment of scientific uncertainty is important at this stage. Consider, for example, the premise: *Killing cowbirds benefits warbler populations.* If the premise is associated with one or more of the types of uncertainties (discussed in Chapter 8) then the premise is inappropriate. In this example, the premise might be made true by replacing "benefits" with "will benefit in some cases," or perhaps by rewriting the premise to begin with: *Though it remains uncertain, there is some reason to expect that killing cowbirds will benefit this particular warbler population.*

Sometimes a premise is true based on accepted scientific information, but not accepted among all stakeholders. Consider, for example, *human activities play an important role in global climate change* as a premise in some argument about climate policy. Climate science has the purview to judge the premise's validity, and climate science indicates the premise is almost certainly true. Nevertheless, some American citizens do not accept the premise as true. These opinions are *not* a basis for judging the premise to be false; however, they are likely a basis for adding a new premise to the argument: *Many Americans do not believe human activities play an important role in global warming.* In cases like these, careful analysis is required to understand how the two premises interact to affect the argument's conclusion. Recall that the purpose of ethical discourse is not so much to assess whether a policy would be politically difficult to pursue but to judge what policy would be ethically justified.

In cases where science and public perception are in conflict, the argument likely will require a premise that speaks to the conflict. Consider, for example, the validity of a premise such as this: *Agencies with the authority to act without widespread public support should pursue policy based on scientists' perceptions of scientific claims (not perceptions of the general public), but they should pursue such policies in a manner that is sensitive to public perception.* This premise suggests public perception affects how a policy should be pursued but not whether it should be pursued. Now consider this premise: *Agencies should pursue policies only when they receive widespread public support.* The word "should" in each of these premises indicates that both premises are ethical premises. These premises might even be thought of as conclusions to an ethical argument that itself requires articulation and analysis.

Assessing ethical premises. The evaluation of some ethical premises is relatively straightforward: they are widely accepted (or rejected) for good reasons that are well-understood. In such cases, the value (and hard work) of ethical discourse is when it exposes how a reasonable-sounding conclusion is actually supported only when one accepts ethical premises that are clearly inappropriate. For examples of this circumstance, see Vucetich and Nelson (2007) and Nelson and Vucetich (2009).

In some cases, however, the appropriateness of an ethical premise is difficult to judge. Sometimes an ethical premise seems to rest on solid reasoning but is not accepted by very many of the stakeholders, or vice versa. Should the lack of support be taken as a sign that the purportedly solid reasons are not actually so solid? In such cases, it may be useful to treat the ethical premise as a conclusion to an argument that requires articulation and evaluation.

Difficulties also arise when two ethical premises seem appropriate but also seem to conflict with one another. Consider, for example, the premises *Kirtland's warbler represents an intrinsically valuable species* and *The lives of individual cowbirds are intrinsically valuable.* One premise suggests cowbirds should be killed if they threaten the viability of Kirtland's warblers, and the other suggests they should not be killed. Using insights from human dimensions inquiry, wildlife professionals might anticipate such an ethical conflict arising among stakeholders. One way to handle this conflict is for professionals to use argument analysis to explain how and why one premise should override the other in this particular circumstance. The result of this process may be to resolve the conflict or it may be to clarify precisely how the stakeholders disagree. Either outcome represents ethical progress. Further inquiry into stakeholder acceptability of management actions, through use of interviews or questionnaires, would then help affirm the outcomes of the argument analyses.

16.6.2.3. Step 3: Synthesis
The result of Step 2 is to judge each argument as being appropriate, inappropriate, or possibly undetermined. Knowing

that a particular argument is inappropriate does not mean that the conclusion of the argument is false. It remains possible that some other argument would justify the conclusion. Consider for example, this conclusion: *We should not drill for oil in Arctic National Wildlife Refuge (ANWR) because it would endanger local caribou populations.* Argument analysis might show that this conclusion (not drilling in ANWR) cannot be supported for that reason. However, the inappropriateness of that conclusion does not mean drilling in ANWR is a good or right thing to do. Another argument—about how exploitation is inconsistent with the principles of a protected area even if the exploitation has minimal effects on the environment—might be able to show that such drilling would be wrong. For reasons such as this, the final step in ethical discourse is to consider the management action in the context of all the arguments that were analyzed (e.g., Nelson and Vucetich 2009).

These three steps of ethical discourse may be implemented in various ways. For example, a research project might be conducted by a few experts and then vetted by peer-review and scientific discourse (e.g., Vucetich and Nelson 2007, Nelson and Vucetich 2009) or by a larger group of people—experts and lay stakeholders alike—engaged in a workshop-like venue. Such workshops are beginning to occur, led by organizations such as the Center for Humans and Nature, the Aldo Leopold Foundation, and the Conservation Ethics Group.

The potential limitation of working with a smaller group of select people is misunderstanding or neglecting reasons that are important but not well-appreciated by that group. The potential limitation of working with a larger group is that larger groups are more likely to include participants who do not appreciate or are not proficient with the principles of ethical discourse. In either case, the success of ethical discourse depends on the participants' collective knowledge of the issue and skill in argument analysis; and, in either case, the legitimacy of ethical discourse is judged by others' ability to find fault with the logic of the analysis.

16.6.3. Ethical and Political Discourse

Despite their similarities, ethical discourse and political discourse differ importantly from one another. Political discourse aims for compromise and concession until all stakeholders agree that the proposed management action is something with which they can live. The purpose of political discourse is to make political progress and avoid civil chaos. Political discourse is typically constrained by the timing of an imminent decision. This circumstance makes participants in political discourse focus on winning rather than on being right. By contrast, ethical discourse aims for basic agreement about an issue. Ethical discourse aims to be "right" rather than to win. Such discourse can make valuable insights during formulation

The Arctic National Wildlife Refuge is critical habitat for a vast herd of caribou (courtesy USFWS)

of objectives (both fundamental and enabling) early in the management process. Recall an early dilemma proposed in Chapter 1, which indicated that if you are not working on the correct things, then the more you try to conduct your work "right," the more wrong you become! Consideration of ethics and engagement in ethical discourse can help reveal what are the right things on which to be working.

SUMMARY

This chapter explored ethics and environmental ethics as a theoretical and practical human dimension of wildlife management. The practical side was largely about constructing and assessing arguments that represent real-world issues in wildlife management. The theoretical side focused on the tool by which those theories (stated as ethical premises) are evaluated. The introduction we provided here was merely a road map to conduct deeper ethical thinking about decisions in wildlife management.

- Although ethical discourse is not a panacea for solving all environmental challenges, wildlife professionals can increase their effectiveness by learning the theory and practice of ethical discourse. It provides an alternative to, or complements, existing political discourse, which (from the perspective of engaged stakeholders) is more about winning and losing than gaining insight into what is right or wrong.
- Ethics is a discipline whose focus is analysis of ethical propositions. The fundamental distinction between ethics and other social sciences is that social science primarily is concerned with the analysis of descriptive propositions about human values, whereas ethics is concerned with the analysis of prescriptive propositions about human values.
- Although there are many ways to categorize the field of environmental ethics, it is centrally concerned with two related topics: (1) *moral considerability;* that is, what sorts of entities deserve membership in the moral community and what justifies that membership? and (2) questions of *moral significance;* that is, how ought humans to behave in an inclusive moral community?
- In this regard, though with an added moral bent of right and wrong, ethical discourse is similar to science in that it follows a systematic and rigorous process of logical thought. Ethical progress and scientific progress are both most likely to occur when the minds involved are open to reason.
- Attending to the ethical dimensions of wildlife management increases the ability of wildlife professionals to have a reliable means of identifying the correct issues to go to work on, to reach the correct conclusions, and to better achieve desired impacts. It will also help managers to be ethically consistent from one situation to another, thereby improving their credibility with peers, partners, stakeholders, and decision makers.

Suggested Readings

Copi, I. M., and C. Cohen. 2005. Introduction to logic. Twelfth edition. Pearson / Prentice Hall, Englewood, New Jersey, USA.

DesJardins, J. R. 2006. Environmental ethics: an introduction to environmental philosophy. Fourth edition. Wadsworth, Belmont, California, USA.

Leopold, A. 1949. A Sand County almanac: and sketches here and there. Oxford University Press, New York, New York, USA.

Moore, K. D. 2004. The Pine Island paradox. Milkweed Editions, Minneapolis, Minnesota, USA.

Pojman, L. P., and P. Pojman, editors. 2008. Environmental ethics: readings in theory and application. Fifth edition. Wadsworth, Belmont, California, USA.

Wolpe, P. R. 2006. Reasons scientists avoid thinking about ethics. Cell 125:1023–1025.

CONTINUING YOUR EDUCATION IN HUMAN DIMENSIONS

LARRY A. NIELSEN AND BARBARA A. KNUTH

The world anticipated the beginning of the twenty-first century with a mixture of excitement and misgiving, largely fueled by the powerful force of globalization on so many issues of human existence: the economy, politics, the food system, and the environment. The connections between global society and wildlife management have become more entangled and explicit. When Dr. Wangari Maathai won the Nobel Peace Prize in 2004 for her work on sustainable development, democracy, and peace in her native Kenya, it highlighted sustainability as a central feature, or at least a desirable goal, of modern life around the globe (Box 17.1). Vice President Al Gore and the United Nations Intergovernmental Panel on Climate Change subsequently received the Nobel Peace Prize in 2007, which vaulted human-induced climate change and global warming onto the policy agenda for all nations. Elinor Ostrom won the 2009 Nobel Prize in Economic Sciences for her pioneering work on common resources (which include most wildlife resources) and the institutions for managing them. During the latter half of the decade, damage to U.S. oil infrastructure due to Hurricane Katrina, and especially the catastrophic impacts on the economy and the environment of British Petroleum's Deep Water Horizon oil spill, caused serious concerns about the ethics and sustainability of the petroleum industry. These natural and manmade catastrophes emphasized the fact that renewable energy is an environmental as well as an economic imperative. Environmental sustainability (and wildlife management as an element of it) has become central to the interests of citizens in developed and developing nations alike.

During the next decades, wildlife managers will encounter opportunities and problems that will test the best minds in the profession. Some challenges will certainly arise from the increasingly complex and sophisticated knowledge of the ecological aspects of wildlife science and management, but more profound changes, in both knowledge and application, can be expected in the human dimensions; that is, describing, understanding, predicting, and incorporating the human component into all levels and stages of wildlife-related governance

and management. Furthermore, wildlife management problems will increasingly be defined in terms of sustainability and environmental quality.

Wildlife managers already are experiencing significant changes. They increasingly are working on teams where each member contributes expertise to the solution of wicked problems. Biologists, statisticians, geographers, educators, botanists, environmental planners, citizens, and politicians assemble to handle wildlife challenges and opportunities. Wildlife managers need to be effective facilitators, problem solvers, team leaders, and team members, as well as biologists or technicians. In fact, the most distinctive quality of a successful wildlife manager in the future may be the capacity to lead or participate effectively as part of a team.

As wildlife managers continue to adapt to deal with new challenges and new understanding of ongoing challenges, the education they receive must also adapt. Universities offer curricula that pertain to the ecological, life, social, physical, and management sciences. Increasingly, they are also providing opportunities for experiential (hands-on) learning, course integration across disciplines, and capstone experiences that encourage teams of students to synthesize and interpret a wide body of evidence. Curricula are becoming more flexible; they require a set of core competencies and specific learning outcomes and also offer specialized training needed in contemporary wildlife management. More than ever before, students are being exposed to the breadth of biological, physical, and social science knowledge needed for effective wildlife management. This book has introduced many of the human dimensions subjects of relevance to wildlife management.

The following sections describe the kinds of human dimensions education that will help wildlife managers launch their careers and continue their professional development. We review components of the human dimensions of wildlife management—many covered in this book and others not—and offer advice about where to find education related to each. By the time you read this, you may be nearing the end of your undergraduate program, so we also suggest some opportuni-

Box 17.1 *WANGARI MAATHAI ON EDUCATION AND ENVIRONMENT*

Dr. Wangari Maathai received the Nobel Peace Prize in 2004 for her grass-roots work to enhance democracy, peace, and environmental quality in her native Kenya. Dr. Maathai began the Green Belt Movement in 1977, which was designed to use the planting of trees as the mechanism to encourage rural Kenyans (particularly women) to invest in themselves, their communities, and their environment. Mrs. Maathai knew that the success of the movement depended on bringing knowledge and empowerment to the people. In her Nobel Prize acceptance speech, she noted the importance of education:

Initially, the work was difficult because historically our people have been persuaded to believe that because they are poor, they lack not only capital, but also knowledge and skills to address their challenges. Instead

they are conditioned to believe that solutions to their problems must come from "outside."

In order to assist communities to understand these linkages, we developed a citizen education program, during which people identify their problems, the causes, and possible solutions. They then make connections between their own personal actions and the problems they witness in the environment and in society.

Those of us who have been privileged to receive education, skills, and experiences, and even power must be role models for the next generation of leadership.

As wildlife professionals, we must use our educational opportunities to learn how to work with the people whose interests we serve—whether in rural Africa, downtown Atlanta, or the wilderness of Alaska.

ties to learn more about human dimensions outside the formal classroom.

17.1. A PART OF THE WHOLE

Human dimensions are present in any conceptual model used to describe the scope of wildlife management, but they may be considered in different ways. A traditional perspective is that the profession involves interactions among animals, habitats, and people. A managerial perspective addresses components that may include substance, processes, and relationships. Relationships form among the people involved, who need to be comfortable with one another and to believe that those making decisions are seriously considering their interests, opinions, and values. Managers need to understand principles of communication, stakeholder engagement, and decision-making processes discussed in earlier chapters of this book. Managers also will benefit from developing skills in conflict resolution and negotiation, and this is a topic of many training sessions and workshops offered around the country.

Ecosystem-based approaches in wildlife management consider the intersection between ecological, sociocultural, and institutional components of managed systems (i.e., coupled systems). The sociocultural and institutional components of ecosystems include the needs, interests, and organizations of people. Ecosystem-based managers are conversant with systems thinking and terminology.

A sub-set of ecosystem management with a human community focus is called community-based management. It incorporates people and their local or regional communities and institutions into management, and it explicitly considers human livelihoods and sociocultural and economic goals along with biological and ecological management goals. For

example, rather than excluding people and their daily activities from a national park, this approach seeks to allow people to live and work within a park in ways that enhance the long-term sustainability of the ecosystem. Regardless of which comprehensive approach to wildlife management is taken, the human dimensions are certain to be core considerations.

These approaches to wildlife management recognize that the variety of ways in which people think, believe, and act are not just external forces that affect the ecosystem, but rather that people and their institutions are important components of these systems. The wildlife profession has adopted that approach for many situations and implementing it will preoccupy science, education, and practice in the field for the foreseeable future.

17.2. HUMAN DIMENSIONS IN PROFESSIONAL EDUCATION

Educational preparation for a career in wildlife conservation and management commonly includes the study of human dimensions described earlier in this book, which ranges from a general survey of the field to in-depth treatment of specializations. Fortunately, there are many opportunities for students to take courses valuable for understanding human dimensions in wildlife management. Some courses that address human dimensions subjects are offered within college wildlife programs and apply social sciences specifically to wildlife-related issues, but courses offered by other departments in the university may delve more deeply into specific social-science disciplines relevant to wildlife management. Courses that deal with natural resources economics and public policy have been available for many years, but many universities also offer courses in public participation, strategic planning, public administration, ethics, governance, and public relations. These courses sometimes include specific applications to conservation or environmental management problems, but typically they have a broader purpose of requiring the student to discover for themselves the applications to wildlife management.

Accrediting and verifying organizations, such as The Wildlife Society, recognize the importance of human dimensions (broadly defined) in their professional certification requirements. These requirements are designed to provide ample opportunity for exploration of social sciences to achieve breadth or depth; however, they should be viewed as minimum requirements. A successful wildlife manager needs a more complete understanding and more skills than these requirements call for. Whatever specific courses one takes, the goal should be to become well-versed in the human component of the systems in which wildlife are managed.

As wildlife management has shifted toward greater stakeholder orientation, professional education in human dimensions has shifted as well. Earlier approaches emphasized certain theories to explain human actions (e.g., economic rationality for decision making or scientific management for organizational structures and operations), but the contemporary vision emphasizes joining with stakeholders to understand their needs, motivations, and interests and to forge acceptable wildlife programs. That requires teamwork, collaboration, and the development of trust, social norms, and shared responsibility. Universities have responded by helping students practice becoming better team members, communicators, and collaborators.

Education in human dimensions, as in other areas, has evolved to include both general and specialized approaches. Educators recognize that every wildlife student and professional needs to understand, appreciate, and incorporate the human component into wildlife management. That need is met in various ways: broadly, through some coverage in each professional course; intensively, in dedicated courses; and experientially, in capstone courses or internships. Wildlife professionals seeking to build their management, communication, or decision-making prowess should study the literature in those areas in more depth than one finds in a module for a wildlife management course or even a survey course on human dimensions.

Careers and hence education are also available for the student or professional seeking to specialize in the human dimensions of wildlife management. Through specialized courses, concentrated use of electives, and focused graduate-level research, students can develop human dimensions expertise and understanding of both the theories and methods of human dimensions disciplines, as well as the applications of human dimensions insight to management. The range of specialties provides exciting opportunities for persons interested in communication and media relations, non-formal adult and youth education, planning and public involvement, demographic and community resource analysis, along with traditional specialties such as economics, law, and policy analysis.

17.3. THE CONTENT OF HUMAN DIMENSIONS EDUCATION

Human dimensions education has three fundamental goals:

1. To help future professionals understand the factors that influence people's values, beliefs, attitudes, decisions, and actions;
2. To help future professionals understand organizations, communities, and institutions; and
3. To build skills that professionals need to work effectively with people, whether they are co-workers, partners, or stakeholders.

This section describes human dimensions educational needs and opportunities. You will not be able to cover all these topics in a typical undergraduate or graduate degree; therefore, you should look for extramural courses and in-service training opportunities in the areas described so you can add important human dimensions knowledge and skills to your professional capabilities.

17.3.1. Decision Making

The decisions people make sometimes seem illogical, arbitrary, contradictory, or hypocritical, but they probably follow predictable patterns, called decision-making styles. At one extreme is rationality: in advance of a decision, the decision maker explicitly sets out to identify all the conditions, rules, and data needed for the decision. At the other extreme is intuition: the decision maker does the processing in his or her head and announces a decision (and those affected might ask for evidence and reasons underlying decisions). Other approaches include incremental, sequential, mixed-scanning, and structured decision-making. Some decision makers use a variant of structured decision-making to handle complex decisions (Chapter 8). Understanding decision-making styles can help wildlife managers choose appropriate styles for particular situations or understand how others arrive at their decisions. This also helps them determine what type of information or evidence might be appropriate for a particular decision-making context.

Risk also influences decisions. Our reactions to unfamiliar, coercive, or potentially catastrophic risks are different from our reactions to common, voluntary, or mild risks. An entire branch of decision making is called risk analysis; it is the basis for much environmental regulation and mitigation. Risk theory has great relevance for understanding human behavior in many wildlife management situations.

Understanding decision making (both the theory and tools available to support decision making) is a very useful area of expertise for a wildlife manager to develop. Degree programs in business management, operations research, public administration, and environmental studies typically offer courses in decision making. Some natural-resource policy and management courses cover decision making in detail and allow practice through group projects that involve role playing.

17.3.2. Stakeholder Involvement

The successful wildlife professional must know how to work with stakeholders as partners in management; this includes understanding of who the stakeholders are, how they can participate in management, and how to work with them effectively. Stakeholder involvement ranges from highly structured activities that must be conducted by specialists to informal activities conducted by managers. Design and implementation of formal stakeholder processes (e.g., task forces and advisory committees) usually needs expert direction. For such activities, managers need to be good facilitators in order to guide the groups to complete their tasks efficiently and effectively.

Increasingly, wildlife agencies rely on stakeholders who volunteer to create habitat, conduct biological surveys, offer education programs, lead tours, and raise funds. Management of volunteers and support organizations requires skills in organization, delegation, relationship building, and promotion—skills that every wildlife professional needs but that they may not come by naturally.

Courses relevant to developing expertise to promote stakeholder involvement are often taught by departments of sociology, psychology, planning, policy, non-formal or extension education, communication, and personnel management. On-the-job experience is also important; it will help a wildlife manager understand the difficulty and importance of the work and therefore be open to further training. Consequently, practicing professionals should seek opportunities to take continuing education courses covering topics such as "Seeking Common Ground," "Facilitation Training," and "Volunteer Management."

Students also can gain experience in stakeholder involvement by participating in groups and activities during their school years (see section 17.5.2). The leadership at most universities is eager to involve students in committees, task forces, study commissions, and other administrative groups. Participatory governance is a core value in universities, and these groups provide excellent exposure to stakeholder involvement. Professional societies (e.g., The Wildlife Society) also present opportunities for participation in student and state chapters or even on committees of the parent society.

17.3.3. Governance

The sum of processes leading to decisions and their implementation in the public sector is referred to as "governance." Good governance involves making decisions and implementing them according to a series of principles including broad participation, consensus, accountability, transparency, responsiveness, efficiency and effectiveness, inclusiveness, and following the rule of law. Wildlife managers are expected to perform their work consistent with these principles. In the context of good governance, the focus shifts from solely valuing the technical expertise of the wildlife professional to also valuing her or his ability to lead people through a collaborative, consultative, and participatory process. Most wildlife professionals, particularly those working for government agencies at local,

state, or national levels, will be affected in some way by what occurs through larger public-policy processes, so knowledge of the policy process will serve the manager well.

Policy processes. Wildlife management occurs within the parameters set by public policies that provide mandates, authority, funding, and limits on wildlife conservation and management programs and personnel, particularly at state and national levels. The process of developing wildlife policy involves a range of participants, each of whom has differing degrees of influence and authority at each stage of the process. Understanding who these people are, what roles they play, and how to influence the decisions and actions that occur at each stage of the wildlife policy process will increase your effectiveness as a wildlife professional. The field of public policy related to wildlife, however, is much richer than simply understanding the set of laws that guide how wildlife management is conducted. Policies aimed primarily at other sectors of society, such as agriculture, rural development, zoning, energy management, and transportation, also have substantial influence on wildlife and its management.

Public policy makers have several sets of tools available to them as they govern, many of which are pertinent for wildlife policy makers. Howlett et al. (2009) characterize these as information-based, authority-based, treasure-based, and organization-based. Information tools include education and communication programs (e.g., information about proper disposal of garbage in bear-proof cans for homeowners in communities located within bear habitat). Authority tools include various types of regulations or permitting systems (e.g., restrictions on hunting area or technique; quotas for number of visitors to wildlife viewing sites on refuges). Treasure tools include financial incentives or disincentives (e.g., payments to landowners to encourage conservation practices). Organization tools refer to the structures used by wildlife agencies to provide services, such as wildlife refuges or recreational access areas. The successful wildlife professional will be familiar with this range of policy tools and understand how they may be applied in wildlife management. The astute professional will know and understand what happens along the path of policy making, who is involved, and how to participate effectively at each stage.

Courses on public policy processes and governance occur in university departments of government, political science, public administration, and city and regional planning. Courses are available in operation of non-governmental organizations (NGOs; often also called non-profit organizations). Social-psychology courses offer valuable insight into the human interaction aspects of working with individuals, boards, and advisory groups. Some university programs in wildlife and natural resources are developing specialty courses focusing on wildlife or natural-resources policy processes.

The general area of leadership development is also relevant to understanding and mastering aspects of public policy and governance. Skills in leadership often are best honed through experience. Universities often offer leadership programs

The Emerging Wildlife Conservation Leaders Program trains students and practicing professionals in a range of skills needed to lead conservation organizations and initiatives. Groups work in teams to produce and deliver conservation initiatives. One group developed a range of materials to support fund raising for the Northern Jaguar Project. (*above*, courtesy of the Emerging Wildlife Conservation Leaders Program; *right*, courtesy of the Melissa Normann/Rainforest Alliance and Northern Jaguar Project)

through their student affairs offices, such as North Carolina State University's "Center for Student Leadership, Ethics and Public Service." Programs like this provide a series of leadership development opportunities and often expose students to experts in the field of leadership and to public leaders with substantial experience. The Wildlife Society also offers leadership training for wildlife professionals.

Law. The codification of policy results in laws. Understanding the legal arena in which wildlife management issues are debated and challenged is a necessity for many wildlife managers. Some of the most important concepts relate to the power and role of government; the protection of constitutional rights; the use of public referenda, court suits, and other mechanisms to stop or prompt actions; private property rights; relationships among different governments, including local, state, national, international, and tribal; and law enforcement and regulations (Bean and Rowland 1997).

Legal concepts are not covered comprehensively in this book but are important at all levels—local, state, national, and international. Local governments use zoning and ordinances to influence land development–uses and activity restrictions, respectively. These can range from extensive environmental review of development proposals, to firearm discharge ordinances that effectively prevent hunting within town limits, to wildlife feeding restrictions that prevent food-conditioning of wildlife. Although it is seldom addressed in wildlife curricula, local authorities (such as county commissions and town planning boards) may have the most direct and important influence on wildlife habitat and use through their local zoning and permitting laws (Meffe et al. 2002). Of course, legal issues at other levels of government need to be understood. Regulatory statutes, tax codes, constitutional issues, and wildlife laws apply at the state level. Federal governments use similar legal

devices supplemented by programs that fund wildlife management activities. International treaties (e.g., Convention on International Trade in Endangered Species) have long been important legal tools with direct influences (e.g., migratory bird and marine mammal treaties, endangered species rules) and indirect influences (e.g., trade policies, debt-for-nature swaps, global climate treaties).

Students can find law courses in government, public administration, business, economics and regional planning departments. Law schools may offer courses about environmental, land-use, or water law to non-major students. Natural resources departments sometimes offer law courses that focus on wildlife, water, or natural resource law. Especially helpful are courses that help students to understand local governance (because that is where so much of the land-use action occurs) as well as the praxis of community-based management of wildlife.

17.3.4. Planning

Planning has become increasingly important in public agencies, businesses, and NGOs. Planning became an imperative in wildlife management when states were given the opportunity to receive their federal wildlife restoration funds as block grants rather than on a project-by-project basis (as long as they had a published strategic plan for their programs). Given the difficult economic conditions that continually affect almost all state governments and NGOs, we can expect planning and the oversight of it by stakeholders (i.e., voters, clients, customers, organization members, donors, and staff) to be an ongoing component of wildlife management. Understanding the techniques of planning—the subtleties and skills within that framework—is essential for wildlife professionals to become effective and efficient in their own work and in guiding the work of their agency or organization.

A related and equally important part of planning not discussed earlier in this book is the organization and execution of individual activities, which is called project planning and budgeting. The stages include setting an objective, planning the work, computing the budget and time schedule, gathering the needed resources, conducting the work, monitoring the progress, and reporting on the project's completion and outcomes. As with the larger organizational planning process, project planning relies on proven conceptual approaches and technical skills that can guide the wildlife professional through a successful project. For example, various techniques of project scheduling (e.g., critical path analysis, Program Evaluation Review Technique charts) can help anticipate where delays in one step are likely to slow the project as a whole and how to avoid them.

Courses in planning, under various titles, are typically offered by departments of public administration, urban and regional planning, and operations research. It is beneficial to have some job experience before taking such courses because planning often seems artificial unless it can be applied to a real situation. Consequently, many continuing education courses in planning are offered to practicing professionals, and in-service training is offered by most employers.

17.3.5. Program Evaluation

Increasing emphasis is being placed on accountability in professional endeavors, especially those in the public sector. It is a professional imperative to produce evidence that a program has worked effectively and efficiently. The federal government operates under a congressional mandate that requires all agencies to describe the outcomes and impacts of their activities. States have followed suit by becoming more explicit and transparent about all their activities. NGOs are accountable to their members and those who provide their funding. No matter where it occurs, wildlife management is subject to these and many other accountability expectations.

Program evaluation concepts and techniques allow wildlife managers to respond to these demands for accountability. Evaluation concepts help managers think about professional practice, and some evaluation techniques help evaluate how a wildlife management program is designed and implemented, what benefits it produces, who receives those benefits, and how much it costs. Evaluation is a key element of many planning and policy process models; it provides the "feedback loops" that allow the next plan or policy to be better than the previous.

Courses in program evaluation may be scattered across the university disciplines and departments. Because the field of evaluation arose in response to the growth of large federal programs in education, health, and welfare, courses are typically located in human ecology, social service, and education curricula. Business schools, regional planning, and public administration departments also often offer evaluation courses.

17.3.6. Ethics

Ethical implications of decisions and activities related to wildlife are receiving increasing attention. Ethical issues have always been present in wildlife management, but currently are given more consideration in management decision making than had been the case historically, in part because people are raising complex ethical questions about the appropriate relationship between humans and wildlife. In addition, science has led to technologies that bring in new ethical questions. For example, is it acceptable for homeowners or farmers to kill nuisance animals, or is it better to develop contraceptives that alter an animal's fundamental physiology and behavior but allow it to live? What is the definition of "nuisance" animal and who has the right to characterize wildlife in this way? New ecosystem-based management models have raised complicated questions about the role of humans in relation to other creatures and within the system. Concerns about human stewardship of wildlife include explicit normative questions about the "right" way for people to interact with and manage the world around them.

An understanding of ethics can help wildlife managers recognize that a clear demarcation does not exist between science and values, and perhaps more importantly, that science alone does not answer the question, "What should we do?" Ethical training provides a basis for applying philosophical and moral arguments to a host of wildlife management challenges, including resolving conflicts, analyzing and communicating risk, encouraging and managing public participation, and assessing the appropriateness or likely acceptability of proposed management actions and outcomes.

Courses in ethics are usually found in philosophy and religion departments. They are also offered in humanities programs and interdisciplinary programs such as those focusing on science, technology, and society (e.g., discussions about ethics of biotechnology are relevant to wildlife management). Some universities offer courses or special seminars on environmental ethics and similar topics. One ethics topic of growing concern is environmental justice with respect to who bears the burden of wildlife and who benefits from it. This class of concerns in wildlife management may be most familiar in cases in developing countries, but if one looks closely it also is evident in North America.

17.3.7. Communication

Probably no aspect of professional practice is more important than communication (Chapter 12), which means that education about communication is equally important. Virtually all employers of wildlife management professionals insist that communication skills are crucial for success in their organizations.

Remember that communication is a multi-way process, so the wildlife manager will usually be a receiver and a sender simultaneously; consequently, good listening skills are crucially important to success in the field. Stakeholders want to know they are heard and understood; good listening skills can demonstrate this to stakeholders and also assure that the wildlife manager truly gets the message.

The ability to develop clear, concise, and coherent messages is vital to effective communication. Written materials (letters to stakeholders, reports to supervisors, evaluations of employees, proposals to secure funding for programs, draft legislation, contracts, work plans, formal publications, press releases) should all get the points across accurately and precisely. The same is true for oral presentations. Wildlife professionals routinely make presentations to committees, advisory boards, regulatory groups, public audiences, school groups, peers at professional meetings, and supervisors. The content of these presentations must be tailored to the individual audience and range from technical and scientific to conversational and persuasive.

The most demanding aspect of communication today may be selecting an effective communication channel. New electronic channels are invented constantly. Today e-mail, texting, social networking programs, and the Internet are primary vehicles for communication, and innovations in electronic communication continue to be developed at a staggering pace. A wildlife professional also needs to be able to use electronic conferencing hardware and software that may not be familiar to many students.

Wildlife managers find that, if used well, the telephone remains one of the most important communication devices they have. If you rapidly return a call to a stakeholder, you will send a powerful message that you care about his or her interests and concerns. If you talk with someone directly rather than sending an e-mail, you can avoid errors of interpretation or potential misunderstandings that may occur in other electronic communications, particularly those that are not interactive.

Wildlife professionals rely on their writing and speaking skills to interact effectively with co-workers, partners, and stakeholders. Consequently, having opportunities to practice these skills in many forums is a key to the education of a successful wildlife professional. The courses offered in introductory speaking and writing are just starting points, and most wildlife programs now require more advanced courses and continuing practice in writing (e.g., writing across the curriculum) and speaking (including the nuances of speaking in teleconference and videoconference venues). Communication skills are also at the core of success in capstone courses where student teams analyze actual situations and prepare management plans. The fundamental message, though, is that communicating is a vital part of the wildlife manager's job *every day,* and she or he needs as much training and practice as possible while in school.

17.4. A GLOBAL PERSPECTIVE ON HUMAN DIMENSIONS EDUCATION

New York Times columnist Thomas Friedman described a new era in U.S. life and policy in 2005 when he published *The World is Flat: A Brief History of the Twenty-First Century* (Friedman 2005). The book's premise is that advances in technology have lowered the barriers among nations in all aspects of life, from trade to communication to terrorism. The best-selling book posits that we all live in a global context where political boundaries are increasingly irrelevant.

Such ideas should not shock any student of wildlife management; they deal every day with the reality that nature does not respect political boundaries or ownership deeds. Migration is one of the most spectacular aspects of wildlife life histories, from the impressive journeys of the monarch butterfly to the spawning runs of anadromous fish and the tundra crossings of caribou. Wildlife managers also struggle constantly against the spread of invasive species.

As in all other aspects of modern life, wildlife management is a global profession for both human-caused and natural reasons. Parallel needs can be seen across continents. Land development in South and Central America is just as important to the sustainability of many bird populations as is development in the prairie pothole region of the northern Great Plains. Education is important for conservation in sub-Saharan Africa and North America if one hopes to sustain cherished wildlife and landscapes on either continent.

Every wildlife professional needs educational experiences that contribute to their global understanding and appreciation. Most universities now have general education requirements that include an international or diversity based course (sometimes the definition of diversity is written broadly enough to include international experiences). Most universities require foreign language training in recognition of the globalization of society. Some universities also have created credentials that illustrate a student's breadth and depth in understanding another nation or global region.

Every wildlife student should take advantage of courses and other opportunities to assemble an international portfolio that demonstrates her or his awareness, understanding, and mastery of global topics. If you have a proclivity for languages, pursue that interest as an essential skill and a competitive advantage in the job market. Take advantage of courses that cover a particular nation, region, or culture and also explore resource-specific courses, such as world forestry. Perhaps most importantly, seek opportunities to visit, study, serve, and work in international settings; the experience will change your life and enhance your professional outlook.

17.5. SOURCES OF HUMAN DIMENSION EDUCATION
17.5.1. Coursework

You can find formal classes in human dimensions subjects in two general areas. First, wildlife programs are offering more classes directed at human dimensions. Comprehensive programs (usually housed within colleges and schools offering natural resource degrees) typically offer overview classes in human dimensions, as well as specific courses in policy, planning, public relations, environmental sociology, and integrated or ecosystem-based resource management. Valuable courses such as these are an integral part of your basic professional preparation.

Second, look for courses offered in interdisciplinary programs, such as environmental studies; science, technology, and society; and international development; or in other disciplines, such as sociology, psychology, government, and economics (e.g., resource economics). In such courses, you should approach learning by first absorbing the principles and then actively identifying applications to wildlife management.

17.5.2. Experience and Practice

You should seek experience outside of the classroom or the formal university curriculum. The most direct form of experience is a summer job or internship. You should aim to work in a professional job (paid or unpaid) at every opportunity, including during all summers while attending college. During the school year, consider volunteering your time to work on faculty or graduate-student research projects that relate to human dimensions, or assisting faculty with extension and outreach activities focused on natural resource needs in various types of communities. A resume that is well-stocked with different kinds of experience that includes exposure to the human dimensions of wildlife management provides the beginner with improved odds of obtaining a professional job after graduation.

Many other avenues are open for developing knowledge, skills, and experience applicable to the human dimensions of wildlife management. Human dimensions are essentially about understanding and embracing the range of human attitudes, beliefs, and behaviors. Any activity that provides the chance to expand your experience with different people in different settings is likely to be of great value. For example, participating in university and professional organizations can be a good learning and professional development opportunity. Every good advisor tells students to join student clubs, serve on university committees, attend departmental seminars, and go to professional meetings. You will improve your resume in this way, and you will also learn about the human dimensions of wildlife management because student clubs are analogous to special interest groups, and committees are analogous to citizen advisory boards. Most wildlife professionals eventually work for large institutions (agencies, NGOs, universities), so taking part in institutional activities as a student is excellent preparation for a professional career. Seek opportunities for elected positions as well (e.g., president of a student association) because these provide leadership experience and signal to potential employers that your peer group valued your contributions. Universities generally seek to be "engaged" with their communities of place (Box 17.2) and offer many opportunities to "get involved." Most students find it rewarding to be a con-

tributing member of these communities, and the communities reinforce a "public service" ethic that is so important to successful wildlife managers.

One of the key contributions of human dimensions to the field of wildlife management is that it points out that not everyone values wildlife in the same way. Wildlife managers work in a world that is filled with individuals who possess all kinds of beliefs and expectations about wildlife. Understanding and incorporating the views of a diverse human population is a cornerstone of good governance of wildlife resources. For example, a manager's understanding of the growing Hispanic segment of the increasingly multi-cultural U.S. population will affect their attainment of wildlife management and conservation goals (Lopez et al. 2005). On a personal level, if you gain experience with and understanding of diverse individuals and groups, you will be better prepared for a successful career. Rather than shrinking from opportunities to acquire that experience, seek ways to immerse yourself in it.

Exposure to an unfamiliar culture, either through an international experience or one in-country that immerses you in a different ethnic or socioeconomic community from the one in which you grew up, should be part of every student's college experiences. Service learning, in which students provide service to the local community as part of their coursework, is another opportunity that every student should seek. Finally, make it a goal to think about diversity with respect to understanding, accepting, and embracing people of different ethnicity, race, gender, age, religion, sexual orientation, and physical ability in the workplace and in wildlife-management decision making. Formal courses and informal experiences that provide exposure to the different perspectives, experiences, and skills that different people possess enhance our collective capacity to succeed in wildlife management.

Universities are excellent places to learn to appreciate diversity. Welcoming and supporting a diverse community of students and scholars is a defining value of most universities; therefore, the community includes people from a wide array of backgrounds and heritages. Consequently, universities always have student groups focused on diversity, and they welcome individuals from different backgrounds as members. Universities also provide training programs about diversity, generally through their student affairs or formal diversity programs.

SUMMARY

Wildlife managers will encounter increasingly expansive and global issues in the foreseeable future, which will challenge the best minds in the profession. Challenges will arise relating to both the ecological aspects of the management environment and the human dimensions aspects—describing, understanding, predicting, and incorporating the human component into all levels and stages of wildlife governance and management will require professionals' dedication to life-long learning.

- In any role you assume as a wildlife professional, you will need to incorporate human dimensions into your work.

Box 17.2 *THE ENGAGED UNIVERSITY*

Colleges and universities do not have to be places where students are removed from the "real world" for four or more years. Most schools today recognize that the closer students are linked to their communities during their college years, the more the students grow as professionals and citizens.

To recognize the importance of university engagement with the community, the Carnegie Foundation for the Advancement of Teaching added a new aspect to its recognized system for classifying universities. This new classification evaluates and recognizes universities based on their "community engagement." Community engagement has two components, one relating to teaching programs and the other relating to faculty–staff-led outreach programs.

Universities are judged to be engaged in their teaching to the extent that they offer opportunities for students to be involved in their communities. As stated in Carnegie's materials, "Curricular Engagement refers to teaching, learning, and scholarship that engage faculty, students, and community in mutually beneficial and respectful collaboration. Their interactions address community identified needs, deepen students' civic and academic learning, enhance the well-being of the community, and enrich the scholarship of the institution." (http://classifications.carnegiefoundation.org/descriptions/ce_2008.php)

To judge how committed an institution is to community engagement in its teaching programs, Carnegie is particularly interested in the following:

- Does the university's mission and strategic plan incorporate engagement?
- Are faculty members encouraged and rewarded for participating in engagement activities?
- Are formal service-learning courses available in many curricula and at many levels?
- Do leadership programs incorporate engagement?
- Do cooperative education programs incorporate engagement?
- Are there study abroad opportunities built around community engagement?
- Can students be involved with research in their disciplines that involve engagement?
- Does learning assessment include engagement?

About 200 universities nationwide are classified as "engaged" through their teaching and learning. If you are interested in being part of engagement activities, check to see whether your university has been awarded the engagement classification (check with your Provost's Office), and then inquire about the range of opportunities that exist. You may be surprised at how many ways universities have to involve students in their local community. You may also find engagement to be rewarding and a good way to gain experiential learning valuable to your career in the wildlife profession.

A student at North Carolina State University provides conservation education to a group of young students in the Dominican Republic (courtesy of Roger Winstead, North Carolina State University)

A student at North Carolina State University interacts with seniors during a service learning experience (courtesy of Roger Winstead, North Carolina State University)

Universities and professional societies recognize that need and have revised their requirements to include considerable training in human dimensions.

- Professional education in human dimensions emphasizes understanding the needs, motivations, and interests of stakeholders and working with stakeholders to forge solutions to management problems. That requires teamwork, collaboration, and trust.
- Increasingly, wildlife professionals must understand the implications of their programs within a global context.

International experiences and education on international topics help prepare the wildlife professional to understand the connections between nations and the cultural contexts that influence wildlife management.

- Understanding and embracing the views of a diverse population is crucial for success in wildlife management. We cannot rely on just our own experience or values to guide us as we manage resources for others. Experience in diverse communities, as a student and a professional, is invaluable.

- Perhaps no other aspect of professional practice is as important as communication. No skills are more important to the success of the wildlife professional than the ability to write and speak well and to use electronic communication technologies to their full capacity.

- When learning about human dimensions, coursework is necessary but not sufficient. Other types of experiences (internships, summer jobs, committees and working groups, professional meetings, seminars, service learning, study abroad, independent research) will enhance your education in the human dimensions and prepare you for a career in natural resource management.

Suggested Reading

Covey, S. R. 1989. The seven habits of highly effective people. Fireside, New York, New York, USA.

Johnson, S. 1998. Who moved my cheese? an amazing way to deal with change in your work and in your life. G. P. Putnam's Sons, New York, New York, USA.

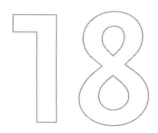

ADAPTIVE VALUE OF HUMAN DIMENSIONS FOR WILDLIFE MANAGEMENT

DANIEL J. DECKER, SHAWN J. RILEY, AND WILLIAM F. SIEMER

In Chapter 1, we observed that people and wildlife affect one another within complex, coupled socioecological systems. We then defined wildlife management as follows:

The guidance of decision-making processes and implementation of practices to influence interactions among people, and between people, wildlife and wildlife habitats, to achieve impacts valued by stakeholders.

We also asserted that human dimensions consideration in wildlife management should focus on understanding public interest in wildlife (impacts of human–wildlife interactions of interest to stakeholders) and the intended and unintended outcomes of management actions. Such considerations reflect the fundamentally human purposes of wildlife management—to create benefits (i.e., net positive impacts) for people, including the conservation of wildlife for the benefit of present and future generations. With this backdrop, other chapters introduced many of the major human dimensions considerations in wildlife management. These consist of core processes such as collaborative governance, stakeholder engagement, planning, situational analysis, decision making, and communication. Chapters also reviewed foundational social sciences (social psychology, sociology, and economics), general human-dimensions research insights, particular foci of research, and applied interests in human dimensions of wildlife management. Toward the end, the importance of wildlife professionals' attention to ethics and continuing education were discussed for improvement in application of human dimensions insights, but the story does not end there—the investigation and application of human dimensions knowledge for wildlife management is ongoing and growing in importance.

By now you recognize that research and experience during the past few decades has gradually uncovered a paradox for wildlife managers. Whereas human dimensions inquiry and stakeholder engagement clarified some of the most difficult aspects of wildlife management, these endeavors also revealed more about the complexity of the sociocultural dimensions of the socioecological system that is pertinent to wildlife management. The sociocultural dimensions of wildlife management discussed in this book, similar to the ecological factors covered in many other books on wildlife management, are ever changing, some nearly imperceptibly, others markedly. Many of these changes are a reflection of the movement toward collaborative governance, described in Chapter 2. Though some tendencies and patterns are pervasive, human behavior often is context-specific. Thus, context-specific knowledge is essential for predicting impacts and stakeholder responses to specific management decisions and actions. That highlights the importance of every manager having at least rudimentary human dimensions knowledge and public engagement skills.

The wildlife management profession is steadily becoming better at anticipating needs and responses of stakeholders, engaging diverse stakeholders and other agencies in collaborative governance of public wildlife resources, and evaluating and adapting to change in both the management environment and in the information base about that environment. This is being accomplished by advances in research, improvements in systems thinking, refinements in decision making, innovations in communication, enhancements in educational preparation of prospective wildlife professionals, and greater use of interdisciplinary management teams. Nevertheless, even with these advances, human dimensions insights do not necessarily make management easier, cheaper, or more efficient. We do believe, however, that the insights provided by human dimensions inquiry and stakeholder engagement improve governance of wildlife resources—they lead to doing the correct things, doing them well, and improving the durability of management decisions and programs arising from them. These positive changes are creating the possibility for wildlife professionals to develop the capacity to be *adaptive;* that is, to learn and adjust. Adaptive managers identify, anticipate, and make changes in enabling objectives and interventions as needed to achieve fundamental objectives of society for wildlife management.

18.1. BEING ADAPTIVE TO REMAIN RELEVANT AND EFFECTIVE IN CHANGING TIMES

Wildlife governance in North America in the twentieth century confronted seemingly insurmountable management obstacles. From the unregulated exploitation of wildlife by "market hunting" in the nineteenth century, to the widespread chemical contamination of the mid-1900s, to urban sprawl and habitat fragmentation of the past several decades, wildlife and those responsible for managing wildlife faced huge challenges. A paradigm emerged from these efforts that reflected the era of exploitation and scarcity that existed throughout much of the world; that is, wildlife management practices centered on protection of wildlife populations against excessive take, encouragement of habitat management practices to promote population growth of some species, and transplant programs to establish species in new areas.

The twenty-first century provides wildlife professionals with challenges of a different nature. The paradigm and practices that worked during the past century need to be evaluated and amended as new expectations for collaborative governance of wildlife resources are taking hold. In North America, circumstances that indicate the need to reassess the adequacy of wildlife management conventions (e.g., as expressed in the North American Model of Wildlife Management [Organ et al. 2010]) include the following: (1) declining numbers of hunters and trappers, who were the traditional base for political support and funding for wildlife management; (2) declining funds and growing demands on state wildlife agencies to provide services to a diverse array and larger population of stakeholders, some of whom we are just beginning to understand; (3) increasing urbanization, suburbanization, exurbanization, and development of rural landscapes combined with growing apathy toward the natural world; (4) the effects of climate change on wildlife habitat; and (5) increasing global-scale social changes and global-scale environmental effects. By nearly every measure and in every place, rates of change in sociocultural–ecological systems are accelerating.

Wildlife agencies, wildlife conservation organizations, and the wildlife profession will need to adapt strategically and productively to meet these challenges. This may seem like the responsibility of veteran wildlife conservation leaders rather than prospective wildlife managers who are still students, but the responsibility for professional adaptation and change belongs to everyone. Novel ideas that enable adaptive behavior at the individual, program team, and organizational level often originate with the newest members of a profession, who have not been acculturated into the conventional thinking and ways of doing things. Ironically, those least invested in an entity's past may be the most inventive in helping it succeed in the future. Every wildlife management student has potential to help wildlife management strategically adapt to sustain its relevancy as a discipline and to positively affect wildlife conservation.

The purpose of this chapter is to provide practical information and guidance for wildlife practitioners to be adaptive leaders, a role they can play regardless of how long they have been working in the wildlife field, what their responsibilities may be, and where they fit in the expansive enterprise of wildlife management. Our intent is to stimulate your thinking about how human dimensions research and application, coupled with stakeholder engagement, can help wildlife agencies and other wildlife organizations adapt as necessary to conserve and manage wildlife into the future. Put simply, we want to help you envision how you can be an adaptive leader and agent of positive change throughout your career, and to think about the ways human dimensions insights and stakeholder engagement can empower you in that pursuit. If you take to heart the suggestions about human dimensions education offered in Chapter 17 (especially in regards to continuing education throughout your career), it will be beneficial to your adaptive strategy.

The remainder of this chapter is organized into three main sections. In the first, we present a conceptual framework for understanding the primary areas where adaptive thinking and action should be applied, and we focus on broadening goals, boundaries, and activities of wildlife management. Human dimensions considerations abound in efforts to achieve such broadening. In the second section, we share a recent synthesis of experiences from state wildlife agencies, relevant literature (not restricted to agencies and applicable to any organization), and the collective experience of some wildlife management scholars and practitioners who suggest best practices to develop adaptive capacity in government wildlife agencies and non-governmental wildlife-conservation organizations. These reflect the breadth and depth of human dimensions considerations required in any change effort. We conclude with some thoughts to help wildlife professionals initiate adaptive change within their organizations, and point out how human dimensions insights and stakeholder engagement activities can aid in this endeavor.

18.2. ORGANIZATIONAL TRANSFORMATION AND ADAPTIVE LEADERSHIP: IMPORTANT CONSIDERATIONS IN COLLABORATIVE GOVERNANCE OF WILDLIFE

The context for wildlife conservation throughout the world has changed in many ways since the inception of the wildlife profession and the establishment of many familiar wildlife management institutions and organizations (Manfredo et al. 2003, Patterson et al. 2003, Jacobson et al. 2010), but traditional thinking about wildlife management in the United States, as described in the North American Model of Wildlife Conservation, remains largely intact today (Geist et al. 2001, Organ et al. 2010). Critics of the current system of wildlife management (e.g., Gill 2004, Nie 2004) are concerned that the model under which wildlife management has operated is not adequately reflecting the diversity of wildlife-related interests that exist in society (Decker et al. 2009a). This concern is one of the driving forces behind the growing interest in human dimensions research and stakeholder engagement. Indicators of public dissatisfaction with the relevancy of wildlife management as it exists today in the United States include the large number of wildlife-related ballot initiatives and referenda, the multitude of wildlife organizations with non-consumptive orientations (e.g., environmental, humane), state and national efforts to find alternative funding sources for government-led wildlife management (i.e., funds not generated directly or indirectly by hunters or trappers), and efforts to reform governance of wildlife (e.g., campaigns to change the composition of wildlife boards and commissions or to end them altogether).

In the United States, state wildlife management agencies (SWAs) have primary legal responsibility for management of most wildlife species, reflecting wildlife laws and other public trust mandates that emerged during the late 1800s. Farmers, ranchers, foresters, hunters, and trappers became established as the primary groups concerned with wildlife management by the early 1900s, and they communicated and lobbied for their interests through organized interest groups. Consistent with the governance approach of the time, SWAs evolved with a focus on these special interest stakeholders (Patterson et al. 2003). The long-term relationship between SWAs and hunters–trappers can be characterized as being highly "path dependent" (Greener 2002).

Path dependency arises from the influence of historical circumstances on organizations and their patterns of behavior, including resistance to adjustment. Further cementing the relationship between consumptive users (i.e., hunters and trappers) and SWAs is an historical dependency on these stakeholders to fund state wildlife management directly with license fees and indirectly with federal excise taxes on firearms, ammunition, and archery equipment (Anderson and Loomis 2006). A resource-dependency perspective on wildlife management (whether in the United States or elsewhere and whether the governance model is facilitated by a government agency [as in the United States] or an NGO) posits that organizational survival is ensured by aligning with other organizations that provide them resources and support (Pfeffer and Salancik 2003). Resource dependency typically affords organizations little opportunity to control their own destinies or to modify dependent relationships. As a result, organizations have less adaptive capacity.

Wildlife conservation organizations (such as SWAs) that break from the historical path, and resource dependencies that expand programs and services to meet the diverse needs of society, encounter many adaptive challenges (Jacobson et al. 2007). Fundamental or transformative change, no matter how sound in principle, often meets resistance from some staff within the organization and from people with whom they have a tradition of interacting (e.g., policy makers, staff of NGOs). Resistance materializes when organizations try to adapt to be more effective given current and anticipated conditions (Putnam 1993:179).

Although transformative change is not easy, avoiding reform when politically active groups are seeking greater collaboration in wildlife resource governance may result in organizations losing relevancy and then legitimacy with society. Legitimacy refers to the extent to which organizations are connected to a broad normative and cultural framework within contemporary society. Organizations need legitimacy to persist. Realigning a wildlife organization's goals, activities, and boundaries to embrace the contemporary needs and expectations of society for wildlife management will promote good governance practices germane to contemporary society. Insight from human dimensions inquiry and stakeholder engagement (if these activities have not been limited to traditional stakeholders) can equip the wildlife manager to guide expansion of a program.

18.2.1. Organizational Change Theory

Small-scale change occurs constantly and often imperceptibly within organizations; however, major changes that involve a break with existing norms and a shift to new norms are uncommon (Aldrich and Ruef 2006). Established bureaucracy typically resists major organizational change, especially when historical path and resource dependencies have long existed (Pfeffer and Salancik 2003). Resistance to change in wildlife organizations, as in all other organizations, is natural.

Organizational transformation involves significant changes in three possible areas: *goals, activities,* and *boundaries* (Aldrich 1999). Two primary elements of goal transformations exist: (1) changes in the breadth of organizational goals, particularly evolution from specialism to generalism (e.g., change from a focus on satisfaction of hunters to a focus on impacts of human–wildlife interactions experienced by diverse stakeholders); and (2) changes in the domain served by an organization (e.g., programs that address a broader range of wildlife values, such as ecological services provided by wildlife; Aldrich 1999).

The second facet of transformation includes changes in activities that have a significant effect on organizational knowledge (Aldrich 1999), such as when an organization initiates social science inquiry to promote more inclusive engagement with non-traditional stakeholders. Such transformation might result in changes in products and services provided by the organization.

The third facet of organizational transformation includes expansion and contraction of boundaries for services and activities. State-level boundaries are delineated by which individuals and organizations (Aldrich 1999) are given opportunity to be active participants in decision making about wildlife management objectives and actions in the state. These boundaries can expand or contract depending upon who SWAs consider to have actionable concerns about wildlife management and valuable insight to bring to decision making. A strategy to minimize dependence on historically narrow relationships, which is consistent with resource dependency theory, is diversification of stakeholders who are invited to be involved in decisions and who receive benefits from wildlife management programs. Impediments to such diversification are usually encountered, and there are different perspectives about how an organization can effectively overcome them.

Organizational change theory has focused on two perspectives (Aldrich 1999). One perspective considers change to be an outcome determined largely by external or "environmental" forces (environmental determinism). The other perspective says organizational change is attributable to internal factors or qualities of organizations (strategic choice).

Environmental determinism refers to how much outside influences control an organization's abilities to make choices about its future (Astley and Van de Ven 1983). In general, a deterministic perspective posits that organizations are highly influenced by or dependent on those organizations or individuals that control the resources necessary for continued existence (Pfeffer and Salancik 2003). The strategic choice perspective holds that organizations are self-directing and able to make and act upon strategic choices about their preferences for change (Astley and Van de Ven 1983). Blending these viewpoints (e.g., Hrebiniak and Joyce 1985, Oliver 1991) leads one to conclude that even though external factors influence organizational behavior, organizational self-interest can be a powerful force for organizational change. With this outlook, organizations have the ability to respond to pressures for change in a strategic manner.

When strategic choice is an option for a wildlife management organization, the importance of having adaptive thinkers in positions throughout an organization becomes readily apparent. Strategic choice and adaptive behaviors go hand in hand. In fact, we argue that whether a wildlife management organization is in a position where strategic choice even is an option depends largely on whether it has adaptive capacity. In order to exercise the strategic choice option, an organization needs enough people on staff who can conceptualize the coupled sociocultural–ecological system in which they work, understand the human dimensions of that system, and envision the kinds of changes–adaptations required to meet wildlife management objectives desired by society.

Some wildlife professionals believe (and it is indeed typical) that senior and mid-level leaders in a wildlife conservation and management organization should be charged with thinking about organizational transformation and adaptive leadership, but we strongly urge and believe that it is preferable to include all staff in such work. We are observing more often than ever before that those agency leaders who are astute about the need for transformative change are recognizing the resource of ideas and energy embodied in their newest employees. In some very important ways, newer members of an organization may have greater capacity to identify the adaptive changes needed to transform and maintain or improve organizational relevance. For example, new staff members are not confined in their thinking by historical path dependencies of the organization. Furthermore, intelligent naiveté on the part of new staff can result in fresh questions, novel observations, and valuable innovations.

If new members of an organization's staff are approached as resources in this respect, asked to participate as peers in deliberations about organizational change, and made to feel safe in responding openly, then they may find themselves in-fluencing the organizations in which they are employed soon after joining them. We argue that new employees with human dimensions "savvy" can lead others toward understanding why and how an organization might adapt to stay in tune with society's expectations with respect to wildlife governance (stakeholder model) as opposed to just the concerns of established interest groups (political model). Senior leaders will welcome that kind of input as they maneuver between keeping traditional stakeholders happy and meaningfully engaging new stakeholders through the process of realigning the organization's goals, activities, and boundaries to embrace the contemporary needs and expectations of society.

Jacobson et al. (2010) found transformative change in wildlife management organizations where (1) leadership promoted cultural change conducive to broadening goals (communication effectiveness); (2) strategies to expand organizational boundaries and grow coalitions included traditional and nontraditional groups (transactional approach to stakeholder engagement); (3) public interest was assessed (e.g., human dimensions inquiry) and accountability was demonstrated (evaluation was conducted to assess the extent to which impacts were produced); and (4) expansion of programs and services occurred.

18.2.2. Envisioning a New Future Is Necessary for Fundamental Change

Wildlife management organizations will improve the likelihood of building and sustaining support by promoting a vision of the future that is robust to the needs and interests of many kinds of stakeholders (Jacobson 2008). This calls for knowledge about diverse stakeholders' needs and expectations as obtained through human dimensions research and stakeholder engagement. If an encompassing vision can be articulated and gain widespread support, then acceptance of the need to diversify programs and services to meet broad societal needs is facilitated. By creating a vision that embraces a broad suite of public interests, reaches out to diverse partners (e.g., wildlife NGOs), and leads to a strategy for change, wildlife management organizations are more likely to increase their conservation capacity via more funding and expertise and to maintain legitimacy with society.

A transformative vision addresses the *who, what, and how* of an organization's mission. This is another way of thinking about Aldrich and Ruef's (2006) three components of change.

1. *Who* are stakeholders and partners? (*major change in boundary domain and depth*)
2. *What* are the desired future conditions and outcomes (impacts) sought through wildlife management that are more encompassing of diverse public values vis-à-vis these resources? (*major change in goals*)
3. *How* are goals set, decisions made, and actions implemented? (*major changes in services and products*)

The framework illustrates many of the human dimensions considerations covered in earlier chapters in this book.

Box 18.1 *KOTTER'S 8-STAGE PROCESS FOR ORGANIZATIONAL TRANSFORMATION*

1. Establishing a sense of urgency
   a. Conducting a market assessment (i.e., a survey to determine who your current stakeholders are and what they want or need).
   b. Communicating the need–opportunity to change in ways that help the organization to maintain relevancy and legitimacy with society.
2. Creating the guiding coalition
   a. Putting together a group with enough power and commitment to lead change.
   b. Getting the group to collaborate and work together as a team.
3. Developing a vision and strategy
   a. Creating a vision to help direct the change effort.
   b. Developing strategies for achieving that vision.
4. Communicating the change vision
   a. Constantly communicating the new vision and strategies.
   b. Having the guiding coalition practice the behavior expected of other staff.
5. Empowering broad-based action
   a. Clearing away obstacles.
   b. Changing systems or structures that undermine the change vision.

   c. Encouraging risk taking and non-traditional ideas, activities, and actions.
6. Generating short-term wins
   a. Planning for visible improvements–wins.
   b. Visibly recognizing and rewarding people who made wins possible.
7. Consolidating change and producing more change
   a. Using increased credibility for change ideas to align the entire organization (e.g., programs, policies) under the new vision for the future.
   b. Hiring, promoting, and developing people who adopt adaptive behaviors.
   c. Reinvigorating the process with new projects, themes, and change agents.
8. Anchoring new approaches in the culture
   a. Creating better performance with respect to stakeholder-oriented behavior, more and better leadership, and more effective management.
   b. Articulating the connections between new adaptive behaviors and organizational success.
   c. Developing means to ensure leadership development and succession.

Adapted from: Kotter (1996)

18.3. ADAPTIVE THINKING AND EFFECTIVE PRACTICES

Leading scholars in the field of organizational behavior have identified preconditions and steps leading to transformative change. Kotter (1996) offers an eight-stage process for creating change in an organization (Box 18.1). We adapted his wording to align better with the circumstances typical of wildlife agencies. You may note the many places where human dimensions considerations (research, evaluation, stakeholder engagement, etc.) are embedded in Kotter's process; this suggests that a wildlife professional who is savvy about human dimensions has a great deal to offer an organization that needs to adopt an adaptive stance, which seems vital for agencies and NGOs alike.

The details leading to adaptive change are situation-dependent. Decker et al. (2011) extracted several points from the literature, case studies, and their experiences that wildlife professionals should keep in mind about adaptive change (Box 18.2). They split their list into points that were primarily people-focused and points that were primarily process-focused, and they suggested that these points should be considered together with an understanding of a particular organization's history, culture, and current context. The adaptive change points identified in Box 18.2 are instructive in that they illustrate the human dimensions considerations inherent

in transformative and adaptive change. A few of these are identified below.

18.3.1. People

Transformative or adaptive leaders do not work in a vacuum; they are constantly analyzing social and political trends. This is essential so that they can draw conclusions about what they will encourage their organization and its stakeholders to transform into or adapt to through time. Some leaders do this analysis by relying on intuition and "gut feelings," but this approach is less common as the world of wildlife management becomes increasingly complex. Alternatively, systematic human dimensions inquiry and stakeholder engagement are being relied upon.

Transformative and adaptive leaders also recognize that adaptive change requires many people (internal and external) to play various kinds of roles. Among the valued traits sought is sociocultural knowledge, which helps groups find common ground, commit to a common purpose, and then cooperate to achieve that purpose. This transformation starts with a vision and related adaptive behavior in the organization's culture, which is a process itself best achieved through a strategy that is informed about the perceptions (beliefs and attitudes) of an organization's staff. One needs to view this as a process whereby a new way of thinking is adopted throughout the organization over a reasonable period of time; change seldom

BOX 18.2 *OBSERVATIONS ABOUT ADAPTIVE CHANGE IN WILDLIFE MANAGEMENT ORGANIZATIONS*

PEOPLE

1. Leadership is needed at multiple levels (even in small teams working on local issues) to effect transformation into an adaptive organization.
2. Change requires that different kinds of players take an adaptive approach to fill different kinds of roles.
3. Cultural change is at the root of successful transformation to an adaptive organization.
4. Commitment to becoming an adaptive organization is essential, but not everyone has to be on board before starting the change process.

PROCESSES

5. Capacity for change often requires novel collaborations and new partnerships—this requires an adaptive approach to avoid reliance only on traditional existing relationships.
6. Strategic vision and planning are vital to turn adversity into opportunity; adaptive behavior can turn opportunity into reality.
7. Stakeholder engagement, transparency, and effective communication are important to build trust.
8. Build and maintain trust with an understanding that demonstrating credibility and accountability is a never-ending responsibility.
9. Communicate the value of adaptive behavior to co-workers, external partners, and old and new stakeholders.
10. Rely on inquiry to confirm intuition.

Source: Decker et al. (2011)

---

happens quickly in organizations, especially in bureaucracies. Strategy for cultural change and innovation adoption–adoption diffusion should be based on social science theory and existing bodies of literature on organizational behavior and social change.

18.3.2. Processes

An initial step in adaptive change is assessing and perhaps building capacity for such change to proceed. That capacity takes several forms, but one valuable component is an adequate knowledge base about the size, characteristics, needs, and preferences of the broader stakeholder population that a wildlife conservation and management organization hopes to serve. These insights are outputs of standard human-dimensions research. In addition, a track record of successful public engagement (trusted, inclusive, and meaningful) is helpful. The value of such a track record is especially apparent when a wildlife management organization experiences difficult budgetary conditions and is expected to "do more with less." Such situations require an adaptive approach where base resources of the organization are used to garner or leverage more resources to apply to high-priority wildlife management issues. Adaptive behavior often involves simultaneously working on two different approaches: (1) forming collaborations with other entities to address specific issues, and (2) building more enduring partnerships around broader, long-term issues. One of the more significant adaptive changes in organizational culture and operation may be developing a philosophy of collaboration and engagement, as presented in this book, and practicing it effectively. Creating unique coalitions is a good way to expand boundaries and stretch thinking about new conservation initiatives, so the inquiry and engagement ideas covered earlier in this book need to be put into operation.

Public input processes should not be considered sufficient to secure public support for adjustments an organization may need to make to adapt to changing circumstances. As detailed in Chapter 11, input is not the same as involvement. It is through meaningful, productive involvement and organization follow-through that stakeholders develop commitment to a vision, build trust in the organization, and support a change strategy. Staff at all levels of an organization can contribute to effective stakeholder engagement by reaching out to a broad spectrum of the public. Opening new and diverse lines of communication with an extensive and inclusive array of stakeholders is one hallmark of an adaptive strategy on a course for success.

The literature on organizational change is replete with references to the essential role of trust in the change equation. Trust is gained through the approach to collaborative governance advocated in this book: authentic effort to seek and use involvement of others, transparent decision making, engagement with stakeholders in all stages of the management process, accountability to stakeholders, and consistent performance. Building trust with new stakeholders requires time and is expedited by a track record that demonstrates how human dimensions research and public involvement practices meaningfully influence decision making. An effective way to build trust and forge support for programs is to take an impacts-management approach (Riley et al. 2003) to wildlife governance that uses human dimensions inquiry to (1) identify impacts to be managed, (2) measure stakeholder reaction to programs, and (3) document the impacts experienced by stakeholders from management of a particular wildlife resource.

Wildlife organizations will be practicing good governance if they strike a balance between these two considerations:

1. Providing stakeholders who have various interests with (a) opportunities to make their needs and concerns known, and (b) meaningful roles in decision making (e.g., through stakeholder engagement processes); and

2. Fulfilling a responsibility to society at large, which includes identifying (e.g., through human dimensions inquiry) and considering some stakeholders who are not organized and are not represented by a formal interest group.

Wildlife management organizations that engage in adaptive change for purposes of addressing a broader swath of public interests in wildlife report that systematic inquiry is needed to identify the variety of wildlife interests that exist and to characterize the stakeholders who have such interests. This may be especially critical for the people who are not direct users of wildlife, but for whom wildlife play a role in their outdoor recreation experiences. Furthermore, inquiry can bring clarity to the expectations that these stakeholders have for wildlife management. The assumption that individual stakeholders are homogenous in the impacts they perceive or desire has been shown repeatedly to be false, even when those stakeholders are members of the same stakeholder group.

18.4. DEVELOPMENTS NEEDED IN HUMAN DIMENSIONS FOR IMPROVED WILDLIFE GOVERNANCE

Continued attention is required in four areas of human dimensions: research, application, integration, and education. Research on human dimensions topics is far behind wildlife managers' requirements for this kind of insight. The topics addressed need to be expanded and the disciplines focused on human dimensions aspects of wildlife management need to be broadened (e.g., more work from anthropological and sociological perspectives could be very useful).

Application of human dimensions insight is enhanced when it occurs in an atmosphere of rigorous management thinking, decision making, and ethical analysis—chapters in this book offer suggestions on how to contribute to such an atmosphere in your work. Closely related to this is the need for greater integration of human and ecological dimensions of wildlife management, of the type encouraged by socioecological systems thinking. Devices like the Manager's Model (described in Chapter 7) and other systems analysis techniques could contribute markedly to this purpose. Finally, it is critical that prospective wildlife professionals be exposed to these ideas during their basic education. The fields reviewed in this book are some, but not all, of the areas of educational preparation needed for the complex enterprise of wildlife management. It should be apparent there is room and need in wildlife management for professionals who bring depth of expertise from a large variety of areas. They all have a place on the multi-disciplinary teams for wildlife management in the twenty-first century.

18.5. ONGOING ISSUES

The shift of wildlife management from concentrating on wildlife population and habitat parameters to producing positive impacts is still underway as professionals explore ways to integrate human and biological dimensions of management in a comprehensive philosophy and approach. A focus on impacts of human–wildlife interactions and innovative ways to manage them will be hallmarks of successful wildlife management. As Decker et al. (2009b) suggest, this is a deceptively simple proposition. Efforts to adopt and apply such a philosophy and approach encounter numerous barriers. We feel that two issues similar to those identified by Decker et al. (2009b), which will affect wildlife management well into the future, are

1. a culture of wildlife management (common among professionals and stakeholders) that typically and uncritically assumes wildlife issues are fundamentally biological or ecological, but that misses the fact that society's interests are in values they associate with wildlife and in impacts of human–wildlife interactions; and
2. governance of wildlife management that (a) clings to an expert-driven model of decision making, (b) privileges certain stakeholder interests and is narrowly funded by them, and therefore (c) resists informed participation of the full breadth of stakeholders.

18.6. RESEARCH, POLICY, AND PRACTICE NEEDS

The approach to wildlife management presented in this book calls for adoption of several research, policy, and practice conventions. Some of these have already occurred. We offer suggestions for research, policy, and practice that reflect, in part, needs identified by Decker et al. (2009b).

RESEARCH NEEDS
- Research should be approached in an integrated fashion that couples ecological and sociocultural inquiry to provide more comprehensive insights about management issues, including the root causes of impacts resulting from human–wildlife interactions.
- Research is needed to improve understanding about the relationship between wildlife abundance (numbers and density) or individual animal behaviors, the extent and nature of interactions between wildlife and humans, and the effects of interactions on perception of impacts.
- Research is needed that improves and updates managers' knowledge about the stakeholders of wildlife management and the human dimensions of the management environment.
- Research is needed that is grounded both in basic social-science theory and actual management contexts, and that builds bridging concepts that address applied needs of wildlife management.
- Research is needed to address evaluation expectations of active and passive adaptive-management approaches.

POLICY NEEDS
- Policies about wildlife-management decision processes should require both scientific data about stakeholders and stakeholder engagement.
- Policies are needed that support co-existence of humans and wildlife, and they should avoid underemphasizing or

overemphasizing negative impacts resulting from human–wildlife interactions.

- Policies are needed that acknowledge jurisdictional limitations and emphasize partnerships across jurisdictions and among diverse organizations (e.g., Landscape Conservation Cooperatives).

PRACTICE NEEDS

- Wildlife professionals and others should develop and apply a science-informed (rather than anecdotal) knowledge base about stakeholders (e.g., their beliefs, attitudes, norms, and behaviors) when engaged in management decision making.
- Manager's Models depicting wildlife-management systems should become standard components of wildlife-issue analysis and program development to ensure systematic, disciplined thinking and to enhance clarity of internal and external communication about a management situation.

SUMMARY

Our hope is that you see the value of adopting an adaptive approach to wildlife management, one in which you continually learn and adjust. We leave you with some concluding thoughts about an adaptive strategy for wildlife management.

- Develop a sound philosophy about the value of an adaptive approach and let it guide your work in wildlife management.
- Demonstrate relevance to society through performance in producing valued benefits and ensuring the public recognizes that these benefits have been created.
- Strive to create a greater awareness of the relationship between the quality of wildlife resources and quality of life for humans through creation of positive economic, aesthetic, psychological, physiological, and recreational impacts.
- Seek innovative, context-specific, capacity-building solutions to wildlife management situations.
- Learn to work effectively in a wildlife governance model that is stakeholder focused; learn how to share power, responsibility, work, and accountability.

Suggested Readings

Badarracco, J. L., Jr. 2002. Leading quietly: an unorthodox guide to doing the right thing. Harvard Business School Press, Cambridge, Massachusetts, USA.

Johnson, S. 1998. Who moved my cheese? an amazing way to deal with change in your work and in your life. G. P. Putnam's Sons, New York, New York, USA.

Kotter, J. P. 1996. Leading change. Harvard Business School Press, Boston, Massachusetts, USA.

APPENDIX
Scientific Names

alligator, American: *Alligator mississippiensis*
antelope, pronghorn: *Antilocapra americana*
banteng: *Bos javanicus*
bass, smallmouth: *Micropterus dolomieu*
bear, black: *Ursus americanus*
bear, brown: *Ursus arctos*
bear, European brown: *Ursus arctos arctos*
bear, grizzly: *Ursus arctos horribilis*
bear, polar: *Ursus maritimus*
beaver, American: *Castor canadensis*
beluga: *Delphinapterus leucas*
bison, American: *Bison bison*
blackbird: family Icteridae
boar, wild: *Sus scrofa*
bullfrog, American: *Lithobates catesbeianus*
butterfly, monarch: *Danaus plexippus*
caribou: *Rangifer tarandus*
catfish: family Ictaluridae
chimpanzee: *Pan* spp.
chinchilla, long-tailed: *Chinchilla lanigera*
chinchilla, short-tailed: *Chinchilla chinchilla*
cod, Atlantic: *Gadus morhua*
condor, California: *Gymnogyps californianus*
cormorant, double-crested: *Phalacrocorax auritus*
cottontail: *Sylvilagus* spp.
cougar: *Puma concolor*
cowbird, brown-headed: *Molothrus ater*
coyote: *Canis latrans*
deer: family Cervidae
deer, Key: *Odocoileus virginianus clavium*
deer, mule: *Odocoileus hemionus*
deer, white-tailed deer: *Odocoileus virginianus*

dolphin: family Delphinidae
duck: *Anas* spp.
eagle: family Accipitridae
eagle, bald: *Haliaeetus leucocephalus*
elephant: family Elephantidae
elephant, African bush: *Loxodonta africana africana*
elephant, African forest: *Loxodonta cyclotis*
elk: *Cervus elaphus*
falcon, aplomado: *Falco femoralis*
fossa: *Cryptoprocta ferox*
frog, Chiricahua leopard: *Lithobates chiricahuensis*
giraffe: *Giraffa camelopardalis*
goat, mountain: *Oreamnos americanus*
goose, Canada: *Branta canadensis*
goose, snow: *Chen caerulescens*
goshawk, northern: *Accipiter gentilis*
grouse, ruffed: *Bonasa umbellus*
gull: *Larus* spp.
heron, black-crowned night: *Nycticorax nycticorax*
human: *Homo sapiens*
iguana: family Iguanidae
jaguar: *Panthera onca*
kangaroo: *Macropus* spp.
lemur: family Lemuridae
leopard, snow: *Uncia uncia*
lion, mountain: *Puma concolor*
lynx: *Lynx* spp.
lynx, Canadian: *Lynx canadensis*
lynx, Eurasian: *Lynx lynx*
mallard: *Anas platyrhynchos*
manatee, Florida: *Trichechus manatus latirostris*
moose: *Alces alces*

mouflon: *Ovis aries*
oryx: *Oryx* spp.
owl, northern spotted: *Strix occidentalis caurina*
panther, Florida: *Puma concolor coryi*
parrot, maroon-fronted: *Rhynchopsitta terrisi*
parrot, thick-billed: *Rhynchopsitta pachyrhyncha*
pheasant, ring-necked: *Phasianus colchicus*
plover, piping: *Charadrius melodus*
prairie dog: *Cynomys* spp.
puffin: *Fraterucla* spp.
raccoon: *Procyon lotor*
racer, black: *Coluber constrictor* subspp.
rattlesnake, eastern massasauga: *Sistrurus catenatus catenatus*
rattlesnake, timber: *Crotalus horridus horridus*
rhinoceros: family Rhinocerotidae
saiga: *Saiga tatarica*
salmon: family Salmonidae
Saola: *Pseudoryx nghetinhensis*
sea turtles: superfamily Chelonioidea

sheep, bighorn: *Ovis canadensis*
sheep, dall: *Ovis dalli*
sparrow, dusky seaside: *Ammodramus maritimus nigrescens*
squirrel: *Sciurus* spp.
tern, Caspian: *Hydroprogne caspia*
thrasher, Le Conte's: *Toxostoma lecontei*
tick: family Ixodidae
tiger: *Panthera* spp.
tortoise, gopher: *Gopherus polyphemus*
turkey, wild: *Meleagris gallopavo*
walrus: *Odobenus rosmarus*
warbler, golden-cheeked: *Dendroica chrysoparia*
warbler, Kirtland's: *Dendroica kirtlandii*
whale: order Cetacea
wolf: *Canis* spp.
wolf, gray: *Canis lupus*
woodpecker, pileated: *Dryocopus pileatus*
zebra, Burchell's: *Equus burchellii*

GLOSSARY

achievement-motivated recreationists People who engage in wildlife-dependent recreation primarily to meet a self-determined standard of performance, such as killing a trophy or shooting the legal limit of game.

active adaptive management A variation of adaptive management that places emphasis on learning about the system. When practicing active adaptive management, managers focus on reduction of uncertainty through the deliberate comparison of different alternatives (e.g., alternatives may be considered different treatments in an experimental design that can be assessed quantitatively).

Adaptive Impact Management A variation on adaptive resource management, adaptive impact management is an approach that focuses on impacts from human–wildlife interactions as fundamental objectives of management, rather than on change in wildlife populations or habitats.

Adaptive Resource Management (ARM) An approach to wildlife management with the purpose of reducing uncertainty about the reason that things happen in a management system, and developing better predictive capability about the effects of perturbations on biological systems. ARM emphasizes management as a learning opportunity. It encourages wildlife agencies to conduct research and management simultaneously, so that managers can reduce uncertainty, improve learning, and improve management.

affect The term *affect* in attitude and norm research refers to a general class of *feeling states* experienced by humans. Emotions are subsumed under this category.

affiliative-motivated recreationists People who engage in wildlife-dependent recreation primarily for the camaraderie and to strengthen personal relationships with others.

agency The capacity of individuals to act independently and to make their own free choices.

appreciative-motivated recreationists People who engage in wildlife-dependent recreation primarily to relax and escape from everyday concerns or to enjoy nature.

attitude A person's favorable or unfavorable evaluation of a person, object, concept, or action. Attitude is an important component to predicting behavior.

behavioral observation A research method wherein the researcher makes direct but unobtrusive observations of subjects during a sample of time periods or during the course of particular events.

beliefs Thoughts about general classes of objects (e.g., all wildlife) or issues (e.g., climate change) that give meaning to the more global cognitions represented in values.

bequest value The value of knowing that something positive, such as wildlife and their habitats, will be available for the enjoyment of future generations.

biological carrying capacity The ability of a habitat or environment to support a population of animals (often conceptualized as the maximum number of animals of a given population supportable by the resources available).

citizen participation Involvement of wildlife management stakeholders in making, understanding, implementing, or evaluating management decisions for improved wildlife management. It may also be referred to as stakeholder engagement.

clients People who pay for professional or expert services (e.g., hunters, who purchase licenses to participate in hunting, may be considered "clients" of the wildlife agency providing those licenses).

cognition The collection of mental processes used in perceiving, remembering, thinking, and understanding, as well as the act of using these processes.

collateral effects The unintended effects that occur during implementation of management actions.

co-management or co-managerial approach Management by two or more entities, which involves shared control and responsibility for a particular wildlife management situation.

communication planning A management activity that ensures communication is linked to program goals.

community-capacity The extent to which a community can take advantage of a new venture, such as restoration of a wildlife species, or successfully mitigate any negative impacts resulting from the same venture.

consequentialism An ethical theory that asserts that the rightness of an action is determined by the consequences of that action.

constituents A group of people who authorize or support the efforts of others to act on their behalf (e.g., licensed trappers who support the efforts of wildlife professionals to act on their behalf may be considered a constituency for furbearer management).

consumer surplus An economic measure of the value of a good or service to consumers. Consumer surplus is calculated by subtracting the amount consumers are willing to pay for a good (e.g., binoculars) or service (e.g., guided birdwatching tour) from its market price. A consumer surplus occurs when the consumer is willing to pay more for a given product or service than the current market price.

content analysis A research method in which text from sources such as books, newspapers, advertisements, or interview transcripts is analyzed to make inferences about messages, their sources, or their intended audiences.

contingent valuation A set of economic methods developed to obtain estimates of the value of changes in the quality of public goods (including wildlife-related recreation) by asking people about their willingness to pay for those changes. Contingent valuation is used especially for non-market goods, such as the value to individual participants of wildlife-viewing trips.

cultural carrying capacity The idea that humans, as societies, have a limit beyond which they will no longer co-habit with or support wildlife.

culture The learned, socially transmitted customs, knowledge, material objects, behavior, and meanings that are shared by a collection of people and expressed in symbols, rituals, stories–narratives, and values–worldviews. Culture is transmitted through the process of socialization, where people learn the attitudes, values, and behaviors appropriate for members of a particular cultural group.

deontology In contrast with consequentialism, it judges an action's rightness by the intention or motivation for action rather than by the results of an action. Examples of deontological perspectives include treating others as you would want to be treated (e.g., the Principle of Ethical Consistency) or respecting the rights of things that possess rights.

direct benefits The benefits that accrue from a personal interaction with wildlife (e.g., wildlife watching, photography, hunting).

economics A social science that studies how society meets competing demands in the face of limited resources.

emotions The term *affect* in attitude and norm research refers to a general class of *feeling states* experienced by humans; *emotions* are subsumed under this category. Although emotional responses are central to personal experiences with wildlife, little human dimensions research has focused on affective responses to wildlife.

enabling objectives Statements that articulate the suite of conditions required to achieve a fundamental objective. Enabling objectives are the focus of management actions with respect to wildlife, habitats, and people.

ethics The discipline whose focus is formal and rigorous analysis of ethical propositions. While other social sciences are concerned with the analysis of descriptive propositions about human values, ethics is concerned with the analysis of prescriptive propositions about human values. Descriptive propositions describe the nature of the world around us, and prescriptive (ethical) propositions are claims about how we ought to behave, value, or relate to the world around us.

exercised values The benefits that are currently accruing to society. Exercised values include direct and indirect benefits to society.

expert authority approach A top-down management approach in which wildlife managers make decisions and take actions without input from stakeholders or partners.

face-to-face interview A data-collection event in which a trained interviewer completes in-person interviews using a carefully designed interview protocol.

focus group A method in which a trained moderator poses a prepared set of questions or topics to a small, relatively homogeneous group of people. Reactions from the group are usually recorded for later analysis. Multiple groups may be convened as part of a single study.

fundamental objective An objective that characterizes the reason for management in terms of desired impacts.

generalizability The breadth of inferences that can be drawn from the findings of a quantitative study; also called external validity.

goals of management Broad statements of agency or organizational intent that are often based on state and federal policies.

governance The set of structures and mechanisms whereby governments and other organizations direct their activities, which includes the processes, laws, rules, and policies that collectively guide actions.

grassroots citizen participation Citizen-initiated involvement, usually at a community or local level (e.g., in wildlife management processes).

impacts The significant beneficial and detrimental effects resulting from events or interactions involving (a) humans and wildlife (including wildlife habitats), (b) wildlife management interventions, and (c) various stakeholders. Impacts are defined and weighted by human values. Impacts are high priority effects for management attention.

indirect expenditures The expenditures generated when the money from direct expenditures is re-spent by merchants to pay salaries and to purchase supplies and services and when employees use their salaries to make purchases.

informational communication Communication intended to distribute information without specific intent to influence how people behave.

inquisitive approach A management approach that actively seeks information about stakeholders and their interests or positions, including knowledge of benefits desired from management.

institution An institution, such as wildlife management, can be an influential organization, or it can be an established set of laws, customs, or practices governing the behavior of society.

intergovernmental organizations (IGOs) Representative bodies consisting of appointed delegates from sovereign states who each represent the governments of those nations. These organizations are formed to address impacts that transcend a single state's borders.

legitimacy (in democratic governance) Legitimacy denotes the legally derived right of a decision-making body (e.g., wildlife agencies), guided by due process, to impose decisions on individuals who were not directly part of a decision process. Acceptance of this right, and deference to authorities by the stakeholders affected by a decision, is linked to public perceptions that a decision-making authority reached their decisions through fair and representative processes.

media planning The process of selecting appropriate channels for messages intended to reach particular stakeholders.

model A theoretical description of a process, situation, or object, expressed in qualitative narrative terms ("soft-system" model) or in quantitative mathematical terms.

motivations Cognitive forces that drive people to achieve particular goals or outcomes.

natural law theory This theory presupposes that what is natural is moral and that what is unnatural is immoral. Natural law theory is similar to *Divine Command Theory,* which presupposes that what is divinely commanded is moral and that what is divinely forbidden is immoral. Christian ethics of environmental stewardship as a directive from God are an example of Divine Command Theory.

net value The value of the benefit received by an individual or group from a product, service, or experience over and above the cost of obtaining it.

nominal group technique A qualitative method in which a trained facilitator convenes a small group of stakeholders or subject matter experts, elicits ideas in writing on a given topic or question, and has group members prioritize the ideas through a voting process.

non-governmental organizations (NGOs) Organizations where the members do not represent governments, although funding for the organizations can come from governments or IGOs.

norms Standards of behavior that guide what people should do or what most people are doing in given circumstances.

objectives Statements that provide measurable definition of the parts of the agency or organizational goal that are expected to be achieved within a particular time frame.

option value The value of maintaining the option to use, experience, or enjoy something in the future (e.g., the opportunity to view some species of wildlife at a particular place at some future time).

participant observation A field research method in which the researcher is part of the context and records observations of subjects in a particular setting.

passive adaptive management A variation of adaptive management that emphasizes effects of management on resources and stakeholders. Within a passive adaptive management approach, learning is regarded as a useful product of management but not as the primary interest of managers.

passive–receptive approach An approach in which managers are alert to, but do not actively and systematically seek out, the concerns of stakeholders.

performative communication Communication conducted to maintain a particular relationship (e.g., a professional relationship, a personal relationship) or to demonstrate a particular trait (e.g., to demonstrate that the communicator is honest or trustworthy).

persuasive communication Communication intended to influence behavior.

Pittman–Robertson Act A federal act in the United States that provides federal assistance to states for acquiring, restoring, and managing wildlife habitat; restoring species; and doing wildlife-related research; it is funded by an excise tax on firearms and ammunition.

pragmatism It is sometimes viewed as another school of consequentialist thought that claims truth or meaning ought to be judged by practical consequences. A pragmatic ethic is therefore judged by its ability to solve ethical problems as we perceive those problems.

public A group of people who are aware of an issue and have mobilized to address it.

Public Trust Doctrine The principle derived through case law that certain natural resources (e.g., wildlife) are preserved for public use, and that the government is required to maintain those resources for the public's reasonable use. Public trust doctrine applies to wildlife in the United States, where wildlife are considered a common property resource held in trust for the people of current and future generations.

reliability The consistency or repeatability of findings from a quantitative study.

representativeness The degree to which the data collected in a quantitative study represent the population of interest.

satisfactions Types of positive outcomes that people seek from their experiences.

secondary data Data that already exist, such as census data, hunting license data, or data from previous surveys.

situation analysis An assessment or analytical (not just descriptive) scoping process, which can be formal or informal, that managers undertake to gain broad understanding of the context for management. Situation analysis identifies relevant impacts and describes the management environment within which impacts occur. The scope of situation analysis usually depends on scale of the management issue and authority of the entity with jurisdiction. The intensity of situation analysis usually depends on what is at stake—the resource involved, the cost of management, and the political consequences of management intervention.

social feasibility assessment A multi-step process of research undertaken to determine how a community will likely be affected socially (and sometimes economically) by a proposed action such as the restoration of a species, and the community's attitudes toward the proposed restoration.

social movement A social movement is a collective, organized attempt to change the social order that acts with some continuity, or as individuals coming together as a group to foster change.

social psychology The scientific study of the way in which people's thoughts, feelings, and behaviors are influenced by their environment.

social structure The enduring, orderly patterns of relationships between elements of society (e.g., roles, institutions, or expectations).

social support A system in which people interested in an activity, such as hunting, are linked with other participants who can encourage and reinforce that interest.

sociology A social science concerned with understanding what people do as members of a group or when interacting with one another. It stresses the social contexts in which people live and how these contexts influence people's lives, and emphasizes how people are influenced by society or social structure and how they, in turn, reshape their society.

sovereign state A geographical area independent from other states and powers, which has a permanent population, a government, and a defined territory on which it exercises internal and external sovereignty.

stakeholder A person or group who significantly affects or is significantly affected by wildlife management.

subsequent effects Unintended effects produced as a consequence of achieving management objectives.

subsidiarity A governing principle that suggests management ought to be handled by the smallest, lowest, or least centralized competent authority.

subsistence use A resource-dependent lifestyle characterized by a pattern of social and economic activity that includes some combination of hunting, fishing, gathering, herding, cultivating, trading, bartering, tool-making, crafting, fuel production, trapping, resource sharing, and household- or village-level consumption; the culture and way of life derived from the land and its resources.

systems A group of interacting or interrelated components that form a unified whole. For example, ecosystems are a system, but so are organisms, organizations, and processes such as wildlife management.

systems thinking The process of developing an understanding of how key factors operating in the situation under study relate to each other. Systems thinking involves the use of various techniques to study systems of many kinds. It involves thinking in a holistic way, rather than using purely reductionist approaches. It aims to gain insights into the whole by understanding the linkages, interactions, and processes between and among the elements that comprise the whole "system."

theory of moral sentiments This theory stresses that reason and emotion are both critical for judging the rightness of an action. For example, in some cases reason is necessary for indicating circumstances where moral attention is required, and emotional sentiments (such as compassion) motivate one to manifest moral attention. Darwin's view on ethics (Chapter 3 of *Descent of Man*) and Leopold's Land Ethic are both related to the Theory of Moral Sentiments (developed philosophically by David Hume and Adam Smith).

transactional approach A management approach in which stakeholders engage each other directly through interactive processes to articulate their values and stakes.

transferability A concept of external validity applied to qualitative studies, which refers to the degree to which salient conditions found in the situation studied match those in another situation of interest.

travel cost method An indirect method of estimating benefits associated with a specific recreation activity, based on the distance traveled to reach the site.

utilitarianism An important form of consequentialism (a form of which dominated American conservation in the twentieth century), which presumes that we ought to act in ways that produce the most utility, happiness, or pleasure for the most people.

validity The accuracy of quantitative data.

value orientations The patterns of direction and intensity characterizing one's basic beliefs.

values Desirable end states, modes of conduct, or qualities of life that humans individually or collectively hold dear. Values are general mental constructs that define what is important to people.

virtue theory A theory that right actions arise from people who are manifestly virtuous, and that moral education ought to focus on identifying precisely which virtues ought to be manifest (e.g., generosity, respect, humility, courage) and how to cultivate such virtues in a person.

wicked problems A term coined to describe problems (especially problems in natural resource management) characterized by scientific uncertainty about cause–effect relationships and social conflicts over goals and management alternatives. Wicked problems may be understood differently by different individuals, depending on the values they associate with the resources.

wildlife management The guidance of decision-making processes and implementation of practices to influence interactions among and between people, wildlife, and habitats to achieve impacts valued by stakeholders.

BIBLIOGRAPHY

Ackoff, R. L. 2001. OR: after the post mortem. Systems Dynamics Review 17:341–346.

Aiken, R. 2009. Net economic values of wildlife-associated recreation: addendum to the 2006 National Survey of Fishing, Hunting and Wildlife-Associated Recreation. U.S. Fish and Wildlife Service Report 2006-5, Washington, D.C., USA.

Ajzen, I., and M. Fishbein. 1980. Understanding attitudes and predicting social behavior. Prentice Hall, Englewood Cliffs, New Jersey, USA.

Aldrich, H. E. 1999. Organizations evolving. Sage, Thousand Oaks, California, USA.

Aldrich, H. E., and M. Ruef. 2006. Organizations evolving. Second edition. Sage, Thousand Oaks, California, USA.

Alkin, M. C. 1990. Debates on evaluation. Sage, London, England, U.K.

Allen, G. M., and E. M. Gould, Jr. 1986. Complexity, wickedness and public forests. Journal of Forestry 84:20–24.

Amante-Helweg, V. 1996. Ecotourists' beliefs and knowledge about dolphins and the development of cetacean ecotourism. Aquatic Mammals 22:131–140.

Anderson, L. E., and D. K. Loomis. 2006. Balancing stakeholders with an imbalanced budget: how continued inequities in wildlife funding maintains old management styles. Human Dimensions of Wildlife 11:455–458.

Anderson, M. W., S. D. Reiling, and G. K. Criner. 1985. Consumer demand theory and wildlife agency revenue structure. Wildlife Society Bulletin 13:375–384.

Applegate, J. E. 1984. Attitudes toward deer hunting in New Jersey: 1972–1982. Wildlife Society Bulletin 12:19–22.

Ariely, D. 2008. Predictably irrational: the hidden forces that shape our decisions. HarperCollins, New York, New York, USA.

Astley, W. G., and A. H. Van de Ven. 1983. Central perspectives and debates in organization theory. Administrative Science Quarterly 28: 245–273.

Austin, J. L. 1962. How to do things with words. Harvard University Press, Cambridge, Massachusetts, USA.

Babbie, E. 2007. The practice of social research. Eleventh edition. Thomson/Wadsworth, Belmont, California, USA.

Bäckstrand, K., and A. K. Jamil Khan. 2010. The promise of new modes of environmental governance. Pages 3–27 in K. Bäckstrand and A. K. Jamil Khan, editors. Environmental politics and deliberative democracy: examining the promise of new modes of governance. Edward Elgar, Northampton, Massachusetts, USA.

Baker, M., and J. Kusel. 2003. Community forestry in the United States. Island Press, Washington, D.C., USA.

Baker, S. V., and J. A. Fritsch. 1997. New territory for deer management: human conflicts on the suburban frontier. Wildlife Society Bulletin 25:404–407.

Ballard, H. L., M. E. Fernandez-Gimenez, and V. E. Sturtevant. 2008. Integration of local ecological knowledge and conventional science: a study of seven community-based forestry organizations in the USA. Ecology and Society 13:37, at www.ecologyandsociety.org/vol13/iss2/art37. Accessed 21 January 2011.

Bangs, E. E., and S. H. Fritts. 1996. Reintroducing the gray wolf to central Idaho and Yellowstone National Park. Wildlife Society Bulletin 24:402–413.

Barker, J. H. 1993. Always getting ready: upterrlainarluta: Yu'ik Eskimo subsistence in southwest Alaska. University of Washington Press, Seattle, USA.

Barrett, L. F., B. Mesquita, K. N. Ochsner, and J. J. Gross. 2007. The experience of emotion. Annual Review of Psychology 58:373–403.

Batcheller, G. R., M. C. Bambery, L. Bies, T. Decker, S. Dyke, D. Guynn, M. McEnroe, M. O'Brien, J. F. Organ, S. J. Riley, and G. Roehm. 2010. The public trust doctrine: implications for wildlife management and conservation in the United States and Canada. The Wildlife Society Technical Review 10-1, Bethesda, Maryland, USA.

Bath, A. J. 1991. Public attitudes in Wyoming, Montana, and Idaho toward wolf restoration in Yellowstone National Park. Transactions of the North American Wildlife and Natural Resources Conference 56:91–95.

Bath, A. J., and T. Buchanan. 1989. Attitudes of interest groups in Wyoming toward wolf restoration in Yellowstone National park. Wildlife Society Bulletin 17:519–525.

Bean, M. J., and M. J. Rowland. 1997. The evolution of national wildlife law. Third edition. Praeger, Westport, Connecticut, USA.

Beard, J. G., and M. G. Ragheb. 1983. Measuring leisure motivation. Journal of Leisure Research 15:219–228.

Bentzen, W. R. 1997. Seeing young children: a guide to observing and recording behavior. Third edition. Delmar, Albany, New York, USA.

Berryman, J. H. 1981. Needed now: an action program to maintain and manage wildlife habitat on private lands. Pages 6–10 in R. T. Dumke, G. V. Burger, and J. R. March, editors. Wildlife management on private lands. The Wildlife Society, Wisconsin Chapter, Madison, USA.

Besley, J. C., and J. Shanahan. 2004. Skepticism about media effects concerning the environment: examining Lomborg's hypotheses. Society and Natural Resources 17:861–880.

Bishop, R. C. 1987. Economic values defined. Pages 24–33 in D. J. Decker and G. R. Goff, editors. Valuing wildlife: economic and social perspectives. Westview, Boulder, Colorado, USA.

Blader, S. L., and T. R. Tyler. 2003a. A four-component model of procedural justice: defining the meaning of a "fair" process. Personality and Social Psychology Bulletin 29:747–758.

Blader, S. L., and T. R. Tyler. 2003b. The group engagement model: procedural justice, social identity, and cooperative behavior. Personality and Social Psychology Review 7:349–361.

Blumer, H. 1969. Symbolic interactionism: perspective and method. Prentice Hall, Englewood Cliffs, New Jersey, USA.

Bogardus, E. S. 1926. The group interview. Journal of Applied Sociology 10:372–382.

Bolen, E. G., and W. L. Robinson. 2003. Wildlife ecology and management. Fifth edition. Prentice Hall, Upper Saddle River, New Jersey, USA.

Borrini-Feyerabend, G., M. T. Farvar, J. C. Nguinguiri, and V. A. Ndangang. 2000. Co-management of natural resources: organizing, negotiating and learning by doing. International Union for Conservation of Nature and Natural Resources, Yaoundé, Cameroon.

Bosnjak, M., and T. L. Tuten. 2003. Prepaid and promised incentives in web surveys: an experiment. Social Science Computer Review 21:208–217.

Bowker, J. M., D. B. K. English, and H. K. Cordell. 1999. Projections of outdoor recreation participation to 2050. Pages 323–350 in H. K. Cordell, editor. Outdoor recreation in American life. Sagamore, Champaign, Illinois, USA.

Boyce, M. 1998. Income, inequality, and pollution: a reassessment of the environmental Kuznets Curve. Ecological Economics 25:147–160.

Bradshaw, B. 2003. Questioning the credibility and capacity of community-based resource management. The Canadian Geographer 47:137–150.

Bradshaw, C. J. A., and B. W. Brook. 2007. Ecological–economic models of sustainable harvest for an endangered but exotic megaherbivore in northern Australia. Natural Resource Modeling 20:129–155.

Brechin, S., P. Wilshusen, C. Fortwangler, and P. West, editors. 2003. Contested nature: promoting international biodiversity with social justice in the twenty-first century. State University Press of New York, Albany, USA.

Bright, A. D., J. F. Lipscomb, and L. Sikorowski. 1997. A cognitive analysis of intergroup attitudes of state wildlife agency personnel and wildlife rehabilitators. Human Dimensions of Wildlife 2:47–67.

Brohn, A. J. 1978. Missouri's "design for conservation." Proceedings of the International Association of Fish and Wildlife Agencies 67:64–67.

Bromley, P. 1997. Nature conservation in Europe. E & FN Spon, London, England, U.K.

Brook, A., M. Zint, and R. de Young. 2003. Landowners' responses to an Endangered Species Act listing and implications for encouraging conservation. Conservation Biology 17:1638–1649.

Brossard, D., B. Lewenstein, and R. Bonney. 2005. Scientific knowledge and attitude change: the impact of a citizen science project. International Journal of Science Education 27:1099–1121.

Broussard Allred, S., and P. J. Smallidge. 2009. Distance learning in natural resources: final report and data files. Cornell University Survey Research Institute, Ithaca, New York, USA.

Brown, T. L., and N. A. Connelly. 1994. Predicting demand for big game and small game hunting licenses: the New York experience. Wildlife Society Bulletin 22:172–178.

Brown, T. L., N. A. Connelly, and D. J. Decker. 1986. First-year results of New York's "Return a Gift to Wildlife" tax checkoff. Wildlife Society Bulletin 14:115–120.

Brown, T. L., C. P. Dawson, and R. L. Miller. 1979. Interests and attitudes of metropolitan New York residents about wildlife. Transactions of the North American Wildlife and Natural Resources Conference 44:289–297.

Brown, T. L., and D. J. Decker. 1982. Identifying and relating organized publics to wildlife management issues: a planning study. Transactions of the North American Wildlife and Natural Resources Conference 47:686–692.

Brown, T. L., D. J. Decker, and C. P. Dawson. 1977. Farmer willingness to tolerate deer damage in the Erie–Ontario Lake Plain. Cornell University, Department of Natural Resources Research and Extension Series Report 8, Ithaca, New York, USA.

Brown, T. L., D. J. Decker, and C. P. Dawson. 1978. Willingness of New York farmers to incur white-tailed deer damage. Wildlife Society Bulletin 6:235–239.

Brown, T. L., D. J. Decker, and D. L. Hustin. 1981. 1978 hunter training course participant study. Cornell University, Department of Natural Resources, Outdoor Recreation Research Unit, Ithaca, New York, USA.

Brown, T. L., D. J. Decker, S. J. Riley, J. W. Enck, T. B. Lauber, P. D. Curtis, and G. F. Mattfeld. 2000a. The future of hunting as a mechanism to control white-tailed deer populations. Wildlife Society Bulletin 28:797–807.

Brown, T. L., D. J. Decker, W. F. Siemer, and J. W. Enck. 2000b. Trends in hunting participation and implications for management of game species. Pages 145–154 in W. Gartner and D. W. Lime, editors. Trends in outdoor recreation, leisure and tourism. CABI, New York, New York, USA.

Brown, T. L., and D. Q. Thompson. 1976. Changes in posting and landowner attitudes in New York State. New York Fish and Game Journal 23:101–137.

Brown Weiss, E. 2000. The five international treaties: a living history. Pages 89–172 in E. Brown Weiss and H. K. Jacobson, editors. Engaging countries: strengthening compliance with international environmental accords. MIT Press, Cambridge, Massachusetts, USA.

Brunson, M. W., and J. J. Kennedy. 1995. Redefining "multiple use": agency responses to changing social values. Pages 143–158 in R. L. Knight and S. F. Bates, editors. A new century for natural resource management. Island Press, Washington, D.C., USA.

Bryan, H. 1977. Leisure value systems and recreational specialization: the case of trout fishermen. Journal of Leisure Research 9:174–187.

Burger, G. V., and J. G. Teer. 1981. Economic and socioeconomic issues influencing wildlife management on private lands. Pages 252–278 in R. T. Dumke, G. V. Burger, and J. R. March, editors. Wildlife management on private lands. The Wildlife Society, Wisconsin Chapter, Madison, USA.

Burgman, M. A. 2005. Risks and decisions for conservation and environmental management. Cambridge University Press, Cambridge, England, U.K.

Butler, J. S., J. Shanahan, and D. J. Decker. 2003. Public attitudes toward wildlife are changing: a trend analysis of New York residents. Wildlife Society Bulletin 31:1027–1036.

Cacioppo, J. T., and W. L. Gardner. 1999. Emotion. Annual Review of Psychology 50:191–214.

Callicott, J. B. 1982. Hume's is/ought dichotomy and the relation of ecology to Leopold's land ethic. Environmental Ethics 4:173–174.

Campa, H., III, S. J. Riley, S. R. Winterstein, T. L. Hiller, S. A. Lischka, and J. P. Burroughs. 2011. Changing landscapes for white-tailed deer in the 21st century: parcelization of landownership and evolving stakeholder values. Wildlife Society Bulletin 35:168–176.

Carpenter, L. H., D. J. Decker, and J. F. Lipscomb. 2000. Stakeholder acceptance capacity in wildlife management. Human Dimensions of Wildlife 5:5–19.

Carrus, G., F. Cini, M. Bonaiuto, and A. Mauro. 2009. Local mass media communication and environmental disputes: an analysis of press communication on the designation of the Tuscan Archipelago National Park in Italy. Society and Natural Resources 22:607–624.

Charnovitz, S. 2006. Accountability of non-governmental organizations in global governance. Pages 21–42 in L. Jordan and P. van Tuijl, editors. NGO accountability: politics, principles, and innovations. Earthscan, Sterling, Virginia, USA.

Chase, L. C., and D. J. Decker. 1998. Citizen attitudes toward elk and participation in elk management: a case study in Evergreen, Colorado. Human Dimensions of Wildlife 3:55–56.

Chase, L. C., T. M. Schusler, and D. J. Decker. 2000. Innovations in stakeholder involvement: what's the next step? Wildlife Society Bulletin 28:208–217.

Chase, L. C., W. F. Siemer, and D. J. Decker. 1999a. Deer management in the village of Cayuga Heights, New York: preliminary situation analysis from a survey of residents. Cornell University, Department of Natural Resources, Human Dimensions Research Unit Publication 99-1, Ithaca, New York, USA.

Chase, L. C., W. F. Siemer, and D. J. Decker. 1999b. Suburban deer management: a case study in the village of Cayuga Heights, New York. Human Dimensions of Wildlife 4:59–60.

Chase, L. C., W. F. Siemer, and D. J. Decker. 2002. Designing stakeholder involvement strategies to resolve wildlife management controversies. Wildlife Society Bulletin 30:937–950.

Christoffel, R. A. 2007. Using human dimensions insights to improve conservation efforts for the eastern massasauga rattlesnake (Sistrurus catenatus catenatus) in Michigan and timber rattlesnake (Crotalus horridus horridus) in Minnesota. Dissertation, Michigan State University, East Lansing, USA.

Christoffel, R. A., D. Hyde, and Y. Lee. 2010. Michigan's eastern massasauga rattlesnake outreach initiative: rattlin' an image. Reptiles and Amphibians 17:130–135.

Cialdini, R. B., C. A. Kallgren, and R. R. Reno. 1991. A focus theory of normative conduct: a theoretical refinement and reevaluation of the role of norms in human behavior. Advances in Experimental Social Psychology 24:201–234.

Clemen, R. T., and T. Reilly. 2004. Making hard decisions with decision tools update. Duxbury, Belmont, California, USA.

Cole, J. S., and D. Scott. 1999. Segmenting participation in wildlife watching: a comparison of casual wildlife watchers and serious birders. Human Dimensions of Wildlife 4:44–61.

Commission on Global Governance. 1995. Our global neighborhood: the report of the Commission on Global Governance. Oxford University Press, New York, New York, USA.

Condon, R. G., P. Collings, and G. Wenzel. 1995. The best part of life: subsistence hunting, ethnicity, and economic adaptation among young adult Inuit males. Arctic 48:3–46.

Connelly, N. A., T. L. Brown, and G. F. Mattfeld. 1995. Interests of urban residents in wildlife and feasibility of wildlife education programs in two urban settings. Cornell University, Department of Natural Resources, Human Dimensions Research Unit Publication 95-7, Ithaca, New York, USA.

Connelly, N. A., D. J. Decker, and S. Wear. 1987. White-tailed deer in Westchester County, New York: public perceptions and preferences. Cornell University, Department of Natural Resources, Human Dimensions Research Unit Publication 87-5, Ithaca, New York, USA.

Conover, M. R. 1997a. Wildlife management by metropolitan residents in

the United States: practices, perceptions, costs, and values. Wildlife Society Bulletin 25:306–311.

Conover, M. R. 1997b. Monetary and intangible valuation of deer in the United States. Wildlife Society Bulletin 25:298–305.

Conover, M. R. 1998. Perceptions of American agricultural producers about wildlife on their farms and ranches. Wildlife Society Bulletin 26:597–604.

Conover, M. R. 2002. Resolving human–wildlife conflicts: the science of wildlife damage management. Lewis Brothers, Boca Raton, Florida, USA.

Conrad, L., and J. Balison. 1994. Bison goring injuries: penetrating and blunt trauma. Journal of Wilderness Medicine 5:371–381.

Conroy, M. J., and J. P. Carroll. 2009. Quantitative conservation of vertebrates. Blackwell, Hoboken, New Jersey, USA.

Cook, M., and S. Mineka. 1989. Observational conditioning of fear to fear-relevant versus fear-irrelevant stimuli in rhesus monkeys. Journal of Abnormal Psychology 98:448.

Copi, I. M., and C. Cohen. 2005. Introduction to logic. Twelfth edition. Pearson / Prentice Hall, Englewood, New Jersey, USA.

Cordell, H. K. 2008. The latest trends in nature-based outdoor recreation. Forest History Today (Spring 2008):4–10.

Cordell, H. K., C. J. Betz, and G. T. Green. 2008. Nature-based outdoor recreation trends and wilderness. International Journal of Wilderness 14:7–13.

Cornelius, R. 1996. The science of emotion: research and tradition in the psychology of emotion. Prentice Hall, Upper Saddle River, New Jersey, USA.

Coser, L. A. 1956. The functions of social conflict. The Free Press, Glencoe, Illinois, USA.

Cowling, R. M., A. T. Knight, S. D. J. Privett, and G. Sharma. 2009. Invest in opportunity, not inventory of hotspots. Conservation Biology 24:633–635.

Cox, K. R., and A. Mair. 1988. Locality and community in the politics of local economic development. Annals of the Association of American Geographers 78:307–325.

Creswell, J. W. 2009. Research design: qualitative, quantitative, and mixed methods approaches. Sage, Thousand Oaks, California, USA.

Curtis, P. D., and J. R. Hauber. 1997. Public involvement in deer management decisions: consensus versus consent. Wildlife Society Bulletin 25:399–403.

Curtis, P. D., R. J. Stout, B. A. Knuth, L. A. Myers, and T. M. Rockwell. 1993. Selecting deer management options in a suburban environment: a case study from Rochester, New York. Transactions of the North American Wildlife and Natural Resources Conference 58:102–116.

Cutlip, S. M., A. H. Center, and G. M. Broom. 1985. Effective public relations. Prentice Hall, Englewood Cliffs, New Jersey, USA.

Czech, B., P. K. Devers, and P. R. Krausman. 2001. The relationship of gender to conservation attitudes. Wildlife Society Bulletin 29:187–194.

Darwin, C. 1871. The descent of man and selection in relation to sex. 1981, Reprint. Princeton University Press, Princeton, New Jersey, USA.

Davenport, M. A., W. T. Borrie, W. A. Freimund, and R. E. Manning. 2002. Assessing the relationship between desired experiences and support for management actions at Yellowstone National Park using multiple methods. Journal of Park and Recreation Administration 20:51–64.

De Boer, W. F., and D. S. Baquete. 1998. Natural resource use, crop damage and attitudes of rural people in the vicinity of the Maputo Elephant Reserve, Mozambique. Environmental Conservation 25:208–218.

Decker, D. J. 1985. Agency image: a key to successful natural resource management. Northeast Section of The Wildlife Society 41:43–56.

Decker, D. J., and T. L. Brown. 1982. Fruit growers' vs. other farmers' attitudes toward deer in New York. Wildlife Society Bulletin 10:150–155.

Decker, D. J., T. L. Brown, B. L. Driver, and P. J. Brown. 1987. Theoretical developments in assessing social values of wildlife: toward a comprehensive understanding of wildlife recreation involvement. Pages 76–95 in D. J. Decker and G. R. Goff, editors. Valuing wildlife: economic and social perspectives. Westview Press, Boulder, Colorado, USA.

Decker, D. J., T. L. Brown, and R. J. Gutierrez. 1980. Further insights into the multiple-satisfactions approach for hunter management. Wildlife Society Bulletin 8:323–331.

Decker, D. J., T. L. Brown, D. L. Hustin, S. H. Clarke, and J. O'Pezio. 1981. Public attitudes toward black bears in the Catskills. New York Fish and Game Journal 28:1–20.

Decker, D. J., T. L. Brown, and G. F. Mattfeld. 1985a. Deer population management in New York: using public input to meet public needs. Pages 185–196 in S. L. Beasom and S. F. Roberson, editors. Game harvest management. Caesar Kleberg Wildlife Research Institute, Kingsville, Texas, USA.

Decker, D. J., T. L. Brown, and W. F. Siemer. 2001. Wildlife management as a process. Pages 77–90 in D. J. Decker, T. L. Brown, and W. F. Siemer, editors. Human dimensions of wildlife management in North America. The Wildlife Society, Bethesda, Maryland, USA.

Decker, D. J., T. L. Brown, J. J. Vaske, and M. J. Manfredo. 2004a. Human dimensions of wildlife management. Pages 187–198 in M. J. Manfredo, J. J. Vaske, B. L. Bruyere, D. R. Field, and P. Brown, editors. Society and natural resources: a summary of knowledge. Modern Litho, Jefferson City, Missouri, USA.

Decker, D. J., and L. C. Chase. 1997. Human dimensions of living with wildlife—a management challenge for the 21st century. Wildlife Society Bulletin 25:788–795.

Decker, D. J., and N. A. Connelly. 1989. Motivations for deer hunting: implications for antlerless deer harvest as a management tool. Wildlife Society Bulletin 17:455–463.

Decker, D. J., and T. Gavin. 1987. Public attitudes toward a suburban deer herd. Wildlife Society Bulletin 15:173–180.

Decker, D. J., and G. R. Goff. 1987. Valuing wildlife: economic and social perspectives. Westview Press, Boulder, Colorado, USA.

Decker, D. J., C. A. Jacobson, and T. L. Brown. 2006a. Situation-specific "impact dependency" as a determinant of management acceptability: insights from wolf and grizzly bear management in Alaska. Wildlife Society Bulletin 34:426–432.

Decker, D. J., C. C. Krueger, R. A. Baer, Jr., B. A. Knuth, and M. E. Richmond. 1996. From clients to stakeholders: a philosophical shift for fish and wildlife management. Human Dimensions of Wildlife 1:70–82.

Decker, D. J., T. B. Lauber, and W. F. Siemer. 2004b. Community-based deer management: a practitioner's guide. Northeast Wildlife Damage Management Research and Outreach Cooperative, Ithaca, New York, USA.

Decker, D. J., and G. F. Mattfeld. 1995. Human dimensions of wildlife management in Colorado: a strategy for developing an agency–university partnership. Journal of Park Recreation Administration 13:25–36.

Decker, D. J., and J. O'Pezio. 1989. Consideration of bear–people conflicts in black bear management for the Catskill region of New York: application of a comprehensive management model. Pages 181–187 in M. Bromley, editor. Bear–people conflicts: proceedings of a symposium on management strategies. Northwest Territories Department of Renewable Resources, Yellowknife, Canada.

Decker, D. J., J. F. Organ, and C. A. Jacobson. 2009a. Why should all Americans care about the North American model of wildlife conservation? Transactions of the 74th North American Wildlife and Natural Resources Conference 74:32–36.

Decker, D. J., R. W. Provencher, and T. L. Brown. 1984. Antecedents to hunting participation: an exploratory study of the social–psycho-logical determinants of initiation, continuation, and desertion in hunting. Cornell University, Department of Natural Resources, Outdoor Recreation Research Unit, Ithaca, New York, USA.

Decker, D. J., and K. G. Purdy. 1988. Toward a concept of wildlife acceptance capacity in wildlife management. Wildlife Society Bulletin 16:53–57.

Decker, D. J., S. J. Riley, J. F. Organ, W. F. Siemer, and L. H. Carpenter. 2011. Applying impact management: a practitioner's guide. Cornell University, Department of Natural Resources, Human Dimensions Research Unit and Cornell Cooperative Extension, Ithaca, New York, USA.

Decker, D. J., S. J. Riley, and W. F. Siemer. In press. Human dimensions of wildlife management. In P. R. Krausman and B. D. Leopold, editors. Essential readings in wildlife management and conservation. Johns Hopkins University Press, Baltimore, Maryland, USA.

Decker, D. J., R. E. Shanks, L. A. Nielsen, and G. R. Parsons. 1991. Ethical and scientific judgments in management: beware of blurred distinctions. Wildlife Society Bulletin 19:523–527.

Decker, D. J., W. F. Siemer, K. L. Leong, S. J. Riley, B. A. Rudolph, and L. H. Carpenter. 2009b. Conclusion: what is wildlife management? Pages 315–327 in M. J. Manfredo, J. J. Vaske, P. J. Brown, D. J. Decker, and E. A. Duke, editors. Wildlife and society: the science of human dimensions. Island Press, Washington, D.C., USA.

Decker, D. J., R. A. Smolka, Jr., J. O'Pezio, and T. L. Brown. 1985b. Social determinants of black bear management for the northern Catskill Mountains. Pages 239–247 in S. L. Beasom and S. F. Roberson, editors. Game harvest management. Caesar Kleberg Wildlife Institute, Kingsville, Texas, USA.

Decker, D. J., M. Wild, S. J. Riley, W. F. Siemer, M. Miller, K. Leong, J. Powers, and J. Rhyan. 2006b. Wildlife disease management: a manager's model. Human Dimensions of Wildlife 11:151–158.

de Groot, R. S., M. A. Wilson, and R. M. J. Boumans. 2002. A typology for the classification, description, and valuation of ecosystem functions, goods and services. Ecological Economics 41:393–408.

Dey, I. 1993. Qualitative data analysis: a user-friendly guide for social scientists. Routledge, New York, New York, USA.

Dick, R. E. 1996. Subsistence economics: freedom from the marketplace. Society & Natural Resources 9:19–29.

Dietz, T., E. Ostrum, and P. C. Stern. 2003. The struggle to govern the commons. Science 302:1907–1912.

Digman, J. M. 1990. Personality structure: emergence of the five-factor model. Annual Review of Psychology 41:417–440.

Dillman, D. A., J. D. Smyth, and L. M. Christian. 2009. Internet, mail, and mixed-mode surveys: the tailored design method. John Wiley & Sons, Hoboken, New Jersey, USA.

Dimopoulos, D., S. Paraskevopoulos, and J. D. Pantis. 2008. The cognitive and attitudinal effects of a conservation education module on elementary school students. Journal of Environmental Education 39:47–62.

Dizard, J. E. 1999. Going wild: hunting, animal rights, and the contested meaning of nature. University of Massachusetts Press, Amherst, USA.

Doig, H. E. 1986. The importance of private lands to recreation. Pages 7–10 in Anonymous, editor. Recreation on private lands: issues and opportunities. President's Commission on Americans Outdoors, Washington, D.C., USA.

Dorn, M. L., and A. G. Mertig. 2005. Bovine tuberculosis in Michigan: stakeholder attitudes and implications for eradication efforts. Wildlife Society Bulletin 33:539–552.

Downs, A. 1972. Up and down with ecology: the issue attention cycle. The Public Interest 28:38–50.

Driver, B. L., H. E. Tinsley, and M. J. Manfredo. 1991. Leisure and recreation experience preference scales: results from two inventories designed to assess the breadth of the perceived benefits of leisure.

Pages 263–287 *in* B. L. Driver, P. J. Brown, and G. L. Peterson, editors. The benefits of leisure. Venture, State College, Pennsylvania, USA.

Duda, M. D., S. J. Bissle, and K. C. Young. 1998. Wildlife and the American mind. Public opinion on and attitudes toward fish and wildlife management. Responsive Management, Harrisonburg, Virginia, USA.

Duda, M. D., and M. Jones. 2009. Public opinion on and attitudes toward hunting. Pages 180–198 *in* Transactions of the Seventy-third North American Wildlife and Natural Resources Conference. Wildlife Management Institute, 23–28 March 2008, Phoenix, Arizona, USA.

Durkheim, E. 1951. Suicide: a study in sociology. Free Press, New York, New York, USA.

Dwyer, J. C., and I. D. Hodge. 1996. Countryside in trust: land management by conservation, recreation, and amenity organizations. John Wiley & Sons, Chichester, West Sussex, England, U.K.

Eliason, S. L. 1999. The illegal taking of wildlife: toward a theoretical understanding of poaching. Human Dimensions of Wildlife 4:27–39.

Ellingwood, M. R., and J. V. Spignesi. 1986. Management of an urban deer herd and the concept of cultural carrying capacity. Transactions of the Northeast Deer Technical Committee 22:42–45.

Ellsworth, P. C., and K. R. Scherer. 2003. Appraisal processes in emotion. Pages 572–595 *in* R. J. Davidson, K. R. Scherer, and H. H. Goldsmith, editors. Handbook of affective sciences. Oxford University Press, Oxford, England, U.K.

Emery, M., and R. E. Purser. 1996. The search conference: a powerful method for planning organizational change and community action. Jossey-Bass, San Francisco, California, USA.

Enck, J. W. 1996. Deer–hunter identity spectrum: a human dimensions perspective for evaluating hunting policy. Dissertation, Cornell University, Ithaca, New York, USA.

Enck, J. W., and T. L. Brown. 2000. Preliminary assessment of social feasibility for reintroducing gray wolves to the Adirondack Park in Northern New York. Cornell University, Department of Natural Resources, Human Dimensions Research Unit Publication Number 00-3, Ithaca, New York, USA.

Enck, J. W., and T. L. Brown. 2002. New Yorkers' attitudes toward restoring wolves to the Adirondack Park. Wildlife Society Bulletin 30:16–28.

Enck, J. W., and D. J. Decker. 1991. Hunters' perspectives on satisfying and dissatisfying aspects of the deer-hunting experience in New York. Cornell University, Department of Natural Resources, Human Dimensions Research Unit Publication 91-4, Ithaca, New York, USA.

Enck, J. W., and D. J. Decker. 1994. Differing perceptions of deer hunting satisfaction among participants and "experts." Transactions of the Northeast Section of The Wildlife Society 51:35–46.

Enck, J. W., and D. J. Decker. 1997. Examining assumptions in wildlife management: a contribution of human dimensions inquiry. Human Dimensions of Wildlife 2:56–72.

Enck, J. W., D. J. Decker, and T. L. Brown. 2000. Status of hunter recruitment and retention in the United States. Wildlife Society Bulletin 28:817–824.

Enck, J. W., D. J. Decker, S. J. Riley, J. F. Organ, L. H. Carpenter, and W. F. Siemer. 2006. Integrating ecological and human dimensions in adaptive management of wildlife-related impacts. Wildlife Society Bulletin 34:698–705.

Enck, J. W., B. L. Swift, and D. J. Decker. 1993. Reasons for decline in waterfowl hunting: insights from New York. Wildlife Society Bulletin 21:10–21.

Ericsson, G., and T. Heberlein. 2003. Attitudes of hunters, locals, and the general public in Sweden now that the wolves are back. Biological Conservation 111:149–159.

Eschenfelder, K. R. 2006. What information should state wildlife agencies provide on their CWD websites? Human Dimensions of Wildlife 11:221–223.

Eubanks, T. L., Jr., J. R. Stoll, and R. B. Ditton. 2004. Understanding the diversity of eight birder subpopulations: socio-demographic characteristics, motivations, expenditures and net benefits. Journal of Ecotourism 3:151–172.

Failing, L., G. Horn, and P. Higgins. 2004. Using expert judgment and stakeholder values to evaluate adaptive management options. Ecology and Society 9:13, at www.ecologyandsociety.org/vol9/iss1/art13. Accessed 29 February 2012.

Fairbank, Maslin, Maullin & Associates. 2006. Sustainable funding for conservation in Iowa: results of a statewide public opinion survey. Iowa Department of Natural Resources, at www.iowadnr.gov/sustainablefunding. Accessed 22 November 2010.

Falk, J. H. 2005. Free-choice environmental learning: framing the discussion. Environmental Education Research 11:265–280.

Fenichel, E. P., and R. D. Horan. 2007. Jointly-determined ecological thresholds and economic trade-offs in wildlife disease management. Natural Resource Modeling 20:511–547.

Fishbein, M., and I. Ajzen. 1975. Belief, attitude, intention, and behavior: an introduction to theory and research. Addison-Wesley, Reading, Massachusetts, USA.

Fitchen, J. M. 1991. Endangered spaces, enduring places. Westview Press, Boulder, Colorado, USA.

Florida Fish and Wildlife Conservation Commission [FWC]. 2007. Florida manatee management plan (*Trichechus manatus latirostris*). December 2007. Florida Fish and Wildlife Conservation Commission, Tallahassee, USA.

Forsyth, C. J., R. Gramling, and G. Wooddell. 1998. The game of poaching: folk crimes in southwest Louisiana. Society and Natural Resources 11:25–38.

Frawley, B. J. 2000. 1999 Michigan deer hunter survey: deer baiting. Michigan Department of Natural Resources Wildlife Division Report 3315, Lansing, USA.

Freeman, M. M. R., and S. R. Kellert. 1992. Public attitudes to whales—results of a six-country survey. Canadian Circumpolar Institute, University of Alberta, Edmonton, Canada.

Friedman, T. 2005. The world is flat: a brief history of the twenty-first century. Farrar, Straus and Giroux, New York, New York, USA.

Frijda, N. H. 1986. The emotions. Cambridge University Press, Cambridge, England, U.K.

Fuller, T. K., W. E. Berg, G. L. Radde, M. S. Lenarz, and G. B. Joselyn. 1992. A history and current estimate of wolf distribution and numbers in Minnesota. Wildlife Society Bulletin 20:42–55.

Gabrey, S. W., P. A. Vohs, and D. H. Jackson. 1993. Perceived and real crop damage by wild turkeys in northeastern Iowa. Wildlife Society Bulletin 21:39–45.

Gashler, K. 2010. Cayuga Heights police training for deer protests. The Ithaca Journal, at www.theithacajournal.com/article/20100519/NEWS01/5190347/Cayuga-Heights-police-training-for-deer-protests. Accessed 19 May 2010.

Gawronski, B., and G. V. Bodenhausen. 2006. Associative and propositional processes in evaluation: an integrative review of implicit and explicit attitude change. Psychological Bulletin 132:692–731.

Geist, V., S. P. Mahoney, and J. F. Organ. 2001. Why hunting has defined the North American model of wildlife conservation. Transactions of the 66th North American Wildlife and Natural Resources Conference 66:175–185.

Geist, V., and J. F. Organ. 2004. The public trust foundation of the North American model of wildlife conservation. Northeast Wildlife 58:9–56.

Gibson, G. C., and S. A. Marks. 1995. Transforming rural hunters into conservationists: an assessment of community-based wildlife management programs in Africa. World Development 23:941–957.

Giddens, A. 1990. The consequences of modernity. Stanford University Press, Stanford, California, USA.

Gigliotti, L. M. 1996. What do our customers want? South Dakota Department of Game, Fish and Parks, Division of Wildlife Report HD-3-96.SAM, Pierre, USA.

Giles, R. H., Jr. 1978. Wildlife management. W. H. Freeman, San Francisco, California, USA.

Gill, R. B. 1996. The wildlife professional subculture: the case of the crazy aunt. Human Dimensions of Wildlife 1:60–69.

Gill, R. B. 2004. Challenges of change: natural resource management professionals engage their future. Pages 35–46 in M. J. Manfredo, J. J. Vaske, B. L. Bruyere, D. R. Field, and P. Brown, Jr., editors. Society and natural resources: a summary of knowledge. Modern Litho, Jefferson City, Missouri, USA.

Gladwell, M. 2006. Blink: the power of thinking without thinking. Little, Brown and Company, New York, New York, USA.

Glahn, J. F., M. E. Tobin, and B. F. Blackwell, editors. 2000. A science-based initiative to manage double-crested cormorant damage to southern aquaculture. U.S. Department of Agriculture Animal and Plant Health Inspection Service 11-55-010, Wildlife Services National Wildlife Research Center, Fort Collins, Colorado, USA.

Glass, R. J., and R. M. Muth. 1989. The changing role of subsistence in Alaska. Transactions of the North American Wildlife and Natural Resources Conference 54:224–232.

Glennon, M. J., and H. E. Kretzer. 2005. Impacts to wildlife from low density, exurban development: information and consideration for the Adirondack Park. Wildlife Conservation Society Technical Paper Series 3, Saranac Lake, New York, USA.

Goedeke, T. L. 2004. In the eye of the beholder: changing social perceptions of the Florida manatee. Society and Animals 12:99–116.

Goffman, E. 1967. The presentation of self in everyday life. Doubleday, New York, New York, USA.

Gore, M. L., and B. Knuth. 2009. Mass media effect on the operating environment of a wildlife-related risk-communication campaign. Journal of Wildlife Management 73:1407–1413.

Gore, M. L., B. A. Knuth, P. D. Curtis, and J. E. Shanahan. 2006. Stakeholder perceptions of risk associated with human–black bear conflicts in New York's Adirondack Park campgrounds: implications for theory and practice. Wildlife Society Bulletin 34:36–43.

Gore, M. L., B. A. Knuth, P. D. Curtis, and J. E. Shanahan. 2007. Campground manager and user perceptions of risk associated with human–black bear conflict: implications for communication. Human Dimensions of Wildlife 12:31–43.

Gore, M. L., B. A. Knuth, C. W. Scherer, and P. D. Curtis. 2008. Evaluating a conservation investment designed to reduce human–wildlife conflict. Conservation Letters 1:136–145.

Gore, M. L., W. F. Siemer, J. E. Shanahan, D. Schuefele, and D. Decker. 2005. Effects on risk perception of media coverage of a black bear–related human fatality. Wildlife Society Bulletin 33:507–516.

Gore, M. L., R. S. Wilson, W. F. Siemer, H. A. Weiczorek Hudenko, C. S. Clarke, S. P. Hart, L. A. Maguire, and B. A. Muter. 2009. Application of risk concepts to wildlife management: special issue introduction. Human Dimensions of Wildlife 14:301–313.

Gray, G. G. 1993. Wildlife and people: the human dimensions of wildlife ecology. University of Illinois Press, Urbana, USA.

Greenbaum, T. L. 1998. The handbook for focus group research. Second edition. Sage, Thousand Oaks, California, USA.

Greene, J. C., V. J. Caracelli, and W. F. Graham. 1989. Toward a conceptual framework for mixed-method evaluation designs. Educational Evaluation and Policy Analysis 11:255–274.

Greener, I. 2002. Theorising path-dependency: how does history come to matter in organisations? Journal of Management History 40:614–619.

Greenwood, D. J., and M. Levin. 1998. Introduction to action research: social research for social change. Sage, Thousand Oaks, California, USA.

Gregory, G. R. 1987. Resource economics for foresters. John Wiley & Sons, New York, New York, USA.

Gregory, R. 2000. Using stakeholder values to make smarter environmental decisions. Environment 42:34–44.

Gregory, R., L. Failing, M. Harstone, G. Long, T. McDaniels, and D. Ohlson. 2012. Structured decision making: a practical guide to environmental management choices. Wiley-Blackwell, Chichester, UK.

Gregory, R., and G. Long. 2009. Using structured decision making to help implement a precautionary approach to endangered species management. Risk Analysis 29:518–532.

Gregory, R., D. Ohlson, and J. Arvai. 2006. Deconstructing adaptive management criteria for applications to environmental management. Ecological Applications 16:2411–2425.

Grovenburg, T. W., J. A. Jenks, R. W. Klaver, K. L. Monteith, D. H. Galster, R. J. Schauer, W. W. Morlock, and J. A. Delger. 2008. Factors affecting road mortality of white-tailed deer in eastern South Dakota. Human–Wildlife Conflicts 2:48–59.

Grunig, J. E. 1983. Communication behaviors and attitudes of environmental publics: two studies. Association for Education in Journalism and Mass Communication, Columbia, South Carolina, USA.

Guba, E. G., and Y. S. Lincoln. 1989. Fourth generation evaluation. Sage, Newbury Park, California, USA.

Gunderson, L. H., C. S. Holling, and S. S. Light. 1995. Barriers and bridges to renewal of regional ecosystems and institutions. Columbia University Press, New York, New York, USA.

Hamaker, E., J. Nesselroade, and P. Molenaar. 2007. The integrated trait–state model. Journal of Research in Personality 41:295–315.

Hamazaki, T., and D. Tanno. 2001. Approval of whaling and whaling-related beliefs: public opinion in whaling and nonwhaling countries. Human Dimensions of Wildlife 6:131–144.

Hammond, J. S., R. L. Keeney, and H. Raiffa. 1999. Smart choices: a practical guide to making better decisions. Harvard Business School Press, Boston, Massachusetts, USA.

Hardin, G. 1968a. The tragedy of the commons. Pages 3–16 in G. Hardin and J. Baden, editors. Managing the commons. W. H. Freeman, San Francisco, California, USA.

Hardin, G. 1968b. The tragedy of the commons. Science 162:1243–1248.

Hart, C. 1998. Doing a literature review. Sage, Thousand Oaks, California, USA.

Hastie, R., and R. M. Dawes. 2001. Rational choice in an uncertain world: the psychology of judgment and decision making. Sage, Thousand Oaks, California, USA.

Hausser, Y., H. Weber, and B. Meyer. 2009. Bees, farmers, tourists and hunters: conflict dynamics around Western Tanzania protected areas. Biodiversity Conservation 18:2679–2703.

Hearne, J. W., and J. Swart. 2000. Modeling: a tool for enhancing the survival prospects of African wildlife. Natural Resource Modeling 13:35–55.

Heberlein, T. A. 2005. Wildlife caretaking vs. wildlife management: a short lesson in Swedish. Wildlife Society Bulletin 33:378–380.

Heberlein, T. A., and R. C. Stedman. 1997. Wildlife in rural America. Pages 772–775 in G. Goreham, editor. Encyclopedia of rural America. ABC-CLIO, Santa Barbara, California, USA.

Heberlein, T. A., and T. Willebrand. 1998. Attitudes toward hunting across time and continents: the United States and Sweden. Game Wildlife 15:1071–1080.

Heerwegh, D. 2006. An investigation of the effect of lotteries on web survey response rates. Field Methods 18:205–220.

Heinen, J. T. 1993. Park–people relations in Kosi Tappu Wildlife Reserve, Nepal: a socio-economic analysis. Environmental Conservation 20:25–34.

Heinen, J. T., and R. Shrivastava. 2009. An evaluation of conservation attitudes and awareness around Kaziranga National Park, Assam,

India: implications for conservation and development. Population and Environment 30:261–275.

Hendee, J. C. 1974. A multiple satisfaction approach to game management. Wildlife Society Bulletin 2:104–113.

Henry, R., N. Tripp, V. Gilligan, E. Smith, G. Pratt, J. Fodge, and L. Berchielli. 2001. Standard operating procedure manual (SOPM) for black bear in New York State. New York State Department of Environmental Conservation, Albany, USA.

Higginbottom, K. 2004. Economics of wildlife tourism. Pages 145–161 in K. Higginbottom, editor. Wildlife tourism: impacts, management and planning. Common Ground, Altona, Victoria, Australia.

Hofer, H., K. L. I. Campbell, and M. L. East. 2000. Modeling the spatial distribution of the economic costs and benefits of illegal game meat hunting in the Serengeti. Natural Resource Modeling 13:151–177.

The Hog Blog. 2010. The day of the condor: part 2, at http://california huntingtoday.com/hogblog/2010/04/14/day-of-the-condor-part-2. Accessed 23 April 2010.

Holling, C. S. 1978. Adaptive environmental assessment and management. John Wiley & Sons, New York, New York, USA.

Horner, S. M. 2000. Embryo, not fossil: breathing life into the public trust in wildlife. University of Wyoming College of Law, Land and Water Law Review 35:1–66.

Hovardas, T., and K. J. Korfiatis. 2008. Framing environmental policy by the local press: case study from the Dadia Forest Reserve, Greece. Forest Policy and Economics 10:316–325.

Howlett, M., M. Ramesh, and A. Perl. 2009. Studying public policy: policy cycles and subsystems. Third edition. Oxford University Press, Oxford, England, U.K.

Hrebiniak, L. G., and W. F. Joyce. 1985. Organizational adaptation: strategic choice and environmental determinism. Administrative Science Quarterly 30:336–349.

Hume, D. 1739. A treatise of human nature. John Noon, London, England, U.K.

Hunt, W. G., W. Burnham, C. N. Parish, K. K. Burnham, B. Mutch, and J. L. Oaks. 2006. Bullet fragments in deer remains: implications for lead exposure in avian scavengers. Wildlife Society Bulletin 34:167–170.

Huntington, H. P. 1992. Wildlife management and subsistence hunting in Alaska. University of Washington Press, Seattle, USA.

Hvenegaard, G. T. 2002. Birder specialization differences in conservation involvement, demographics, and motivations. Human Dimensions of Wildlife 7:21–36.

Hvenegaard, G. T., J. R. Butler, and D. K. Krystofiak. 1989. Economic values of bird watching at Point Pelee National Park, Canada. Wildlife Society Bulletin 17:526–531.

Izard, C. E. 2007. Basic emotions, natural kinds, emotion schemas, and a new paradigm. Psychological Science 2:260–280.

Jacobs, M. H. 2009. Why do we like or dislike animals? Human Dimensions of Wildlife 14:1–11.

Jacobs, M. H., J. J. Vaske, and J. M. Roemer. 2012. Toward a mental systems approach to human relationships with wildlife: the role of emotional dispositions. Human Dimensions of Wildlife 17:4–15.

Jacobson, C. A. 2008. Wildlife conservation and management in the 21st century: understanding challenges for institutional transformation. Dissertation, Cornell University, Ithaca, New York, USA.

Jacobson, C. A., and D. J. Decker. 2006. Ensuring the future of state wildlife management: understanding challenges for institutional change. Wildlife Society Bulletin 34:531–536.

Jacobson, C. A., and D. J. Decker. 2008. Governance of state wildlife management: reform and revive or resist and retrench? Society and Natural Resources 21:441–448.

Jacobson, C. A., D. J. Decker, and L. Carpenter. 2007. Securing alternative funding for wildlife management: insights from agency leaders. Human Dimensions of Wildlife Management 71:2106–2113.

Jacobson, C. A., J. F. Organ, D. J. Decker, G. R. Batcheller, and L. Carpenter. 2010. A conservation institution for the 21st century: implications for state wildlife agencies. Journal of Wildlife Management 74:203–209.

Jett, J. S., B. Thapa, and J. K. Yong. 2009. Recreation specialization and boater speed compliance in manatee zones. Human Dimensions of Wildlife 14:278–293.

Jiménez, J. E. 1996. The extirpation and current status of wild chinchillas Chinchilla lanigera and C. brevicaudata. Biological Conservation 77:1–6.

Jones, J., M. Andriamarovololona, and N. Hockley. 2008. The importance of taboos and social norms to conservation in Madagascar. Conservation Biology 22:976–987.

Kaczensky, P., M. Blazic, and H. Gossow. 2001. Content analysis of articles on brown bears in the Slovenian press, 1991–1998. Forest, Snow, and Landscape Research 761:121–135.

Kahneman, D., P. Slovic, and A. Tversky, editors. 1982. Judgment under uncertainty: heuristics and biases. Cambridge University Press, Cambridge, England, U.K.

Kahneman, D., and A. Tversky. 2000. Choices, values, and frames. Cambridge University, Cambridge, England, U.K.

Kallman, H., editor. 1987. Restoring America's wildlife, 1937–1987: the first 50 years of the Federal Aid in Wildlife Restoration (Pittman–Robertson) Act. U.S. Fish and Wildlife Service, Washington, D.C., USA.

Kaltenborn, B. P., T. Bjerke, and J. Nyahongo. 2006. Living with dangerous animals—self-reported fear of potentially dangerous species in the Serengeti region, Tanzania. Human Dimensions of Wildlife 11:397–408.

Kaltenborn, B. P., T. Bjerke, and J. Vittersø. 1999. Attitudes towards large carnivores among sheep farmers, wildlife managers, and research biologists in Norway. Human Dimensions of Wildlife 4:57–73.

Kant, I. 1930. Lectures on ethics. Translated by Louis Infield. Methuen, London, England, U.K.

Kaplowitz, M. D., T. D. Hadlock, and R. Levine. 2004. A comparison of web and mail survey response rates. Public Opinion Quarterly 68:94–101.

Kasperson, R. E., and J. X. Kasperson. 1992. The social amplification and attenuation of risk. Annals of the American Academy of Political and Social Science 545:95–105.

Kausler, D. H., and B. C. Kausler. 2001. The graying of America: an encyclopedia of aging, health, mind, and behavior. Second edition. University of Illinois Press, Champaign, USA.

Keeney, R. L. 1992. Value-focused thinking: a path to creative decision making. Harvard University Press, Cambridge, Massachusetts, USA.

Keeney, R. L., and R. Gregory. 2005. Selecting attributes to measure achievement of objectives. Operations Research 53:1–11.

Keeney, R. L., and H. Raiffa. 1993. Decisions with multiple objectives: preferences and values trade-offs. Cambridge University Press, Cambridge, England, U.K.

Kellert, S. R. 1979. Phase one: public attitudes toward critical wildlife and natural habitat issues. U.S. Fish and Wildlife Service Report No. PB-80-138332. U.S. Government Printing Office, Washington, D.C., USA.

Kellert, S. R. 1980a. Contemporary values of wildlife in American society. Pages 31–60 in W. W. Shaw and E. H. Zube, editors. Wildlife values. U.S. Department of Agriculture Forest Service, Rocky Mountain Experiment Station, Fort Collins, Colorado, USA.

Kellert, S. R. 1980b. Public attitudes, knowledge and behaviors toward wildlife and natural habitats. Transactions of the North American Wildlife and Natural Resources Conference 45:111–124.

Kellert, S. R., and M. O. Westervelt. 1983. Historical trends in American animal use and perception. International Journal for the Study of Animal Problems 4:133–146.

Kennedy, J. J. 1985. Viewing wildlife managers as a unique professional culture. Wildlife Society Bulletin 13:571–579.

Keohane, R. O., P. M. Haas, and M. A. Levy. 1993. The effectiveness of international environmental institutions. Pages 3–24 in P. M. Haas, R. O. Keohane, and M. A. Levy, editors. Institutions for the earth: sources of effective international environmental protection. MIT Press, Cambridge, Massachusetts, USA.

King, S. T., and J. R. Schrock, editors. 1985. State wildlife regulations. Volume 3. Controlled wildlife: a three-volume guide to U.S. wildlife laws and permit procedures. Association of Systematics Collections, Lawrence, Kansas, USA.

Klein, M. L., S. R. Humphrey, and H. F. Percival. 1997. Effects of ecotourism on distribution of waterbirds in a wildlife refuge. Conservation Biology 9:1454–1465.

Klofstad, C. A., S. Boulianne, and D. Basson. 2008. Matching the message to the medium: results from an experiment on internet survey email contacts. Social Science Computer Review 26:498–509.

Knight, A. T., and R. M. Cowling. 2007. Embracing opportunism in the selection of priority conservation areas. Conservation Biology 21:1124–1126.

Kotchen, M. J., and S. D. Reiling. 2000. Environmental attitudes, motivations, and contingent valuation of nonuse values: a case study involving endangered species. Environmental Economics 32:93–107.

Kotter, J. P. 1996. Leading change. Harvard Business School Press, Boston, Massachusetts, USA.

Kretser, H. E., P. J. Sullivan, and B. A. Knuth. 2008. Housing density as an indicator of spatial patterns of reported human–wildlife interactions in Northern New York. Landscape and Urban Planning 84:282–292.

Krueger, C. C., D. J. Decker, and T. A. Gavin. 1986. A concept of natural resource management: an application to unicorns. Transactions of the Northeast Section of The Wildlife Society 43:50–56.

Krueger, R. A., and M. A. Casey. 2009. Focus groups: a practical guide for applied research. Fourth edition. Sage, Los Angeles, California, USA.

Kuentzel, W. F. 1994. Skybusting and the slob hunter myth. Wildlife Society Bulletin 22:331–336.

Kuhl, A., N. Balinova, E. Bykova, Iu. N. Arylov, A. Esipov, A. A. Lushchikina, and E. J. Milner-Gulland. 2009. The role of saiga poaching in rural communities: linkages between attitudes, socio-economic circumstances and behavior. Biological Conservation 142:1442–1450.

Lachapelle, P. R., S. F. McCool, and M. E. Patterson. 2003. Barriers to effective natural resource planning in a "messy" world. Society and Natural Resources 16:473–490.

Lancia, R. A., C. E. Braun, M. W. Callopy, R. D. Dueser, J. G. Kie, C. J. Martinka, J. D. Nichols, T. D. Nudds, W. R. Porath, and N. G. Tilghman. 1996. ARM! For the future: adaptive resource management in the wildlife profession. Wildlife Society Bulletin 24:436–442.

Langenau, E. E., E. J. Flegler, and H. R. Hill. 1985. Deer hunters' opinion survey, 1984. Michigan Department of Natural Resources Wildlife Division Report 3012, Lansing, USA.

Lasswell, H. 1948. The structure and function of communication in society. Pages 37–51 in L. Bryson, editor. The communication of ideas: a series of addresses. Harper, New York, New York, USA.

Lauber, T. B., and T. L. Brown. 2006. Learning by doing: policy learning in community-based deer management. Society and Natural Resources 19:411–428.

Lauber, T. B., and B. A. Knuth. 1997. Fairness in moose management decision-making: the citizen's perspective. Wildlife Society Bulletin 25:776–787.

Lauber, T. B., R. C. Stedman, D. J. Decker, and B. A. Knuth. 2011. Linking knowledge to action in collaborative conservation. Conservation Biology 25:1186–1194.

Lauber, T. B., E. J. Taylor, and D. J. Decker. 2009. Factors influencing membership of federal wildlife biologists in The Wildlife Society. Journal of Wildlife Management 73:980–988.

Lauber, T. B., E. J. Taylor, D. J. Decker, and B. A. Knuth. 2010. Challenges of professional development: balancing the demands of employers and professions in federal natural resource agencies. Organization and Development 23:446–464.

Lee, J., and D. Scott. 2004. Measuring birding specialization: a confirmatory factor analysis. Leisure Sciences 26:245–260.

Lee, K. N. 1993. The compass and gyroscope: integrating science and politics for the environment. Island Press, Washington, D.C., USA.

Lemelin, R. H., and E. C. Wiersma. 2007. Perceptions of polar bear tourists: a qualitative analysis. Human Dimensions of Wildlife 12:45–52.

Leonard, J. 2007. Fishing and hunting recruitment and retention in the U.S. from 1990 to 2005: addendum to the 2001 National Survey of Fishing, Hunting, and Wildlife-Associated Recreation. U.S. Government Printing Office Report 2001-11, Washington, D.C., USA.

Leong, K. M. 2010. The tragedy of becoming common: landscape change and perceptions of wildlife. Society and Natural Resources 23:111–127.

Leong, K. M., D. J. Decker, T. B. Lauber, D. B. Raik, and W. F. Siemer. 2009a. Overcoming jurisdictional boundaries through stakeholder engagement and collaborative governance: lessons learned from white-tailed deer management in the U.S. Pages 221–247 in K. Andersson, E. Eklund, M. Lehtola, and P. Salmi, editors. Beyond the rural–urban divide: cross-continental perspectives on the differentiated countryside and its regulation. Emerald Group, Bingley, England, U.K.

Leong, K. M., J. F. Forester, and D. J. Decker. 2009b. Moving public participation beyond compliance: uncommon approaches to finding common ground. George Wright Forum 26:23–39.

Leong, K. M., K. A. McComas, and D. J. Decker. 2007. Matching the forum to the fuss: using co-orientation contexts to address the paradox of public participation in natural resource management. Environmental Practice 9:1–12.

Leong, K. M., K. A. McComas, and D. J. Decker. 2008. Formative co-orientation research: a tool to assist with environmental decision making. Environmental Communication: A Journal of Nature and Culture 2:257–273.

Leopold, A. 1933. Game management. Charles Scribner's Sons, New York, New York, USA.

Leopold, A. 1949. A Sand County almanac: and sketches here and there. Oxford University Press, New York, New York, USA.

Lerner, J. S., and D. Keltner. 2000. Beyond valence: toward a model of emotion-specific influences on judgment and choice. Cognition & Emotion 14:473–493.

Lewis, D., G. B. Kaweche, and A. Mwenya. 1990. Wildlife conservation outside protected areas—lessons from an experiment in Zambia. Conservation Biology 4:171–180.

Lindblom, C. E. 1959. The science of muddling through. Public Administration Review 19:79–88.

Lindsey, P. A., R. Alexander, L. G. Frank, A. Mathieson, and S. S. Romanach. 2006. Potential of trophy hunting to create incentives for wildlife conservation in Africa where alternative wildlife-based land uses may not be viable. Animal Conservation 9:283–291.

Lindsey, P. A., P. A. Roulet, and S. S. Romanach. 2007. Economic and conservation significance of the trophy hunting industry in sub-Saharan Africa. Biological Conservation 134:455–469.

Lischka, S. A., S. J. Riley, and B. A. Rudolph. 2008. Effects of impact perception on acceptance capacity for white-tailed deer. Journal of Wildlife Management 72:502–509.

Liu, J., T. Dietz, S. R. Carpenter, C. Folke, M. Alberti, C. L. Redman, S. H. Schneider, E. Ostrom, A. N. Pell, J. Lubchenco, W. W. Taylor, Z. Ouyang, P. Deadman, T. Kratz, and W. Provencher. 2007. Coupled human and natural systems. Ambio 36:639–649.

Liu, J. H., B. Bonzon-Liu, and M. Pierce-Guarino. 1997. Common fate between humans and animals? The dynamic systems theory of groups and environmental attitudes in the Florida Keys. Environment and Behavior 29:87–122.

Loker, C. A., and D. J. Decker. 1995. Colorado black bear hunting referendum: what was behind the vote? Wildlife Society Bulletin 23:370–376.

Loker, C. A., D. J. Decker, R. B. Gill, T. D. I. Beck, and L. H. Carpenter. 1994. The Colorado black bear controversy: a case study in contemporary wildlife management. Cornell University, Department of Natural Resources, Human Dimensions Research Unit Publication 94-4, Ithaca, New York, USA.

Lomborg, B. 2001. The skeptical environmentalist. Cambridge University Press, Cambridge, England, U.K.

Lopez, R. R., A. Lopez, R. N. Wilkins, C. Torres, R. Valdez, J. G. Teer, and G. Bowser. 2005. Changing Hispanic demographics: challenges in natural resource management. Wildlife Society Bulletin 33:553–564.

Low, B., S. R. Sundaresan, I. R. Fischhoff, and D. I. Rubenstein. 2009. Partnering with local communities to identify conservation priorities for endangered Grevy's zebra. Biological Conservation 142:1548–1556.

Lybecker, D., B. L. Lamb, and P. D. Ponds. 2002. Public attitudes and knowledge of the black-tailed prairie dog: a common and controversial species. BioScience 52:607–613.

Manfredo, M. J. 1988. Second-year analysis of donors to Oregon's nongame tax checkoff. Wildlife Society Bulletin 16:221–224.

Manfredo, M. J., editor. 2002. Wildlife viewing: a management handbook. Oregon State University Press, Corvallis, USA.

Manfredo, M. J. 2008. Who cares about wildlife? Springer, New York, New York, USA.

Manfredo, M. J., B. L. Driver, and M. A. Tarrant. 1996. Measuring leisure motivation: a meta-analysis of the recreation experience preference scales. Journal of Leisure Research 28:188–213.

Manfredo, M. J., D. C. Fulton, and C. L. Pierce. 1997. Understanding voter behavior on wildlife ballot initiatives: Colorado's trapping amendment. Human Dimensions of Wildlife 2:22–39.

Manfredo, M. J., and B. Haight. 1986. Oregon's nongame tax checkoff: a comparison of donors and nondonors. Wildlife Society Bulletin 14:121–126.

Manfredo, M. J., T. L. Teel, and A. D. Bright. 2003. Why are public values toward wildlife changing? Human Dimensions of Wildlife 8:287–306.

Manfredo, M. J., T. L. Teel, and A. D. Bright. 2004. Applications of the concepts of values and attitudes in human dimensions of natural resource research. Pages 271–282 in M. J. Manfredo, J. J. Vaske, B. L. Bruyere, D. R. Field, and P. Brown, editors. Society and natural resources: a summary of knowledge. Modern Litho, Jefferson City, Missouri, USA.

Manfredo, M. J., T. L. Teel, and K. L. Henry. 2009a. Linking society and environment: a multi-level model of shifting wildlife value orientations in the western U.S. Social Science Quarterly 90:407–427.

Manfredo, M. J., J. J. Vaske, P. J. Brown, D. J. Decker, and E. A. Duke. 2009b. Wildlife and society: the science of human dimensions. Island Press, Washington, D.C., USA.

Mangun, W. R., editor. 1992. American fish and wildlife policy: the human dimension. Southern Illinois University Press, Carbondale, USA.

Mangun, W. R., and W. W. Shaw. 1984. Alternative mechanisms for funding nongame wildlife conservation. Public Administration Review 44:407–413.

Manring, N. J. 1998. Alternative dispute resolution and organizational incentives in the U.S. Forest Service. Society and Natural Resources 11:67–80.

Marchini, S., and D. W. Macdonald. 2012. Predicting ranchers' intention to kill jaguars: case studies in Amazonia and Patanal. Biological Conservation 147:213–221.

Marks, S. A. 1991. Hunting culture in black and white: nature, history, and ritual in a Carolina community. Princeton University Press, Princeton, New Jersey, USA.

Martin, S. R. 1997. Specialization and differences in setting preferences among wildlife viewers. Human Dimensions of Wildlife 2:1–18.

Marx, K., and F. Engels. 1986. Selected works. International, New York, New York, USA.

Mastro, L. L., M. R. Conover, and S. N. Frey. 2008. Deer–vehicle collision prevention techniques. Human–Wildlife Conflicts 2:80–92.

Mattfeld, G. F., D. J. Decker, and T. L. Brown. 1984. Developing human dimensions in New York's wildlife research program. Transactions of the North American Wildlife and Natural Resources Conference 49:54–65.

Mattfeld, G. F., G. R. Parsons, T. L. Brown, and D. J. Decker. 1998. Integrating human dimensions in wildlife management: experiences and outlooks of an enduring partnership. Transactions of the North American Wildlife Natural Resources Conference 63:244–256.

McCall, G. J., and J. L. Simmons, editors. 1969. Issues in participant observation: a text and reader. Addison-Wesley, Reading, Massachusetts, USA.

McCleery, R. A., R. B. Ditton, J. Sell, and R. R. Lopez. 2006. Understanding and improving attitudinal research in wildlife sciences. Wildlife Society Bulletin 34:537–541.

McFarlane, B. 1994. Specializations and motivations of birdwatchers. Wildlife Society Bulletin 22:361–370.

McNaught, D. A. 1987. Wolves in Yellowstone?—park visitors respond. Wildlife Society Bulletin 15:518–521.

Meares, T. L. 2000. Norms, legitimacy and enforcement. Oregon Law Review 79:391–408.

Meffe, G. K., L. A. Nielsen, R. L. Knight, and D. A. Schenborn. 2002. Ecosystem management: adaptive, community-based conservation. Island Press, Washington, D.C., USA.

Meine, C. 1988. Aldo Leopold: his life and work. University of Wisconsin Press, Madison, USA.

Mertig, A. G., and R. E. Dunlap. 2001. Environmentalism, new social movements, and the new class: a cross-national investigation. Rural Sociology 66:113–136.

Messmer, T. A. 2009. Human–wildlife conflicts: emerging challenges and opportunities. Human–Wildlife Conflicts 3:10–17.

Messmer, T. A., and C. E. Dixon. 1997. Extension's role in achieving hunter, landowner, and wildlife agency objectives through Utah's big game posted hunting unit program. Transactions of the North American Wildlife and Natural Resources Conference 62:47–57.

Messmer, T. A., and S. A. Schroeder. 1996. Perceptions of Utah alfalfa growers about wildlife damage to their hay crops: implications for managing wildlife on private land. Great Basin Naturalist 56:254–260.

Michaelidou, M. 2002. Moving beyond colonial perspectives in conservation: sustaining forests, wildlife and mountain villages in Cyprus. Dissertation, Cornell University, Ithaca, New York, USA.

Michaelidou, M., and D. J. Decker. 2002. Challenges and opportunities facing wildlife conservation and cultural sustainability in the Pafos Forest, Cyprus: historical overview and contemporary perspective. European Journal of Wildlife Research 48:291–300.

Michaelidou, M., and D. J. Decker. 2003. European Union policy and local perspectives: nature conservation and rural communities in Cyprus. The Cyprus Review 15:121–145.

Michaelidou, M., and D. J. Decker. 2005. Incorporating local values in European Union conservation policy: the Cyprus case. Human Dimensions of Wildlife Journal 10:1–12.

Miles, M. B., and A. M. Huberman. 1994. Qualitative data analysis: an expanded sourcebook. Second edition. Sage, Thousand Oaks, California, USA.

Millar, M. M., A. C. O'Neil, and D. A. Dillman. 2009. Are mode preferences real? Washington State University, Social and Economic Sciences Research Center (SESRC) Technical Report 09-003, Pullman, USA.

Mills, C. W. 1959. The sociological imagination. Oxford University Press, New York, New York, USA.

Minnis, D. L. 1998. Wildlife policy-making by the electorate: an overview of citizen-sponsored ballot measures on hunting and trapping. Wildlife Society Bulletin 26:75–83.

Minnis, D. L., R. Holsman, L. Grice, and R. B. Peyton. 1997. Focus groups as a human dimensions research tool: three illustrations of their use. Human Dimensions of Wildlife 2:40–49.

Minnis, D. L., and R. B. Peyton. 1995. Cultural carrying capacity: modeling a notion. Pages 19–34 in J. B. McAninch, editor. Urban deer: a manageable resource? Proceedings of the 1993 Symposium of the North Central Section. The Wildlife Society, St. Louis, Missouri, USA.

Minton, S. A. 1987. Poisonous snakes and snakebite in the U.S. Northwest Scientist 61:130–137.

Mishra, C. 1997. Livestock depredation by large carnivores in the Indian trans-Himalaya: conflict perceptions and conservation prospects. Environmental Conservation 24:338–343.

Mitchell, J. M., G. J. Pagac, and G. R. Parker. 1997. Informed consent: gaining support for removal of overabundant white-tailed deer on an Indiana state park. Wildlife Society Bulletin 25:447–450.

Moore, C. M. 1987. Group techniques for idea building. Sage, Newbury Park, California, USA.

Moore, K. D. 2004. The Pine Island paradox. Milkweed Editions, Minneapolis, Minnesota, USA.

Morgan, D. L., editor. 1993. Successful focus groups: advancing the state of the art. Sage, Newbury Park, California, USA.

Morgan, D. L. 1996. Focus groups. Annual Review of Sociology 22:129–152.

Mortenson, K. G., and R. S. Krannich. 2001. Wildlife managers and public involvement: letting the crazy aunt out. Human Dimensions of Wildlife 6:277–290.

Morzillo, A. T., A. G. Mertig, N. Garner, and L. Jianguo. 2007. Resident attitudes toward black bears and population recovery in East Texas. Human Dimensions of Wildlife 12:417–429.

Moscovici, S. 1985. Social influence and conformity. Pages 347–412 in G. Lindzey and E. Aaronson, editors. The handbook of social psychology. Third edition. Erlbaum, New York, New York, USA.

Moss, M. B., J. D. Fraser, and J. D. Wellman. 1986. Characteristics of nongame fund contributors vs. hunters in Virginia. Wildlife Society Bulletin 14:107–114.

Muter, B. A. 2009. Risk perception, social networks, and media frames associated with human–cormorant interactions in the Great Lakes. Thesis, Michigan State University, East Lansing, USA.

Muth, R. M., and J. F. Bowe, Jr. 1998. Illegal harvest of renewable natural resources in North America: toward a typology of the motivations for poaching. Society and Natural Resources 11:9–24.

Muth, R. M., D. E. Ruppert, and R. J. Glass. 1987. Subsistence use of fisheries resources in Alaska: implications for Great Lakes fisheries management. Transactions of the American Fisheries Society 116:510–518.

Myers, K. H., G. R. Parsons, and P. E. T. Edwards. 2010. Measuring the recreational use value of migratory shorebirds on the Delaware Bay. Marine Resource Economics 25:247–264.

Natural Resource Modeling. 2000. Conserving African wildlife: modeling issues (introduction to special issue). Natural Resource Modeling 13:5–11.

Naughton-Treves, L., R. Grossberg, and A. Treves. 2003. Paying for tolerance: rural citizens' attitudes toward wolf depredation and compensation. Conservation Biology 17:1500–1511.

Nelson, D. 1992. Citizen task forces on deer management: a case study. New England Wildlife 49:92–96.

Nelson, F. 2009. Developing payments for ecosystem services approaches to carnivore conservation. Human Dimensions of Wildlife 14:381–392.

Nelson, M. P., and J. A. Vucetich. 2009. On advocacy by environmental scientists: what, whether, why, and how. Conservation Biology 23:1090–1101.

Nelson, R. K. 1983. Make prayers to the raven: a Koyukon view of the northern forest. University of Chicago Press, Chicago, Illinois, USA.

Nelson, R. K., K. H. Mautner, and G. R. Bane. 1982. Tracks in the wildland: a portrayal of Koyukon and Nunamiut subsistence. University of Alaska, Cooperative Park Studies Unit, Anthropology and Historic Preservation, Fairbanks, USA.

Nesslage, G. M., M. Wilberg, and S. J. Riley. 2006. Dynamics of range collapse of two large carnivore species under a bounty system. Intermountain Journal of Science 12:63–76.

Newmark, W. D., N. L. Leonard, H. I. Sariko, and D. G. M. Gamassa. 1993. Conservation attitudes of local people living adjacent to five protected areas in Tanzania. Biological Conservation 63:177–183.

New York State Department of Environmental Conservation [NYSDEC]. 2003. A framework for black bear management in New York. New York State Department of Environmental Conservation, Albany, USA.

Ng, J. W., C. Nielsen, and C. C. St. Clair. 2008. Landscape and traffic factors influencing deer–vehicle collisions in an urban environment. Human–Wildlife Conflicts 2:34–47.

Nichols, J. D., P. J. Heglund, M. G. Knutson, M. E. Seamans, J. E. Lyons, J. M. Morton, M. T. Jones, G. S. Boomer, and B. K. Williams. 2011. Climate change, uncertainty, and natural resource management. Journal of Wildlife Management 75:6–18.

Nie, M. 2004. State wildlife policy and management: the scope and bias of political conflict. Public Administration Review 64:221–233.

Nye, J. S., Jr., P. D. Zelikow, and D. C. King. 1997. Why people don't trust government. Harvard University Press, Cambridge, Massachusetts, USA.

O'Brien, D. J., S. M. Schmitt, S. D. Fitzgerald, D. E. Berry, and G. J. Hickling. 2006. Managing the wildlife reservoir of Mycobacterium bovis: the Michigan, USA, experience. Veterinary Microbiology 112:313–323.

O'Brien, D. J., S. M. Schmitt, B. A. Rudolph, and G. Nugent. 2011. Recent advances in the management of bovine tuberculosis in free-ranging wildlife. Veterinary Microbiology 151:23–33.

Oldfield, T. E. E., R. J. Smith, S. R. Harrop, and N. Leader-Williams. 2003. Field sports and conservation in the United Kingdom. Nature 423:531–533.

Oliver, C. 1991. Strategic responses to institutional processes. The Academy of Management Review 16:145–179.

Organ, J. F., and G. R. Batcheller. 2009. Reviving the public trust doctrine as a foundation for wildlife management in North America. Pages 161–171 in M. J. Manfredo, J. J. Vaske, P. J. Brown, D. J. Decker, and E. A. Duke, editors. Wildlife and society: the science of human dimensions. Island Press, Washington, D.C., USA.

Organ, J. F., L. H. Carpenter, D. Decker, W. F. Siemer, and S. J. Riley. 2006. Thinking like a manager: reflections on wildlife management. Wildlife Management Institute, Washington, D.C., USA.

Organ, J. F., D. J. Decker, S. J. Riley, J. E. McDonald, Jr., and S. P. Mahoney. 2012. Adaptive management in wildlife conservation. Pages 43–54 *in* N. Silvy, editor. The wildlife techniques manual. Volume 2. Seventh edition. Johns Hopkins University Press, Baltimore, Maryland, USA.

Organ, J. F., and M. R. Ellingwood. 2000. Wildlife stakeholder acceptance capacity for black bears, beavers, and other beasts in the east. Human Dimensions of Wildlife 5:63–75.

Organ, J. F., and E. K. Fritzell. 2000. Trends in consumptive recreation and the wildlife profession. Wildlife Society Bulletin 28:780–787.

Organ, J. F., S. P. Mahoney, and V. Geist. 2010. Born in the hands of hunters: the North American model of wildlife conservation. The Wildlife Professional 4:22–27.

Osherenko, G. 1988. Can co-management save Arctic wildlife? Environment 30:6–34.

Otto, D., D. Monchuk, K. Jintanakul, and C. Kling. 2007. The economic value of Iowa's natural resources. Commissioned by the Sustainable Funding for Natural Resources Study Committee, Iowa General Assembly. Iowa Department of Natural Resources, at www.iowadnr.gov/sustainablefunding. Accessed 22 November 2010.

Palmer, B. C., M. R. Conover, and S. N. Frey. 2010. Replication of a 1970s study on domestic sheep losses to predators on Utah's summer rangelands. Journal of Range Ecology and Management 63:689–695.

Parkes, C., and J. Thornley. 2000. Deer: law and liabilities. Swan Hill Press, Shrewsbury, England, U.K.

Parry, D., and B. Campbell. 1992. Attitudes of rural communities to animal wildlife and its utilization in Chobe Enclave and Mababe Depression, Botswana. Environmental Conservation 23:207–217.

Pate, J., M. Manfredo, A. Bright, and G. Tishbein. 1996. Coloradans' attitudes toward reintroducing the gray wolf into Colorado. Wildlife Society Bulletin 24:421–428.

Patterson, M. E., J. M. Montag, and D. R. Williams. 2003. The urbanization of wildlife management: social science, conflict, and decision making. Urban Forestry & Urban Greening 1:171–183.

Patton, M. Q. 2002. Qualitative evaluation and research methods. Third edition. Sage, Newbury Park, California, USA.

Pease, M. L., R. K. Rose, and M. J. Butler. 2005. Effects of human disturbance on the behavior of wintering ducks. Wildlife Society Bulletin 33:103–112.

Pelstring, L., J. Shanahan, and B. Perry. 1997. The press and citizen participation: a content analysis. Proceedings of the Eastern Wildlife Damage Management Conference 8:149–160.

Peters, B. G. 2000. Globalization, institutions, and governance. Pages 29–57 *in* B. G. Peters and D. J. Savoie, editors. Governance in the twenty-first century: revitalizing the public service. McGill-Queen's University Press, Montreal, Quebec, Canada.

Peters, E. 2006. The functions of affect in the construction of preferences. Pages 454–463 *in* S. Lichtenstein and P. Slovic, editors. The construction of preference. Cambridge University Press, Cambridge, England, U.K.

Peterson, M. N., T. R. Peterson, and M. J. Peterson. 2002. Cultural conflict and the endangered Florida Key deer. Journal of Wildlife Management 66:947–968.

Peyton, R. B., and E. E. Langenau. 1985. A comparison of attitudes held by BLM biologists and the general public towards animals. Wildlife Society Bulletin 13:117–120.

Pfeffer, J., and G. R. Salancik. 2003. The external control of organizations: a resource dependence perspective. Stanford University Press, Stanford, California, USA.

Pierre, J. 2000. Understanding governance. Pages 1–10 *in* J. Pierre, editor. Debating governance: authority, steering, and democracy. Oxford University Press, New York, New York, USA.

Pinchot, G. 1947. Breaking new ground. Island Press, Washington, D.C., USA.

Polasky, S. 2008. Why conservation planning needs socioeconomic data. Proceedings of the National Academy of Sciences of the United States of America 105:6505–6506.

Pollan, M. 2006. The omnivore's dilemma: a natural history of four meals. University of California Press, Berkeley, USA.

Posavac, E. J., and R. G. Carey. 2006. Program evaluation. Seventh edition. Prentice Hall, Upper Saddle River, New Jersey, USA.

Prell, C., K. Hubacek, and M. Reed. 2009. Stakeholder analysis and social network analysis in natural resources management. Society and Natural Resources 22:501–518.

Primm, S. A. 1996. A pragmatic approach to grizzly bear conservation. Conservation Biology 10:1026–1035.

Primm, S. A., and T. Clark. 1996. Making sense of the policy process for carnivore conservation. Conservation Biology 10:1036–1045.

Purdy, K. G., and D. J. Decker. 1986. A longitudinal investigation of the social–psychological influences on hunting participation in New York: study one (1983–1985). Cornell University, Department of Natural Resources, Human Dimensions Research Unit Series Publication 86-7, Ithaca, New York, USA.

Purdy, K. G., D. J. Decker, and T. L. Brown. 1985. New York's 1978 hunter training course participants: the importance of social–psychological influences on participation in hunting from 1978–1984. Cornell University, Department of Natural Resources, Human Dimensions Research Unit Series Publication 85-7, Ithaca, New York, USA.

Purdy, K. G., D. J. Decker, and T. L. Brown. 1989. New York's new hunters: influences from beginning to end. Cornell University, Department of Natural Resources, Human Dimensions Research Unit Series Publication 89-3, Ithaca, New York, USA.

Putnam, R. O. 1993. Making democracy work: civic traditions in modern Italy. Princeton University Press, Princeton, New Jersey, USA.

Radtke, T. M., and C. D. Dieter. 2010. Selection of pathways to foraging sites in crop fields by flightless Canada geese. Human–Wildlife Interactions 4:202–206.

Raik, D. B., W. F. Siemer, and D. J. Decker. 2005. Intervention and capacity considerations in community-based deer management: the stakeholders' perspective. Human Dimensions of Wildlife 10:259–272.

Rasmussen, L. 2001. Letters, notes, and comments: research note. Journal of Religious Ethics 29:201–206.

Reading, R. P., B. J. Miller, and S. R. Kellert. 1999. Values and attitudes toward prairie dogs. Anthrozoos 12:43–52.

Redpath, S. M., B. E. Arroyo, F. M. Leckie, P. Bacon, N. Bayfield, R. J. Gutierrez, and S. J. Thirgood. 2004. Using decision modeling with stakeholders to reduce human–wildlife conflict: a raptor–grouse case study. Conservation Biology 18:350–359.

Regan, H. M., M. Colyvan, and M. A. Burgman. 2002. A taxonomy and treatment of uncertainty for ecology and conservation biology. Ecological Applications 12:618–628.

Responsive Management. 2006. Sportsmen's attitudes. Unpublished survey about various hunting and fishing issues. Responsive Management, Harrisonburg, Virginia, USA.

Responsive Management/National Shooting Sports Foundation [RM/NSSF]. 2008. The future of hunting and the shooting sports: research-based recruitment and retention strategies. Responsive Management/National Shooting Sports Foundation, produced for the U.S. Fish and Wildlife Service under Grant Agreement CT-M-6-0, Harrisonburg, Virginia, USA.

Riley, S. J., and D. J. Decker. 2000. Wildlife stakeholder acceptance capacity for cougars in Montana. Wildlife Society Bulletin 28:931–939.

Riley, S. J., D. J. Decker, L. H. Carpenter, J. F. Organ, W. F. Siemer, G. F. Mattfeld, and G. Parsons. 2002. The essence of wildlife management. Wildlife Society Bulletin 30:585–593.

Riley, S. J., W. F. Siemer, D. J. Decker, L. H. Carpenter, J. F. Organ, and L. T. Berchielli. 2003. Adaptive impact management: an integrative approach to wildlife management. Human Dimensions of Wildlife 8:81–95.

Rittel, H. W. J., and M. M. Webber. 1973. Dilemmas in a general theory of planning. Policy Sciences 4:155–169.

Robinson, J. G., and K. H. Redford, editors. 1991. Neotropical wildlife use and conservation. University of Chicago Press, Chicago, Illinois, USA.

Rogers, E. M. 1995. Diffusion of innovations. Fourth edition. Free Press, New York, New York, USA.

Rokeach, M. 1973. The nature of human values. Free Press, New York, New York, USA.

Rosas-Rosas, O. C., and R. Valdez. 2010. The role of landowners in jaguar conservation in Sonora, Mexico. Conservation Biology 24:366–371.

Rowe, G., and L. J. Frewer. 2005. A typology of public engagement mechanisms. Science, Technology & Human Values 30:251–290.

Rudolph, B. A., S. J. Riley, G. H. Hickling, B. J. Frawley, M. S. Garner, and S. R. Winterstein. 2006. Regulating hunter baiting for white-tailed deer in Michigan: biological and social considerations. Wildlife Society Bulletin 34:314–321.

Ruys, K. I., and D. A. Stapel. 2008. The secret life of emotions. Psychological Science 19:385–391.

Salatiel, J., and L. R. Irby. 1998. Perceptions of game damage in Montana by resource agency personnel and agricultural producers. Wildlife Society Bulletin 26:84–91.

Sali, M. J., D. M. Kuehn, and L. Zhang. 2008. Motivations for male and female birdwatchers in New York State. Human Dimensions of Wildlife 13:187–200.

Sander, D., D. Grandjean, and K. R. Scherer. 2005. A systems approach to appraisal mechanisms in emotion. Neural Networks 18:317–352.

Sandrey, R. A., S. T. Buccola, and W. G. Brown. 1983. Pricing policies for antlerless elk hunting permits. Land Economics 59:432–443.

Sandström, C., J. Pellika, O. Ratamäki, and A. Sande. 2009. Management of large carnivores in Fennoscandia: new patterns of regional participation. Human Dimensions of Wildlife 14:37–50.

Sax, J. L. 1970. The public trust doctrine in natural resource law: effective judicial intervention. Michigan Law Review 68:471–566.

Scarce, R. 1998. What do wolves mean?—conflicting social constructions of Canis lupus in "Bordertown." Human Dimensions of Wildlife 3:26–45.

Scherer, K. R. 1999. Appraisal theory. Pages 637–663 in T. Dalgleish and M. Power, editors. Handbook of cognition and emotion. John Wiley & Sons, London, England, U.K.

Schuett, M. A., D. Scott, and J. O'Leary. 2009. Social and demographic trends affecting fish and wildlife management. Pages 18–30 in M. J. Manfredo, J. J. Vaske, P. J. Brown, D. J. Decker, and E. A. Duke, editors. Wildlife and society: the science of human dimensions. Island Press, Washington, D.C., USA.

Schusler, T. M., and D. J. Decker. 2002. Engaging local communities in wildlife management area planning: an evaluation of the Lake Ontario Islands Search Conference. Wildlife Society Bulletin 30:1226–1237.

Schweitzer, A. 1923. The philosophy of civilization II. Page 254 in J. Naish, translator. Civilization and ethics. A & C Black, London, England, U.K.

Scott, D., S. Baker Menzel, and K. Chulwon. 1999. Motivations and commitments among participants in the great Texas birding classic. Human Dimensions of Wildlife 4:50–67.

Shanks, R. E. 1992. The role of environmental paradigms in communication between New York State Bureau of Wildlife and New York State County Legislators. Thesis, Cornell University, Department of Natural Resources, Ithaca, New York, USA.

Sherwood, D. 2002. Seeing the forest for the trees: a manager's guide to applying systems thinking. Nicholas Brealey, London, England, U.K.

Shiota, M. N., D. Keltner, and O. P. John. 2006. Positive emotion disposi-

tions differentially associated with Big Five personality and attachment style. Journal of Positive Psychology 1:61–71.

Shrader-Frechette, K., and E. D. McCoy. 1994. How the tail wags the dog: how value judgments determine ecological science. Environmental Values 3:107–120.

Siemer, W. F. 2009. Toward a practice of impacts management: insights from an exploratory case study. Dissertation, Cornell University, Ithaca, New York, USA.

Siemer, W. F., G. R. Batcheller, R. J. Glass, and T. L. Brown. 1994. Characteristics of trappers and trapping participation in New York. Wildlife Society Bulletin 22:100–111.

Siemer, W. F., and D. J. Decker. 2006. An assessment of black bear impacts in New York. Cornell University, Department of Natural Resources, Human Dimensions Research Unit Series Publication 06-6, Ithaca, New York, USA.

Siemer, W. F., D. J. Decker, P. Otto, and M. L. Gore. 2007a. Working through black bear management issues: a practitioners' guide. Northeast Wildlife Damage Management Research and Outreach Cooperative, Ithaca, New York, USA.

Siemer, W. F., D. J. Decker, and J. Shanahan. 2007b. Media frames for black bear management stories during issue emergence in New York. Human Dimensions of Wildlife 12:89–100.

Siemer, W. F., P. Hart, D. Decker, and J. Shanahan. 2009. Factors that influence concern about human–black bear interactions in residential settings. Human Dimensions of Wildlife 14:185–197.

Siemer, W. F., S. A. Jonker, and T. L. Brown. 2004. Attitudes toward beaver and norms about beaver management: insights from baseline research in New York. Cornell University, Department of Natural Resources, Human Dimensions Research Unit Series Publication 04-5, Ithaca, New York, USA.

Silverman, D. 2004. Qualitative research: theory, method and practice. Second edition. Sage, London, England, U.K.

Simon, H. A. 1990. Invariants of human behavior. Annual Review of Psychology 41:1–19.

Singer, P. 1990. Animal liberation. Second edition. New York Review of Books, New York, New York, USA.

Skonhoft, A., and C. W. Armstrong. 2005. Conservation of wildlife. A bioeconomic model of a wildlife reserve under the pressure of habitat destruction and harvesting outside the reserve. Natural Resource Modeling 20:511–547.

Slovic, P., E. Peters, M. Finucane, and D. MacGregor. 2005. Affect, risk and decision making. Health Psychology 24:S35-S40.

Smith, A. 1759. The theory of moral sentiments. Second edition. Printed for A. Millar, London, England, U.K.

Smith, C. A. 2011. The role of state wildlife professionals under the public trust doctrine. Journal of Wildlife Management 75:1539–1543.

Smith, E. R., and J. DeCoster. 2000. Dual-process models in social and cognitive psychology: conceptual integration and links to underlying memory systems. Personality and Social Psychology Review 4:108–131.

Smith, F. E. 1980. The public trust doctrine: instream flows and resources. A discussion paper prepared by the California Water Policy Center. U.S. Fish and Wildlife Service, Sacramento, California, USA.

Snow, D. A., and P. E. Oliver. 1995. Social movements and collective behavior: social psychological dimensions and considerations. Pages 571–600 in K. S. Cook, G. A. Fine, and J. S. House, editors. Sociological perspectives on social psychology. Allyn and Bacon, London, England, U.K.

Sorice, M. G., and J. R. Conner. 2010. Predicting private landowner intentions to enroll in an incentive program to protect endangered species. Human Dimensions of Wildlife 15:77–90.

Sovoda, F. J. 1980. Professionalism and wildlife management on private lands. Wildlife Society Bulletin 8:96–97.

Stedman, R. C. 1993. Expanding the concepts of hunters and hunting: a social world analysis. Thesis, Cornell University, Ithaca, New York, USA.

Stedman, R. C., and D. J. Decker. 1993. What hunting means to nonhunters: comparing hunting-related experiences, beliefs, and benefits reported by hunters and nonhunters. Cornell University, Department of Natural Resources, Human Dimensions Research Unit Series Publication 93-10, Ithaca, New York, USA.

Stedman, R. C., and D. J. Decker. 1996. Illuminating an overlooked hunting stakeholder group: non-hunters and their interest in hunting. Human Dimensions of Wildlife 1:29–41.

Stedman, R. C., and T. A. Heberlein. 2001. Hunting and rural socialization. Rural Sociology 66:599–617.

Stewart, D. W. 1984. Secondary research: information sources and methods. Sage, Beverly Hills, California, USA.

Stout, R. J., D. J. Decker, and B. A. Knuth. 1992. Evaluating citizen participation: creating a communication partnership that works. Transactions of the North American Wildlife and Natural Resources Conference 57:135–140.

Stout, R. J., D. J. Decker, and B. A. Knuth. 1994. Public involvement in deer management decision-making: comparison of three approaches for setting deer population objectives. Cornell University, Department of Natural Resources, Human Dimensions Research Unit Publication 94-2, Ithaca, New York, USA.

Stout, R. J., D. J. Decker, B. A. Knuth, J. C. Proud, and D. H. Nelson. 1996. Comparison of three public-involvement approaches for stakeholder input into deer management decisions: a case study. Wildlife Society Bulletin 24:312–317.

Stout, R. J., R. C. Stedman, D. J. Decker, and B. A. Knuth. 1993. Perceptions of risk from deer-related vehicle accidents: implications for public preferences for deer herd size. Wildlife Society Bulletin 21:237–249.

Stryker, S. 1980. Symbolic interactionism: a social structural version. Benjamin/Cummings, Menlo Park, California, USA.

Sun, L., G. C. van Kooten, and G. M. Voss. 2005. Demand for wildlife hunting in British Columbia. Canadian Journal of Agricultural Economics 53:25–46.

Susskind, L., and J. L. Cruikshank. 1987. Breaking the impasse: consensual approaches to resolving public disputes. Basic Books, New York, New York, USA.

Tamietto, M., and B. de Gelder. 2010. Neural bases of the non-conscious perception of emotional signals. Nature Reviews Neuroscience 11:697–709.

Tanner, G., and R. W. Dimmick. 1983. An assessment of farmers' attitudes toward deer and deer damage in west Tennessee. Proceedings of the Eastern Wildlife Damage Control Conference 1:195–199.

Taylor-Powell, E., S. Steele, and M. Douglah. 1996. Planning a program evaluation. University of Wisconsin Cooperative Extension G3658-1, Madison, USA.

Tchamba, M. N. 1996. History and present status of the human/elephant conflict in the Waza–Logone region, Cameroon, West Africa. Biological Conservation 75:35–41.

Teel, T. L., and M. J. Manfredo. 2010. Understanding the diversity of public interests in wildlife conservation. Conservation Biology 24:128–139.

Templeton, J. F. 1994. The focus group. Probus, Chicago, Illinois, USA.

Thompson, J. G. 1993. Addressing the human dimensions of wolf reintroduction: an example using estimates of livestock predation and costs of compensation. Society and Natural Resources 6:165–179.

TRAFFIC Southeast Asia. 2007. A matter of attitude: the consumption of wild animal products in Ha Noi, Vietnam. A TRAFFIC report. Greater Mekong Programme, Ha Noi, Vietnam.

Treves, A., R. L. Jurewicz, L. Naughton-Treves, and D. S. Wilcove. 2009. The price of tolerance: wolf damage payments after recovery. Biodiversity Conservation 18:4003–4021.

Treves, A., R. B. Wallace, L. Naughton-Treves, and A. Morales. 2006. Co-managing human–wildlife conflicts: a review. Human Dimensions of Wildlife 11:383–396.

Trouteaud, A. R. 2004. How you ask counts: a test of internet-related components of response rates to a web-based survey. Social Science Computer Review 22:385–392.

Trumbull, D. J., R. Bonney, D. Bascom, and A. Cabral. 2000. Thinking scientifically during participation in a citizen-science project. Science Education 84:265–275.

Trumbull, D. J., R. Bonney, and N. Grudens-Schuck. 2005. Developing materials to promote inquiry: lessons learned. Science Education 89:879–900.

Tuten, T. L., M. Galesic, and M. Bosnjak. 2004. Effects of immediate versus delayed notification of prize draw results on response behavior in web surveys. Social Science Computer Review 22:377–384.

Tyler, T. R. 1998. Public mistrust of the law: a political perspective. University of Cincinnati Law Review 66:847–875.

Tyler, T. R. 2000. Social justice: outcome and procedure. International Journal of Psychology 35:117–125.

U.S. Department of Agriculture [USDA]. 2003. National resources inventory 2001 annual NRI. Urbanization and development of rural land. U.S. Department of Agriculture, Natural Resources Conservation Service, at www.nrcs.usda.gov/technical/NRI/2001/nri0.urban.pdf. Accessed 1 September 2010.

U.S. Fish and Wildlife Service [USFWS]. 1987. Northern Rocky Mountain wolf recovery plan. U.S. Fish and Wildlife Service, Denver, Colorado, USA.

U.S. Fish and Wildlife Service [USFWS]. 1988. 1985 national survey of fishing, hunting, and wildlife-associated recreation. U.S. Government Printing Office, Washington, D.C., USA.

U.S. Fish and Wildlife Service [USFWS]. 1993. 1991 national survey of fishing, hunting, and wildlife-associated recreation. U.S. Government Printing Office, Washington, D.C., USA.

U.S. Fish and Wildlife Service [USFWS]. 2007. 2006 national survey of fishing, hunting, and wildlife-associated recreation. U.S. Government Printing Office, Washington, D.C., USA.

van Vliet, N. 2010. Participatory vulnerability assessment in the context of conservation and development projects: a case study of local communities in southwest Cameroon. Ecology and Society 15:6, www.ecologyandsociety.org/vol15/iss2/art6. Accessed 17 May 2011.

Vaske J. J. 2008. Survey research and analysis: applications in parks, recreation, and human dimensions. Venture, State College, Pennsylvania, USA.

Vaske, J. J., and D. Whittaker. 2004. Normative approaches to natural resources. Pages 283–294 in M. J. Manfredo, J. J. Vaske, B. L. Bruyere, D. R. Field, and P. Brown, editors. Society and natural resources: a summary of knowledge. Modern Litho, Jefferson City, Missouri, USA.

Vaske, J. J., D. Whittaker, B. Shelby, and M. J. Manfredo. 2002. Indicators and standards: developing definitions of quality. Pages 143–171 in M. J. Manfredo, editor. Wildlife viewing in North America: a management planning handbook. Oregon State University Press, Corvallis, USA.

Von Winterfeldt, D., and W. Edwards. 1986. Decision analysis and behavioral research. Cambridge University Press, Cambridge, England, U.K.

Vucetich, J. A., and M. P. Nelson. 2007. What are 60 warblers worth? Killing in the name of conservation. Oikos 116:1267–1278.

Vucetich, J. A., and M. P. Nelson. 2010. Sustainability: virtuous or vulgar? BioScience 60:539–544.

Wagner, K. K., R. H. Schmidt, and M. R. Conover. 1997. Compensation programs for wildlife damage in North America. Wildlife Society Bulletin 25:312–319.

Wallace, M. S., H. L. Stribling, and H. A. Clouts. 1989. Factors influencing land access selection by hunters in Alabama. Transactions of the North American Wildlife and Natural Resources Conference 54:183–189.

Walters, C. J. 1986. Adaptive management of renewable resources. McGraw-Hill, New York, New York, USA.

Walters, C. J., and C. S. Holling. 1990. Large-scale management experiments and learning by doing. Ecology 71:2060–2068.

Washburn, B. E., S. C. Barras, and T. W. Seamans. 2007. Foraging preferences of captive Canada geese related to turfgrass mixtures. Human–Wildlife Conflicts 2:214–223.

Webb, E. J., D. T. Campbell, R. D. Schwartz, and L. Sechrest. 1966. Unobtrusive measures: nonreactive research in the social sciences. Rand McNally, Chicago, Illinois, USA.

Weber, M. 1958. Max Weber: essays in sociology. H. Gerth and C. W. Mills, editors and translators. Galaxy, New York, New York, USA.

Weber, R. P. 1990. Basic content analysis. Second edition. Sage, Newbury Park, California, USA.

Weladji, R. B., R. M. Stein, and P. Vedeld. 2003. Stakeholder attitudes towards wildlife policy and the Bénoué Wildlife Conservation Area, North Cameroon. Environmental Conservation 30:334–343.

Whittaker, D., J. J. Vaske, and M. J. Manfredo. 2006. Specificity and the cognitive hierarchy: value orientations and the acceptability of urban wildlife management actions. Society and Natural Resources 19:515–530.

Wholey, J. S., H. P. Hatry, and K. E. Newcomer, editors. 2010. Handbook of practical program evaluation. Third edition. Jossey-Bass, San Francisco, California, USA.

Wigley, T. B., and M. A. Melchiors. 1987. State wildlife management programs for private land. Wildlife Society Bulletin 15:580–584.

Wildlife Management Institute [WMI]. 1973. North American wildlife policy 1973 and the American game policy 1930. North American Wildlife and Natural Resources Conference 38:152–181.

Wilkinson, K. 1991. The community in rural America. Social Ecology Press, Middleton, Wisconsin, USA.

Willemsen, A., editor. 2008. The dignity of living beings with regard to plants: moral consideration of plants for their own sake. Federal Ethics Committee on Non-Human Biotechnology ECNH, Berne, Switzerland.

Williams, B. K., R. C. Szaro, and C. D. Shapiro. 2007. Adaptive management: the US Department of the Interior technical guide. U.S. Department of the Interior, Adaptive Management Working Group, Washington, D.C., USA.

Williamson, S. J. 1998. Origins, history, and current use of ballot initiatives in wildlife management. Human Dimensions of Wildlife 3:51–59.

Williamson, S. J., S. Adair, K. L. Brown, and J. Turner. 2001. Contributions made by hunters toward conservation of the North American landscape. Transactions of the 66th North American Wildlife and Natural Resources Conference 66:255–269.

Wilson, M. A. 1997. The wolf in Yellowstone: science, symbol, or politics? Deconstructing the conflict between environmentalism and wise use. Society and Natural Resources 10:453–468.

Winter, S. C., and P. J. May. 2001. Motivation for compliance with environmental regulations. Journal of Policy Analysis and Management 20:675–698.

Witter, D. J., and W. W. Shaw. 1979. Beliefs of birders, hunters, and wildlife professionals about wildlife management. Transactions of the 44th North American Wildlife and Natural Resources Conference 44:298–305.

Witter, D. J., and S. L. Sherriff. 1983. Obtaining constituent feedback: implications for conservation programs. Transactions of the North American Wildlife and Natural Resources Conference 48:42–49.

Witter, D. J., D. L. Tylka, and J. E. Werner. 1981. Values of urban wildlife in Missouri. Transactions of the 46th North American Wildlife and Natural Resources Conference 46:424–431.

Wolch, J. R., A. Gullo, and U. Lassiter. 1997. Changing attitudes toward California's cougars. Society and Animals 5:9–116.

Woodward, R. T., and R. C. Bishop. 1999. Optimal-sustainable management of multi-species fisheries: lessons from a predator–prey model. Natural Resource Modeling 12:355–377.

Wright, B. A., R. A. Kaiser, and N. D. Emerald. 2001. A national trend assessment of hunter access problems: perceptions of state wildlife administrators, 1984–1997. Human Dimensions of Wildlife 6:145–146.

Yaffee, S. L. 1994. The wisdom of the spotted owl: policy lessons for a new century. Island Press, Washington, D.C., USA.

Young, O. R. 1994. International governance: protecting the environment in a stateless society. Cornell University Press, Ithaca, New York, USA.

Zajonic, R. B. 2000. Feeling and thinking. Pages 31–58 in J. P. Fargas, editor. Feeling and thinking: the role of affect in social cognition. Cambridge University Press, Paris, France.

Zeiler, H., A. Zedrosser, and A. Bath. 1999. Attitudes of Austrian hunters and Vienna residents toward bear and lynx in Austria. Ursus 11:193–200.

Zinn, H. C., and W. F. Andelt. 1999. Attitudes of Fort Collins, Colorado, residents toward prairie dogs. Wildlife Society Bulletin 27:1098–1106.

Zinn, H. C., M. J. Manfredo, and J. J. Vaske. 2000. Social psychological bases for stakeholder acceptance capacity. Human Dimensions of Wildlife 5:20–23.

INDEX

food conditioning, *187*
formative program evaluation, 133
frameworks: jurisdictional management, 31, *33*; legal, 191–92; social amplification of risk, 160, 168. *See also* PrOACT framework
framing: messages, 171; social movements, 66
functionalist perspective in sociology, 59–60
fundamental objectives. *See* objectives, fundamental
funding: building support for new sources of, 217–18; for compensation for negative impacts, 201; for state wildlife agencies, 207

game management, definition of, 204
game species: compensation programs for damage from, 186; restoration of, 194
Geer v. Connecticut, 6
gender and hunting, 63
general attitudes, 45
generalizability: from focus group interviews, 124; of research findings, 123
genetic rescue of wolves, 223, 227
global governance, 15, 22
globalization, 66–67, 238, 244
goal-interference models, 54
goals: of communication, 162; in community-based management, 239; complexity of, and communication, 157, 161; in decision making, 103, 140; emotions and, 52; ethics and, 224; of human dimensions education, 240; Landscape Conservation Cooperatives and, 18, 19; in management process, 89–90, 92; motivation and, 53, 54, 211; in organizational transformation, 250, 251; in program evaluation, 133–34; of public trust, 5; of wildlife management, 9, 43. *See also* conflict among stakeholders; messy problems in wildlife management; wicked problems in wildlife management
goal-seeking systems, 88
gopher tortoise management in Florida, 148, *148*
Gore, Al, 238
governance: co-management model of, *16*, 16–17, 135–36, 149–51, 161; decentralized, 16–17; definition of, 4, 15; developments needed to improve, 254; education in, 241–42; empirical, 15; evolution of structures of, 15; global, 15, 22; normative, 15, 23; overview of, 1; Public Trust Doctrine, 4–5, 6–7, 26, 64; regional, 16–18, 24; top-down, 27–29; wildlife management as act of, 15–16. *See also* collaborative governance
government jurisdictions in wildlife management, 181–82
Grand River Grasslands, prairie restoration in, 152
grassroots participation, 152–53
gray wolf. *See* wolf management
Greater Yellowstone Ecosystem: gray wolf restoration in, 189–91, 197–98; grizzly bear management in, 181–82
Great Lakes Fishery Commission, 165

grizzly bear management in Yellowstone ecosystem, 181–82
group influences, 62
Gulf of Mexico, oil spill in, 238

habitat loss with urbanization and exurban development, 209–10
harm to wildlife, economic value of, 72
harvest policies: bioeconomic modeling to inform, 77–79, *78*, *79*, 82; overexploitation and, 80–81; structured decision-making and, *104*; telephone interviews and, 128; values and, 72
health: abundant wildlife and, 12, 158; chronic wasting disease and, 162; of environment, 179, 224, 229; as impact, 3, 6, 27, 167, 177, 187, 191, 206–7; TB infection and, 18; of wildlife, 69, 70, 110, 223
hedonic pricing method, 73–74
heuristics in decision making, 102
history: overexploitation of wildlife in, 80; of stakeholder engagement, 33; of use management, 203–5
human dimensions inquiries: attributes of, 115–16; benefits of, 248; combining methods for, 134–35; context of, 117–21; determination of need for, 205; methods of, 122–24; overview of, 115; qualitative methods of, 124–26; quantitative methods of, 126–34; questions to ask about, 116–17; resources for, 136–38; sample questions and matrix, 119. *See also* information needed from human dimensions inquiries
human dimensions of wildlife management: defined, 3; developments needed to improve, 254; education in, 239–44; fundamental concepts of, 3–7; governance and, 4; as informing decision processes, 10; social sciences and, 7. *See also* human dimensions inquiries; impacts; stakeholders
Human Dimensions of Wildlife Management in North America (Decker, Brown, and Siemer, eds.), vii
human welfare ethics, 226
Human-Wildlife Conflict Collaboration Conservation Conflict Resolution training, 34–35, *35*
hunting: attitudes toward, 216–17; biological-ecological impacts of, 213–14; conflict perspective and, 61; consumer surplus from, 71; declining participation in, 208, 210; encouraging, 183–84; functionalist perspective on, 60; hunters as stakeholders, 27–28; in national parks, 34; of overabundant wildlife, 182; in Pennsylvania, 214, *214*; on private land, 184; socialization into culture of, 62–63; social network of, 212–13; social psychological perspective on, 59; sociocultural and demographic influences on participation trends, 60; sociological perspective on, 58, 59; studies of, 12; subsistence value of, 71; symbolic interactionist perspective on, 61; use of bait in, 21; by youth, 214
Hurricane Katrina, 238

identities, social construction of, 62
IGOs. *See* intergovernmental organizations
image of agency and communication, 163, *163*
immunocontraception, 158
impact dependency, 8
impact management approach to black bears, 96–99, *98–99*
impacts: balancing positive and negative, 180–81; biological-ecological, 213–14; concept of, 3; in cost-benefit analysis, 76; of deer, in Michigan, 179; of elk viewing, 51; forms of, 6; as fundamental objectives, 90–91, 93; identifying and managing with wildlife use and users, 213–15; issues of scale in managing, 181–82; as linking social and ecological systems, 4; of overabundant wildlife, managing, 182–87; relationship between populations and, 179–80, *180*; of scarce wildlife, managing, 191–92; stakeholders and, 6–7, 31, *33*
implicit attitudes, 44
incentives: to increase prey base for predators, 201; for Web-based survey completion, 129
inclusive moral theories, 229
inclusiveness of research, 123
incrementalism in decision making, 102
independent survey firms or consultants, 117
indicators, 50–51
indirect and induced effects in regional economic impact analysis, 77
inequality in wildlife management outcomes, 64
influence, collaboration and pooling of, 18–20
information and education approach to stakeholders, 153, 159, 160–61
information base in management process, 96, 140
information-by-objective matrix, 119
information needed from human dimensions inquiries: building support for new sources of funding, 217–18; documenting activity trends and patterns, 207–8; identifying and managing impacts, 213–15; overview of, 118, 207; understanding demographic influences, 208–9; understanding public support for use activities, 216–17; understanding wildlife recreation as process, 210–13
informative communication: example of, *164*; media planning for, 164; message design for, 164–67, *166*, *167*; overview of, 163–64
innate emotional dispositions, 52
input from stakeholders: seeking, 11, *11*, 153, 172; unsolicited, 38. *See also* engagement of stakeholders
inquisitive approach to stakeholder engagement, 144–46
inquisitive communication, 161, 162
institutional sphere of influence, 30, *31*
instrumental models of behavior, 21
instrumental value, 225, 226
interactional approach to community, 64
interconnectivity of global population, 66–67
intergovernmental organizations (IGOs): in collaborative governance, 22; in wildlife management, 15, 19–20